EVALUATED KINETIC DATA ON GAS PHASE ADDITION REACTIONS: REACTIONS OF ATOMS AND RADICALS WITH ALKENES, ALKYNES AND AROMATIC COMPOUNDS.

EVALUATED KINETIC DATA ON GAS PHASE ADDITION REACTIONS: REACTIONS OF ATOMS AND RADICALS WITH ALKENES, ALKYNES AND AROMATIC COMPOUNDS.

J. A. Kerr and M. J. Parsonage

Department of Chemistry, University of Birmingham, Edgbaston, Birmingham.

BUTTERWORTHS

THE BUTTERWORTH GROUP

ENGLAND
Butterworth & Co (Publishers) Ltd
London: 88 Kingsway, WC2B 6AB

AUSTRALIA
Butterworth & Co (Australia) Ltd
Sydney: 586 Pacific Highway, Chatswood, NSW 2067
Melbourne: 343 Little Collins Street, 3000
Brisbane: 240 Queen Street, 4000

CANADA
Butterworth & Co (Canada) Ltd
Toronto: 14 Curity Avenue, 374

NEW ZEALAND
Butterworth & Co (New Zealand) Ltd
Wellington: 26-28 Waring Taylor Street, 1

SOUTH AFRICA
Butterworth & Co (South Africa) (Pty) Ltd
Durban: 152-154 Gale Street

First published 1972

ISBN 0 408 70403 9

Printed in England by Camelot Press, Southampton, Hants

Preface

The tremors of the literature explosion have been felt as much in chemical kinetics as in any other branch of physical science. The days when monographs could contain comprehensive lists of known kinetic parameters for whole areas of the subject are gone. There is just too much data to be covered adequately in a single volume. This situation has led to the development of kinetic data compilations and evaluations. The initiative in this direction was taken by the National Bureau of Standards in Washington D.C. with the National Standard Reference Data System. Subsequent efforts have included kinetic compilations and evaluations sponsored by the Office for Scientific and Technical Information in the U.K. and we are grateful for the financial support we have received from OSTI for the preparation of the present evaluation.

Caught in the interregnum between c.g.s. and SI units, we have opted for a dual system of energy units, kcal mol^{-1} (kJ mol^{-1}), mainly for the benefit of transatlantic readers who have yet to experience the full impact of the SI system. The line has been drawn, however, before the conversion of pressure units to the SI system, the only other unit in experimental kinetics to be affected, and we have persisted with units of torr. For those readers who have faithfully re-calibrated their manometers and their thinking to SI units, we remind them that 1 torr = 133 Nm^{-2}. The solidus convention of expressing symbols and units, e.g. A/s^{-1}, has been used except for ratios of rate constants, where the symbols are followed by the units in brackets to avoid confusion with a series of solidi.

An extensive literature search was made until the end of 1970, and it is hoped that all quantitative data on gas-phase addition reactions are included for this period. Some data published during the first half of 1971 are also included. A few qualitative studies are mentioned for reactions which were studied as part of a series of unsaturated substrates with a particular reactant.

If we have learned anything from the present work, it is that data evaluation is a slow business. Our original aim was to deal with liquid-phase reactions as well, but this proved to be impossible in the year spent preparing the manuscript of the gas-phase data, plus four months preparing the camera-ready copy for printing.

Acknowledgements

We thank Dr. David Garvin and his colleagues of the Kinetic Data Center at the National Bureau of Standards for providing us with a bibliography. To all colleagues who kindly supplied missing information or preprints of papers, we extend our sincere thanks. We are considerably indebted to Mrs. Patricia Kay for her endless patience and perseverance, extending over many rejected layouts, typescripts lost in the post and authors' errors, in the typing or rather type-setting of this book.

Table of Contents

List of Figures

List of Tables

1. INTRODUCTION

1.1 GENERAL BACKGROUND

In 1963 the U.S. Government inaugurated the National Standard Reference Data System to critically evaluate and compile quantitative data in physical science. This programme has been administered by the National Bureau of Standards, Washington, D.C. As part of this effort the National Bureau of Standards has sponsored compilations and evaluations of rate data in chemical kinetics. The former consist of logically arranged tables of standardised kinetic results, sometimes with short comments or brief notes, while the latter attempt to provide critical assessment of reported kinetic parameters for a limited number of reactions or for a particular class or type of reaction [1]. The first NSRDS-NBS publication in the field of gas kinetics was the compilation of Trotman-Dickenson and Milne [2] entitled Tables of Bimolecular Gas Reactions, and this has been followed by two evaluations, the first by Johnston [3], Gas Phase Reaction Kinetics of Neutral Oxygen Species, and the second by Benson and O'Neal [4], Kinetic Data on Gas Phase Unimolecular Reactions. None of the above publications has dealt with gas phase addition reactions of atoms and radicals to multiple bonds and this publication is intended to fill that gap.

Kondratiev's Handbook of Kinetic Constants of Gaseous Reactions was published in 1970 [5]. This excellent work includes reactions of the type considered here, but it is essentially a compilation, while we hope to present an evaluation of considerably fewer reactions. To a certain extent the compilations of Trotman-Dickenson and Milne [2] and the supplement of Ratajczak and Trotman-Dickenson [6] overlap with the handbook of Kondratiev [5]. In an attempt to avoid such duplication in the future, the International Council of Scientific Unions, through its Committee on Data for Science and Technology (CODATA), has set up a Task Force in chemical kinetics under the chairmanship of Dr. S. W. Benson to co-ordinate data programmes in chemical kinetics.

1.2 SCOPE AND LIMITATIONS

This evaluation deals with gas phase addition reactions of atoms and radicals to alkenes, alkynes and aromatic compounds. Ideally it would be desirable to evaluate liquid phase data at the same time, but this was not possible in the time available. By addition reactions we mean the reaction of an atom or radical, in its ground electronic state unless otherwise indicated, with a multiple bond to form a single product, i.e. an adduct radical.

The reactions of oxygen atoms have been omitted, since they are to appear in a separate evaluation in the NSRDS-NBS series [7], but with this exception we hope to have considered all the quantitative kinetic data for this type of gas phase reaction which have appeared in the literature up to the end of 1970. Also included are relevant papers which were published in the principal journals containing kinetic papers during the first half of 1971, but our literature search for this period was incomplete. Many of the early qualitative publications have been omitted where they are adequately summarised in a more recent paper or review. The starting point for our literature search was the compilation of Kondratiev [5]. We are grateful to Dr. David Garvin and his collaborators at the Kinetic Data Center of the National Bureau of Standards for providing us with a bibliography. We have attempted to evaluate all published data, although it is inevitable that some quantitative data have been missed in this evaluation. We would appreciate hearing about any such omissions. Also, we appeal to authors who measure rate parameters of reactions subsidiary to the main theme of their papers to give some indication of all their quantitative data in the abstract of the paper.

The evaluation of gas phase unimolecular reactions by Benson and O'Neal [4] has served both as a model and as a source of inspiration for the present effort, and we have tried to approach the ideals of data evaluation as formulated by Garvin [1].

1.3 GLOSSARY OF SYMBOLS AND ABBREVIATIONS

Symbol	Meaning	Value or Unit(s)
k_1	rate constant of addition reaction under consideration	$\ell\ mol^{-1}s^{-1}$
k_{1+2}	composite rate constant of addition reaction with two possible pathways	$\ell\ mol^{-1}s^{-1}$
k_{olefin}	rate constant for addition of species under consideration to olefin (usually relative rate constant)	$\ell\ mol^{-1}s^{-1}$
k_{RH}	rate constant for H-transfer reaction of substrate RH with species under consideration, e.g. $\dot{R}' + RH \longrightarrow R'H + \dot{R}$	$\ell\ mol^{-1}s^{-1}$
k_c	rate constant for radical combination reaction	$\ell\ mol^{-1}s^{-1}$
k_d	rate constant for radical disproportionation reaction	$\ell\ mol^{-1}s^{-1}$
k	Boltzmann constant	1.3805×10^{-16} erg molecule^{-1}K^{-1}
A	A-factor of Arrhenius equation	same units as corresponding rate constant
E	activation energy of Arrhenius equation	kcal mol^{-1}(kJ mol^{-1})
K_p	equilibrium constant in pressure units	atm$^{\Delta n}$
K_c	equilibrium constant in concentration units	$(mol\ \ell^{-1})^{\Delta n}$
ΔH_f°	standard enthalpy of formation, ideal gas, 1 atm	kcal mol^{-1}(kJ mol^{-1})
ΔH°	standard enthalpy change for reaction	kcal mol^{-1}(kJ mol^{-1})
S°	standard entropy, ideal gas, 1 atm	cal mol^{-1}K^{-1}(J mol^{-1}K^{-1})
ΔS°	standard entropy change for reaction	cal mol^{-1}K^{-1}(J mol^{-1}K^{-1})
$\Delta S^\ddagger$	entropy of activation	cal mol^{-1}K^{-1}(J mol^{-1}K^{-1})
C_p°	standard heat capacity, ideal gas, 1 atm	cal mol^{-1}K^{-1}(J mol^{-1}K^{-1})
ΔC_p°	standard heat capacity change for reaction	cal mol^{-1}K^{-1}(J mol^{-1}K^{-1})
$\langle \Delta C_p^\circ \rangle$	mean heat capacity change over a temperature range	cal mol^{-1}K^{-1}(J mol^{-1}K^{-1})
$(\ )_p$	indicates the pressure standard state of ideal gas, 1 atm	
$(\ )_c$	indicates the concentration standard state, 1 mol ℓ^{-1}	
$(\ )_T$	data refer to temperature T,K	
T	temperature	K
Δn	change in number of moles for chemical reaction	
[ ]	Square brackets are used in three ways; the context will indicate which meaning is intended: (i) concentration (ii) an estimated thermodynamic quantity which involves errors in excess of ±2 cal mol^{-1}K^{-1} or ±2 kcal mol^{-1} (iii) a tentative kinetic result, or a kinetic result calculated from transition state theory	
R	gas constant	1.987 cal mol^{-1}K^{-1} 8.314 J mol^{-1}K^{-1}
h	Planck constant	6.625×10^{-27}erg s
RRKM	Rice-Ramsperger-Kassel-Marcus theory of unimolecular reactions	
ℓ	denotes units of litre in rate constant units	
log	logarithm to the base 10	
ℓn	logarithm to the base e	
Oℓ	olefin	

1.4 DATA SHEET DETAILS

Filing Order

The reactant atoms and radicals are arranged under four headings:
(a) atoms,
(b) alkyl radicals,
(c) halogenated radicals and
(d) other radicals.

Under each heading the attacking species are arranged in order of increasing formula weight. The substrates are arranged in the following order for each attacking species:
monoalkenes; monoalkenes containing halogen, N and O atoms; dialkenes; cycloalkenes; aromatic compounds and alkynes.

Reaction Thermodynamic Data

Following the reaction heading there is a summary of the thermodynamic properties of each of the reactants and product, i.e. enthalpy of formation, entropy and heat capacity, from which the listed enthalpy, entropy and heat capacity changes are calculated. The thermodynamic data are taken from the most up-to-date sources available or are estimated by the group additivity method, as discussed in Section 1.9. The values of ΔH_f° and ΔH° are in kcal mol^{-1}(kJ mol^{-1}) and S°, C_p°, ΔS° and ΔC_p° are in cal $mol^{-1}K^{-1}$(J $mol^{-1}K^{-1}$). ΔH_f° and ΔH° with estimated errors in excess of ± 2 kcal mol^{-1} and S°, C_p°, ΔS° and ΔC_p° with estimated errors in excess of ± 2 cal $mol^{-1}K^{-1}$ are given in brackets.

Kinetic Results

The experimental temperature coefficients of the rate constants are expressed in terms of the logarithmic form of the Arrhenius equation, $\log k = \log A - E/2.3RT$, where k and A are in ℓ $mol^{-1}s^{-1}$, E is in kcal mol^{-1}(kJ mol^{-1}) and the gas constant is defined in kcal mol^{-1}(kJ mol^{-1}). Rate constants at a single temperature are followed by the temperature in brackets or, for single temperature measurements, by the temperature given in the conditions column.

Conditions

This column lists the temperature or temperature range of the experiments, and here it should be pointed out that this is the range of temperatures to which the Arrhenius parameters refer, and not the complete experimental range, which is occasionally larger.

Also listed is the pressure or pressure range of the study which refers to the total pressure of the system.

References and Experimental Methods

The references are numbered sequentially as they occur in the text, and thereafter the same reference retains that number in the remainder of the text. Where the same authors appear to have reported the same experimental results in two publications, these publications are given under a single reference number. To avoid excessive repetition, the principal experimental details are given only the first time the reference occurs, and for subsequent occasions it is necessary to refer back to the first appearance of the reference. This generally occurs in the first system of any attacking species, i.e. in the ethylene system. The following information is contained in the experimental method section:
(i) some indication if the quoted values of kinetic parameters differ from those in the original paper, e.g. if the rate constant selected for the reference reaction is different from that used by the original authors;
(ii) the following types of information: the nature of the system; the source of the attacking species; the method of measuring the rate or the rate constant; analytical procedure; any reference reaction.
It was not possible to standardise this information owing to the wide range of methods, and consequently the details given vary according to the system.
For detailed descriptions of the various experimental techniques involved in studying addition reactions, the reader should consult the reviews listed in Table I.

TABLE I SUMMARY OF REVIEWS ON ADDITION REACTIONS

Author and reference	Title	Species discussed
R.J.Cvetanović. Advan.Photochem., 1, 115 (1963)	"Addition of Atoms to Olefins in the Gas Phase"	O, H, Halogen, C, N, S
B.A.Thrush. Prog.Reaction Kinetics, 3, 65 (1965)	"Reactions of Hydrogen Atoms in the Gas Phase"	H
G.Chiltz, P.Goldfinger, G.Huybrechts, G.Martens and G.Verbeke. Chem.Rev., 63, 355 (1963)	"Atomic Chlorination of Simple Hydrocarbon Derivatives in the Gas Phase"	Cℓ
F.S.Dainton and P.B.Ayscough. Photochemistry and Reaction Kinetics, Cambridge University Press, p.64 (1965)	"Gaseous Photochlorination"	Cℓ
H.E.Gunning and O.P.Strausz. Advan.Photochem., 4, 143 (1966)	"The Reactions of Sulphur Atoms"	S
S.W.Benson and W.B.De More. Ann.Rev.Phys.Chem., 16, 397 (1965)	"Gas Kinetics"	Various atoms and radicals
J.A.Kerr. Ann.Rep.Chem.Soc., 64, 73 (1967)	"Gas Kinetics"	Various atoms and radicals
(i) M.A.A.Clyne (ii) J.A.Kerr Ann.Rep.Chem.Soc., 65, 165 (1968)	"Gas Kinetics"	Various atoms and radicals
J.A.Kerr and A.F.Trotman-Dickenson. Prog.Reaction Kinetics, 1, 107 (1961)	"The Reactions of Alkyl Radicals"	CH_3, C_2H_5, n-C_3H_7, i-C_3H_7, n-C_4H_9, t-C_4H_9
J.M.Tedder and J.C.Walton. Prog.Reaction Kinetics, 4, 39 (1967)	"Reactions of Halogenomethyl Radicals"	$CCℓ_3$, CF_3
D.D.Drysdale and A.C.Lloyd. Oxid.and Comb.Rev., 4, 157 (1970)	"Gas Phase Reactions of the Hydroxyl Radical"	OH
L.I.Avramenko and R.V.Kolesnikova. Advan.Photochem., 2, 25 (1964)	"Mechanisms and Rate Constants of Elementary Gas Phase Reactions involving Hydroxyl and Oxygen Atoms"	OH, O
A.C.Lloyd. Int.J.Chem.Kinet., in press	"Evaluated and Estimated Kinetic Data for the Gas Phase Reactions of the Hydroperoxyl Radical"	HO_2

Preferred Rate Parameters

The preferred rate constants, expressed by the Arrhenius equation or the value at a single temperature, are those assessed to be most consistent with the criteria given below. Tentative or suspect results are given in square brackets.

Comments

The reasons for selecting the preferred rate parameters are discussed here, along with any other mechanistic or relevant points related to the reaction. The criteria upon which the evaluations have been made are discussed in Section 1.5.

1.5 CRITERIA FOR EVALUATING KINETIC PARAMETERS

Where possible the rate constant for the addition reaction is compared with that for the reverse radical decomposition reaction via the equilibrium constant and the thermodynamic data. Thus, for the reaction

$$\dot{X} + O\ell \xrightarrow{1} X\dot{O}\ell$$

the equilibrium constant is given by $K_1 = k_1/k_{-1}$, and from the thermodynamic relations $\Delta G_1^\circ = -RT\ell nK_1 = \Delta H_1^\circ - T\Delta S_1^\circ$, and the Arrhenius equations for the forward and reverse reactions, it can be shown [8] that $\Delta H_1^\circ = E_1 - E_{-1}$ and $\Delta S_1^\circ = R\ell n(A_1/A_{-1})$.

Usually ΔH_1° can be calculated or estimated, and if the activation energy of the radical decomposition reaction (E_{-1}) is known, it is then possible to calculate E_1. Likewise, from ΔS_1° and A_{-1} it is possible to calculate the value of A_1 to compare with the experimental result. Following the usual convention, all the thermodynamic data quoted in this evaluation refer to a standard state of 1 atm, whereas the kinetic parameters refer to a standard state of 1 mol ℓ^{-1}. The thermodynamic data can be converted to the latter standard state by use of the equations

$$\Delta H^\circ\ (1\ \text{mol}\ \ell^{-1}) = \Delta H^\circ\ (1\ \text{atm}) - \Delta nRT$$

$$\Delta S^\circ\ (1\ \text{mol}\ \ell^{-1}) = \Delta S^\circ\ (1\ \text{atm}) - \Delta nR[1 + \ell n(0.0821\ T)]$$

Temperature corrections to ΔS° and ΔH° were made by the use of the following equations:

$$\Delta H_T^\circ = \Delta H^\circ_{298} + \langle\Delta C_p^\circ\rangle(T - 298)$$

$$\Delta S_T^\circ = \Delta S^\circ_{298} + \langle\Delta C_p^\circ\rangle\ell n(T/298)$$

Many of the kinetic data on the radical decomposition reactions are taken from the evaluation of Benson and O'Neal [4]. We have tried to avoid the situation of using their preferred data to assess the kinetic parameters of addition reactions if the Benson and O'Neal data are based on the addition kinetic parameters and the thermodynamic data.

An example of the calculation, comparing the experimental rate parameters of the addition reaction with those calculated from the parameters for the reverse reaction together with the thermodynamic data, is shown in Table II.

TABLE II THERMODYNAMIC DATA FOR $C_2H_5 + C_2H_4 \xrightarrow{1} n\text{-}C_4H_9$

X	C_2H_5	C_2H_4	$n\text{-}C_4H_9$	$\Delta X^\circ_{p(300)}$	$\Delta X^\circ_{c(300)}$	$\Delta X^\circ_{c(500)}$
$\Delta H^\circ_{f(300)}$	26.0	12.5	15.8	-22.7	-22.1	-21.6
$S^\circ_{(300)}$	59.8	52.4	78.3	-33.9	-25.5	-24.3
$C^\circ_{p(300)}$	11.1	10.3	23.2	1.8	$\langle\Delta C_p^\circ\rangle = 2.4$	
$C^\circ_{p(500)}$	16.3	14.9	34.2	3.0		

The experimental rate constant for the reverse radical decomposition reaction (-1) is given by reference [9]

$$\log\ (k_{-1}/s^{-1}) = 13.57 - 28.7/2.3RT$$

and in conjunction with the above thermochemical data it follows that:

$\Delta H_1^\circ = E_1 - E_{-1}$, hence $E_1 = 28.7 - 21.6$

$\Delta S_1^\circ/4.575 = \log A_1 - \log A_{-1}$, hence $\log A_1 = 13.57 - 5.31$

This yields the calculated rate constant $\log (k_1/\ell\ mol^{-1}s^{-1}) = 8.26 - 7.1/2.3RT$ at 500 K, which compares very well with the preferred parameters, $\log k_1 = 8.20 - 7.3/2.3RT$ measured at about 400 K.

The application of transition-state theory to addition reactions is discussed in Section 1.8. It suffices to say here that the calculations are difficult and the results are open to question owing to the difficulty of defining the activated complex. It is possible, however, to set a lower limit to the A-factor for an addition reaction by considering the A-factor for the reverse radical decomposition and the entropy change for the system. From transition-state theory the A-factor for the unimolecular decomposition of a radical is given by

$A = (ekT/h)\exp(\Delta S^\ddagger/R)\ s^{-1}$

For bond-breaking reactions it is reasonable to assume that the entropy of the activated complex will be larger than that of the reactant, i.e. $\Delta S^\ddagger > 0$. For the decompositions of radicals, however, it has been shown [4,10,11] that the complex is quite "tight" with a structure resembling that of the reactant, so that $\Delta S^\ddagger \geqslant 0$ is a reasonable approximation, from which it follows that $A_{-1}/s^{-1} \geqslant ekT/h$ is likewise reasonable. At 600 K this corresponds to $\log (A_{-1}/s^{-1}) \geqslant 13.53$. Together with the ΔS° for the reaction, this sets a lower limit to the A-factor of the addition reaction given by $\log A \geqslant 13.53 + \Delta S_c^\circ/4.575$. For example, for the reaction $\dot{C}C\ell_3 + CH_2{=}CH_2 \longrightarrow C C\ell_3 CH_2\dot{C}H_2$, the entropy change is estimated to be -34.4 cal $mol^{-1}K^{-1}$ and, hence, the minimum A-factor is $10^{7.8}$, considerably higher than the original experimental value.

The third method of evaluating the kinetic data for addition reactions involves comparisons between sets of data for different species adding to common substrates. This idea has been applied where there is no kinetic information on the reverse reaction and it has not been possible to estimate ΔS° with any certainty. As a general rule-of-thumb, the A-factors for the additions of atoms to unsaturates lie in the range $10^{10\pm1}\ \ell\ mol^{-1}s^{-1}$, while those for radicals are about $10^{8\pm1}\ \ell\ mol^{-1}s^{-1}$. Experimental A-factors much outside these limits can be viewed with suspicion.

A further comparison which is occasionally useful is that for non-electrophilic radicals; the relative rate constants for addition to a given multiple bond follow much the same pattern as those for the H transfer reactions involving attack of the radicals on a weak X-H bond, say a tertiary C-H bond in an alkane. Since there are considerably more data available on the H transfer reactions [2,5,6], this can be used to gain a rough indication of the addition rate constants.

In the few cases where there are several determinations of the rate constant which are assessed to be of equal reliability, the values have been averaged to yield the preferred data.

1.6 REFERENCE REACTIONS

Apart from the reactions of NF_2 and $CC\ell$, all the rate constants of radical addition reactions are derived from ratios of rate constants and assumed or measured rate constants of the reference reaction. An attempt has been made to use evaluated data as far as possible for the reference reactions.

For most of the radicals the reference reaction is the combination reaction of the radical. The few rate constants of radical combination reactions which have been measured are mainly from experiments involving the rotating-sector technique and are subject to large errors. Considerable doubt surrounds the rate constants of the larger alkyl radicals, as the combination data cannot be reconciled with the kinetic data for the reverse decompositions of the dimer molecules [10,12]. It appears that the combination reactions of the larger radicals could involve significant activation energies. Until this difficulty is resolved the k_c values remain in doubt. In the meantime there is little alternative to assuming zero activation energy for these combination reactions. The selected A-factors are either experimental values or estimated values [8], some of which were calculated from the empirical correlation between the combination rate constants and the molecular weight of the radical [13].

1.7 EXPERIMENTAL ERRORS

Atomic and free radical addition reactions are among the most difficult reactions to study kinetically. To the usual problems of dealing with a reactive attacking species must be added the problem that the product is a radical. Thus, in following the rate of the reaction either by the rate of disappearance of the attacking species (commonly used for atomic additions) or by the rate of formation of the products from the adduct radical (used for many radical additions) there is more than the usual scope for the occurrence of secondary reactions which introduce systematic errors. For radical additions there is another source of systematic error in that the concentrations of the radicals cannot be directly measured, and consequently it is necessary to employ a reference reaction, often the radical combination reaction, for which the rate constants may not be accurately known.

When these types of error are compounded with all the usual sources of errors in kinetic measurements [4,14], it is apparent that rate constants of the addition reactions involve large errors, even by kinetic standards. There are no absolute rate constants for addition reactions which are known to better than about ±10 to ±15%, and errors of ±50 to ±100% are not uncommon.

The errors involved in the Arrhenius parameters depend upon the temperature range over which the reaction has been studied, and for these reactions a range of 75-100° is fairly typical. As a rough guide it is estimated that the minimum errors involved in determining the activation energy for an addition reaction (2 to 8 kcal mol^{-1}) are in the range ±1 to ±2 kcal mol^{-1} with an associated error in log ($A/\ell\ mol^{-1}s^{-1}$) of ±(0.6 to 1.2) at 350 K.

It has been pointed out by Cvetanović that, although the errors in the absolute rate constants of the addition reactions are large, those for the relative rate constants for atoms or radicals adding to the series of methyl-substituted ethylenes should be considerably smaller, and he has suggested errors of a few percent in these relative rate constants [15]. Recent measurements of the relative rate constants of H atoms adding to the series of olefins tend to support this estimate.

1.8 GENERAL FEATURES OF ADDITION REACTIONS

Position of Addition to Asymmetrical Alkenes

There are surprisingly few quantitative data on the relative rates of addition of atoms at the different sites in asymmetrical olefins in the gas phase. For additions to asymmetrical terminal olefins it has long been recognised [15-17] that addition occurs predominantly at the terminal C atom, and this is usually ascribed to the formation of the less reactive, or thermodynamically more stable, radical. In other words, the more exothermic reaction is the faster.

The recent experiments of Falconer and Sunder [18] on H atom additions to propene, butene-1 and isobutene appear to be the only really definitive studies for additions of atoms to olefins. The H atoms were generated by the mercury-photosensitised decomposition of H_2 and the relative rates of addition at the different sites of olefins were determined by careful analyses of the products ensuing from the adduct radicals. The percentages of non-terminal additions were found to be 5.7 ± 0.1, 5.7 ± 0.1 and 0.48 ± 0.03 for propene, butene-1 and isobutene, respectively, at room temperature.

Quantitative data on the positions of addition of free radicals to asymmetrical olefins in the gas phase are more plentiful. It has been shown from the analyses of the products of the reactions of the adduct radicals that about 14% of CD_3 radicals add non-terminally to butene-1 [19]. Likewise, Miyoshi and Brinton [20] have reported that approximately 10% of CH_3 radicals add non-terminally to propene at $\sim$ 400 K. Additions of alkyl radicals to alkynes appear to follow the same pattern with about 10% of CH_3, i-C_3H_7 and t-C_4H_9 radicals adding non-terminally to propyne at 400 K [21].

Tedder and co-workers [17,22] have reported extensively on the relative rates of addition of $CC\ell_3$ [17,22], n-C_3F_7 [23], SF_5 [24] and CF_2Br [25] radicals mainly to asymmetrical halogenated olefins. Although there has recently been a re-interpretation of the $CC\ell_3$ addition results, leading to much higher Arrhenius parameters than previously reported [22,26], the conclusions regarding the positions of addition to the asymmetrical olefins are unaffected. All of this work involved product analysis and is unambiguous with regard to the positions of addition. Tedder and co-workers have also measured the Arrhenius parameters for the different positions of addition to the asymmetrical olefins, from which the general conclusion appears to be that the differences in reactivity of the two sites in the olefin arise largely from differences in activation energy, the A-factors remaining essentially constant.

A summary of the relative rates of $CC\ell_3$ addition to the two positions in some asymmetrical olefins in the gas phase is given in Table III.

TABLE III RELATIVE RATE CONSTANTS FOR ADDITION OF $CC\ell_3$ TO THE ALTERNATIVE SITES IN ASYMMETRICAL OLEFINS IN THE GAS PHASE AT 423 K [27]

CH_2=CHF 1 : 0.077	CF_2=CH_2 1 : 85	$CC\ell_2$=CF_2 1 : 5
CH_2=$CHCH_3$ 1 : 0.071	CF_2=CHF 1 : 3.5	$CC\ell_2$=$CHC\ell$ 1 : 30
CH_2=CF_2 1 : 0.012	CF_2=$CC\ell_2$ 1 : 0.2	
CH_2=$CFCH_3$ 1 : 0.0071	CF_2=$CHC\ell$ 1 : 0.04	
	CF_2=CFBr 1 : 0.03	

With the exceptions of the olefins CH_2=CHF, CH_2=CF_2, CHF=CF_2 and possibly CH_2=$CFCH_3$, the relative rates of attack of $CC\ell_3$ at the two possible sites are as predicted on the basis of the stability of the adduct radical. The enthalpies of formation of most of the fluoro-radicals are not known accurately, but there seems little doubt that the sequence runs ΔH_f° ($R\dot{C}C\ell_2$) < ΔH_f° ($R\dot{C}H_2$) < ΔH_f° ($R\dot{C}HF$) < ΔH_f° ($R\dot{C}F_2$). The anomalous behaviour of CF_2=CH_2 has been confirmed for CF_3 addition in the gas phase [28], where the adduct is exclusively $CF_3CH_2\dot{C}F_2$, and the same pattern of reactivity has been observed in solution [29]. Similarly, for the addition of heptafluoropropyl radicals to the fluoroethylenes, CH_2=CHF, CH_2=CF_2 and CHF=CF_2, there is preferential addition at the carbon atom with the fewer F substituents, in contrast to the prediction of the more stable radical being favoured [23]. On the other hand, it has been shown that CF_2Br radicals add with approximately equal facility to both carbon atoms in CFH=CF_2 over the temperature range 330-440 K [25].

Similar results have been obtained for addition of SF_5 radicals to CH_2=CF_2 and CHF=CF_2, with attack predominantly at the less fluorinated carbon atom [24].

Another interesting exception, where addition gives rise to the thermodynamically less stable radical, is for the addition of alkyl radicals to allene, which occurs exclusively at the terminal carbon atom [30], although addition at the central atom would give rise to an allyl radical instead of a vinyl radical from terminal addition. The explanation in this case appears to be in the unique geometry of the allene molecule, in which the π-bonds are mutually perpendicular. In the transition state the incipient radical from centre attack resembles a primary rather than an allylic radical, since the orbital of the odd electron does not overlap with the π-orbital of the remaining double bond.

To sum up, it appears that for atom or radical addition to asymmetrical unsaturates the reactions yield predominantly the more stable adduct radical, with the exception of additions to fluoro-olefins. At present there is no satisfactory qualitative explanation of these effects, and so there can be no meaningful quantitative description of them.

Flannery [31] concluded that the correlation between atom localisation energies in asymmetric olefins with experimental activation energies for CH_3, CF_3 and $CC\ell_3$ radical addition, supports the idea that the direction of addition is determined primarily by the relative stabilities of the adduct radicals. This conclusion would appear to be premature when the correlation is carried out with the evaluated data of this work.

It is sometimes necessary to estimate the Arrhenius parameters for addition at the unfavoured site in asymmetrical olefins from the relative rate constants for addition at both atoms. Two extreme positions have been adopted on this point: (i) it is assumed that the difference in reactivities is due entirely to differences in A-factors with E remaining constant [4,10], and (ii) it is assumed that the difference in reactivities is due entirely to differences in E with the A-factors remaining constant [18]. The latter assumption derives from the fact that the $\Delta H°$ for addition to the terminal position in the asymmetrical olefin is always more favourable. At the same time there is some direct support for the second idea from the measurements of Tedder and co-workers [17] of the differences in Arrhenius parameters for additions of various radicals to asymmetrical olefins where the significant changes are in the activation energies and not the A-factors.

Nevertheless it seems unlikely that the A-factors will remain exactly constant, since it has been shown that for the addition of CH_3 and NF_2 radicals to a series of substituted ethylenes there is a marked decrease in the A-factor as the olefin becomes more crowded (in keeping with the $\Delta S°$ of the reactions). On this basis one would expect a smaller A-factor for addition at the more substituted carbon atom of an asymmetrical olefin. It is therefore probable that both A and E will be different for additions at the two sites, although changes in E should have the greater effect upon the rate constant. Accordingly, in our calculations of this type we have assumed that the A-factor is constant and assigned the total difference in rate constants to a difference in E.

Finally, as regards the position of addition of atoms and radicals to asymmetrical olefins, we should mention an alternative suggestion to the interpretation of the formation of the more stable adduct radical, giving rise to the faster rate of addition. This involves the idea that "the orientation of radical addition is determined principally by the relative strengths of the two possible bonds initially formed" [17,32], by which is presumably meant that terminal addition is favoured since the C-C bond formed between the attacking radical and the terminal carbon atom is stronger than the alternative bond involving the non-terminal carbon atom. This would appear to offer nothing new, since the strength of the bond formed, which is an alternative way of stating the enthalpy change of the reaction, must be directly related to the stability of the adduct radical, i.e. the stronger the bond the more stable the radical.

A-Factors for Addition Reactions

There have been few attempts at calculating A-factors for atom and radical addition reactions *a priori* by transition-state theory. As in any application of this theory to bimolecular reactions of radicals, the difficulty lies in predicting the properties of the activated complex, which simply have to be guessed.

The calculations of Knox and co-workers [33] on the additions of $C\ell$ atoms to chloroethylenes, $\dot{C}\ell + O\ell \xrightarrow{1} C\ell\dot{O}\ell$, are among the most detailed and illustrate the general approach to the problem. From transition-state theory the rate constants can be calculated from

$$k_1 = (kT/h)(Q^{\ddagger}/Q_{C\ell}Q_{O\ell})\exp(-\Delta E_0^{\circ\ddagger}/RT)$$

where $Q^{\ddagger}$, $Q_{C\ell}$ and $Q_{O\ell}$ are the partition functions measured from the lowest vibrational state of the molecules for the activated complex, $C\ell$ and the olefin; $\Delta E_0^{\circ\ddagger}$ is the difference in energy between the lowest vibrational state of the reactants and that of the complex ($C\ell\dot{O}\ell^{\ddagger}$). The partition functions can then be expressed in terms of the masses, moments of inertia, symmetry numbers and vibrational frequencies of the complex, $C\ell$ and $O\ell$, at which point it is necessary to specify the properties of the complex, the properties of the reactants usually being well characterised. For the above reactions Knox and co-workers assumed that the structure of the complex $C\ell\dot{O}\ell$ was identical with that of the final adduct radical $C\ell\dot{O}\ell$ and assigned vibrational frequencies on the basis of those measured for chloroethanes.

The results of these calculations are shown in Table IV.

TABLE IV COMPARISONS OF CALCULATED AND EXPERIMENTAL A-FACTORS FOR THE REACTION $\dot{C}\ell + O\ell \xrightarrow{1} C\ell\dot{O}\ell$

Olefin	log A_1(350 K) calculated from T.S.T [33]	log A_1 preferred experimental	log A_1(350 K) minimum from $\Delta S°$
CH_2=CH_2	10.63	10.65	10.21
CH_2=$CHC\ell$	9.64	10.64	10.12
$CC\ell_2$=CH_2	9.86	[10.5]	9.36
$CHC\ell$=$CHC\ell$(cis)	9.78	10.78	9.41
$CHC\ell$=$CHC\ell$(trans)	9.97	10.60	9.41
$CC\ell_2$=$CHC\ell$	9.61	10.54	9.32
$CC\ell_2$=$CC\ell_2$	10.28	[10.6][a]	9.86

[a] calculated from log k_1 = [9.5] at 298 K and assuming E_1 = 1.5 kcal mol^{-1}.

Also included in Table IV are A-factors calculated on the basis of the over-all entropy change for the reaction and a minimum A-factor for the reverse reaction (-1) given by $A_{-1} \geqslant ekT/h$. Thus log $A_1 \geqslant \Delta S^o_c/4.575 + ekT/h$. In these calculations we have neglected the small temperature corrections to ΔS^o_c. There is a substantial body of experimental evidence on alkyl radical decompositions [4,10,11] which indicates that the entropies of activation are small for these reactions, so the above assumption is a reasonable approximation.

On the whole, the comparison between the calculated and experimental A-factors in Table IV is rather disappointing. Apart from the data for ethylene, the preferred experimental A-factors are all considerably higher than the values calculated from transition-state theory, and, assuming that the experimental values are reliable, this can only mean that the model selected for the activated complex is too "tight". Unfortunately, there is no way of predicting this a priori.

The minimum A-factors of the $C\ell$ additions to the olefins, set from the lower limit A-factor of the reverse reactions, are about 0.4 log units lower than those calculated directly from transition-state theory. Again, assuming that the $\Delta S°$ values are reliable, one must conclude that the activated complex for both the forward and reverse reactions is considerably looser than anticipated on the basis of the results for hydrocarbon radical decompositions.

Reactivity of Alkenes and Molecular Structure

Following the examples of organic and polymer chemists, gas phase kineticists have sought an appropriate molecular property as an index for the reactivity of unsaturated molecules towards the addition of atoms and free radicals. Rate parameters for CH_3 + aromatics [34], H + olefins [35,36], CH_3 and C_2H_5 + olefins [37], $CC\ell_3$ + halogeno-olefins [17,38], CF_3 + olefins [39,40] and NF_2 + olefins [39] have been correlated with various molecular parameters of the substrates, such as atom localisation energies, maximum free valency, singlet-triplet excitation energies, electron affinities, ionisation energies, incremental stabilisation energies of the reacting species in the transition state, and the Alfrey-Price intrinsic reactivity parameter.

Correlations between these properties and the rate constants of addition, which imply a constant A-factor of the reactions, are clearly not as satisfactory as correlations with activation energies. The results of the present evaluation show that while there is not a large variation in the A-factors for different attacking radicals adding to the same olefin, there can be significant variations in the A-factors for the same species adding to a progressively more substituted olefin centre.

For most attacking species the best correlation is obtained between the activation energy and the atom localisation energy, the energy required to disengage one electron from the system and localise it at the selected carbon atom. The trend in these correlations is for the increase in activation energy to parallel the increase in localisation energy. The localisation energy represents the electron density pattern at a stage in the reaction later than the transition state and consequently only an approximate correlation would be expected. The fact that such a correlation exists at all indicates that the transition state in addition reactions bears more resemblance to the final product than to the reactants, a conclusion already supported by transition-state theory calculations.

While these types of correlation are obviously useful for predictive purposes and give some idea of the sorts of considerations necessary in calculating rate parameters from molecular properties, they fall a long way short of an a priori approach to explaining the trends in the rate constants and Arrhenius parameters.

Nature of the Transition State

Considerable discussion has taken place on the natures of the transition states and intermediates in the reactions of atoms and free radicals with unsaturated molecules. Much of the argument has centred around what may be regarded as two limiting structures of the transition state: (i) a σ-complex, in which a well-defined primary valence bond is developed between the radical and the substrate and which consequently has a structure essentially that of the product radical; (ii) a π-complex, or charge-transfer complex, in which the bonding arises from the mutual perturbation of the dispersion fields of the reactants.

On the whole, the idea of a σ-complex seems to be favoured over that of the π-complex, but the matter is still the subject of lively debate for several reactions. Stefani and co-workers [39,40] have given an excellent summary of the present position regarding this question, following earlier reviews of Szwarc [41] and Cvetanović [15].

As an example of the difficulty in pin-pointing the transient species in these systems, the reader should consult recent discussions on the nature of the intermediate in the addition of oxygen atoms to olefins [42,43].

1.9 SOURCES OF THERMODYNAMIC DATA

Enthalpy Data of Attacking Atoms and Radicals

With the few exceptions given below, the enthalpies of formation of the attacking species are taken either from Benson's book [8] or from the estimates of Benson and O'Neal [4,12]. $\Delta H_f^\circ(n\text{-}C_5H_{11})$ was estimated by the group additivity method [12]. $\Delta H_f^\circ(CH_2F)$ and $\Delta H_f^\circ(n\text{-}C_3F_7)$ were calculated from the bond dissociation energies $D(CH_2F\text{-}H) = 101$ [44] and $D(n\text{-}C_3F_7\text{-}H) = 104$ kcal mol^{-1} [45]. $\Delta H_f^\circ(CF_2C\ell)$ is listed by Kerr and Trotman-Dickenson [45]. The only two attacking species for which meaningful estimates of the enthalpy of formation could not be made were the radicals CF_2Br and SF_5.

Entropy and Heat Capacity Data of Attacking Atoms and Radicals

Here again the vast majority of the data are from the compilations of Benson and co-workers [4,8,12]. For the $n\text{-}C_5H_{11}$ radical $S°$ and C_p° were estimated by group additivity [12] and for CH_2F estimates were based on CH_3F as a model [12]. $S°(CF_2Br)$ and $C_p^\circ(CF_2Br)$ were calculated by the method of bond additivity.

Enthalpy Data of Substrate Molecules

The enthalpies of formation of alkenes, alkynes and aromatic compounds not containing a halogen atom were either taken directly from compilations [8,46] or estimated by the group additivity methods as described by Benson [8,46].

The properties of molecules can be considered to be made up of additive contributions from the individual atoms or bonds in the molecule. An extension of this is to assume the property as being composed of contributions due to groups. A group is defined as a polyvalent atom with its ligands, and the nomenclature used is to identify first the atom and then its ligands. Thus, $[C - (H)_3(C)]$ represents a C atom connected to three H atoms and a C atom, and $[C_B - (F)]$ represents a C atom on a benzene ring connected to an F atom. The property to be calculated is written before the bracket, e.g. $\Delta H_f^\circ[C_B - (F)]$.

An example of a typical group additivity calculation is as follows for pentafluorobromobenzene:

ΔH_f° 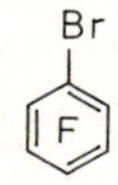

$$= 5\ \Delta H_f^\circ[C_B - (F)] + \Delta H_f^\circ[C_B - (Br)] + 4 \text{ ortho (F)(F) correction} + 2 \text{ ortho (Br)(F) correction}$$
$$= -214.0 + 10.7 + 20.0 + 0$$
$$= -183.3 \text{ kcal mol}^{-1}$$

The enthalpies of formation of many of the substrates containing halogen atoms are not very accurately known. We have opted to use the experimental data as listed by Cox and Pilcher [47] as far as possible, with the exceptions of $\Delta H_f^\circ(CH_2{=}CCl_2)$, $\Delta H_f^\circ(CH_2{=}CF_2)$ and $\Delta H_f^\circ(CH_2{=}CHF)$, which are the estimates of Benson [8]. For other unsaturates for which the enthalpies of formation are not listed in reference 47, we have made estimates from group additivity rules [8,46].

Entropy and Heat Capacity Data of Substrate Molecules

Entropy and heat capacity data were taken either from the tabulations of Benson and co-workers [8,46] or were calculated by group additivity methods [8,46]. An example of a calculated entropy is shown below:

To estimate S° (hexyne-3):

$$S^\circ (CH_3CH_2C{\equiv}CCH_2CH_3) = 2\ S^\circ[C - (C)(H)_3] + 2\ S^\circ[C - (C_t)(C)(H)_2] + 2\ S^\circ[C_t - (C)] - 2R\ln 3(\text{symmetry of } CH_3)$$
$$= (2 \times 30.41) + (2 \times 10.30) + (2 \times 6.35) - 4.4$$
$$= 60.82 + 20.60 + 12.70 - 4.4$$
$$= 94.12 - 4.4 = 89.7 \text{ cal mol}^{-1}\text{s}^{-1}$$

Enthalpy Data of Adduct Radicals

Experimental data are available for only a few of the radicals [8,12]. Most of the enthalpies were calculated either directly from group additivity rules [12] or from the known or estimated bond dissociation energy D(R-H) in conjunction with the experimental or estimated value of $\Delta H_f^\circ(RH)$, i.e.

$$\Delta H_f^\circ(R) = D(R\text{-}H) + \Delta H_f^\circ(RH) - \Delta H_f^\circ(H)$$

A typical group additivity calculation is as follows:

To estimate ΔH_f°(2-methylbutyl-2 radical):

$$\Delta H_f^\circ(CH_3CH_2\dot{C}(CH_3)_2) = 3\ \Delta H_f^\circ[C - (C)(H)_3] + \Delta H_f^\circ[C - (\dot{C})(C)(H)_2] + \Delta H_f^\circ[\dot{C} - (C)_3]$$
$$= -30.24 - 4.95 + 37.0$$
$$= 1.8 \text{ kcal mol}^{-1}$$

Most of the enthalpies of formation of adducts formed by addition to halogen-containing substrates were calculated from the list of bond dissociation energies given in Table V

TABLE V BOND DISSOCIATION ENERGIES FOR ESTIMATING ENTHALPIES OF FORMATION OF ADDUCT RADICALS

Bond	D(R-H)[a]	Bond	D(R-H)[a]	Bond	D(R-H)[a]
RCH_2-H	98[b]	CH_2:CHCH(R)-H	83[b]	RCHF-H	[100[c]]
R_2CH-H	95[b]	CH_2:CHC(R_2)-H	[80[c]]	RCF_2-H	103[b]
R_3C-H	92[b]	C_6H_5CH(R)-H	78[b]	R_2CF-H	[97[c]]
R_2C:CH-H	[108[b]]	C_6H_5C(R_2)-H	[73[c]]	RCHCl-H	[97[c]]
R_2C:CR-H	[108[c]]	CH_2:C(C_6H_5)-H	[95[c]]	$RCCl_2$-H	95[b]

[a] kcal mol^{-1}. [b] reference 45. [c] estimate based on data in reference 45.

The enthalpies of formation of the parent molecules used for calculating the ΔH_f° of the halogenated adduct radicals were either taken from Cox and Pilcher [47] or, when no experimental value was available, calculated from group additivity rules [12]. Thus, for example, $\Delta H_f^\circ(C_2C\ell_5)$ is estimated as follows:

$$\Delta H_f^\circ(C_2C\ell_5) = D(C_2C\ell_5 - H) + \Delta H_f^\circ(C_2C\ell_5H) - \Delta H_f^\circ(H)$$
$$= [95] - 34.1 - 52.1$$
$$= [9] \text{ kcal mol}^{-1}$$

Entropy and Heat Capacity Data of Adduct Radicals

These were calculated by group additivity rules or from model compounds as outlined by O'Neal and Benson [12]. A sample calculation is as follows:

To estimate S°(pentyl-2 radical):

$$S^\circ(CH_3\dot{C}HCH_2CH_2CH_3) = S^\circ[C - (\dot{C})(H)_3] + S^\circ[\dot{C} - (C)_2(H)] + S^\circ[C - (\dot{C})(C)(H)_2] + [C - (C)_2(H)_2]$$
$$+ [C - (C)(H)_3] - 2R\ell n3(CH_3 \text{ groups})$$
$$= 30.41 - 10.74 + 9.42 + 9.42 + 30.41 - 4.4$$
$$= 86.0 \text{ cal mol}^{-1}\text{K}^{-1}$$

REFERENCES FOR SECTION 1.

[1] Evaluation of Rate Data, David Garvin, The Combustion Institute, Specialist Session on Chemical Kinetic Calculations, October 30, 1968, University of California, Berkeley, California.

[2] Tables of Bimolecular Gas Reactions, A.F.Trotman-Dickenson and G.S.Milne, NSRDS-NBS 9, 1967.

[3] Gas Phase Reaction Kinetics of Neutral Oxygen Species, H.S.Johnston, NSRDS-NBS 20, 1968.

[4] Kinetic Data on Gas Phase Unimolecular Reactions, S.W.Benson and H.E.O'Neal, NSRDS-NBS 21, 1970.

[5] Handbook of Kinetic Constants of Gaseous Reactions, V.N.Kondratiev, "Nauka", Moscow, 1970.

[6] Supplementary Tables of Bimolecular Gas Reactions, E.Ratajczak and A.F.Trotman-Dickenson, Publications Department, University of Wales Institute of Science and Technology, Cardiff, Wales.

[7] Evaluation of Gas Phase Reactions of Oxygen Atoms, J.T.Herron, to be published in NSRDS-NBS series.

[8] Thermochemical Kinetics, S.W.Benson, J.Wiley and Sons, New York, 1968.

[9] W.E.Morganroth and J.G.Calvert, J.Amer.Chem.Soc., 88, 5387 (1966).

[10] H.M.Frey and R.Walsh, Chem.Rev., 69, 103 (1969).

[11] J.A.Kerr and A.C.Lloyd, Quart.Rev., 22, 549 (1968).

[12] H.E.O'Neal and S.W.Benson, Int.J.Chem.Kinet., 1, 221 (1969).

[13] L.Bertrand, G.R.De Maré, G.Huybrechts, J.Olbregts and M.Toth, Chem.Physics Letters, 5, 183 (1970).

[14] Foundations of Chemical Kinetics, S.W.Benson, McGraw-Hill, New York, 1960.

[15] R.J.Cvetanović, Advan.Photochem., 1, 115 (1963).

[16] J.I.G.Cadogan and D.H.Hey, Quart.Rev., 8, 308 (1954).

[17] J.M.Tedder and J.C.Walton, Prog.Reaction Kinetics, 4, 37 (1967).

[18] W.E.Falconer and W.A.Sunder, Int.J.Chem.Kinet., 3, 395 (1971).

[19] J.R.McNesby and A.S.Gordon, J.Amer.Chem.Soc., 79, 5902 (1957).

[20] M.Miyoshi and R.K.Brinton, J.Chem.Phys., 36, 3019 (1962).

[21] R.R.Getty, J.A.Kerr and A.F.Trotman-Dickenson, J.Chem.Soc.(A), 1360 (1967).

REFERENCES

[22] H.W.Sidebottom, J.M.Tedder and J.C.Walton, Int.J.Chem.Kinet., in press.

[23] J.Gibb, M.J.Peters, J.M.Tedder, J.C.Walton and K.D.R.Winton, Chem.Comm., 978 (1970).

[24] H.W.Sidebottom, J.M.Tedder and J.C.Walton, Trans.Faraday Soc., 66, 2038 (1970).

[25] J.M.Tedder and J.C.Walton, Trans.Faraday Soc., 66, 1135 (1970).

[26] J.A.Kerr and M.J.Parsonage, Int.J.Chem.Kinet., in press.

[27] D.P.Johari, H.W.Sidebottom, J.M.Tedder and J.C.Walton, J.Chem.Soc.(B), 95 (1971).

[28] S.E.Braslavsky, F.Casas and O.Cifuentes, J.Chem.Soc.(B), 1059 (1970).

[29] R.N.Haszeldine, J.Chem.Soc., 2504 (1952);
R.Gregory, R.N.Haszeldine and A.E.Tipping, J.Chem.Soc.(C), 3020 (1968).

[30] R.R.Getty, J.A.Kerr and A.F.Trotman-Dickenson, J.Chem.Soc.(A), 979 (1967).

[31] J.B.Flannery, J.Phys.Chem., 70, 3707 (1966).

[32] J.M.Tedder and J.C.Walton, Proc.Chem.Soc., 420 (1964).

[33] J.H.Knox, Trans.Faraday Soc., 62, 1206 (1966);
P.C.Beadle, J.H.Knox, F.Placido and K.C.Waugh, Trans.Faraday Soc., 65, 1571 (1969).

[34] M.Szwarc and J.H.Binks, Theoretical Organic Chemistry, Kekulé Symposium, London, p.262, 1959.

[35] K.R.Jennings and R.J.Cvetanović, J.Chem.Phys., 35, 1233 (1961).

[36] K.Yang, J.Amer.Chem.Soc., 84, 3795 (1962).

[37] J.E.Bloor, A.C.R.Brown and D.G.L.James, J.Amer.Chem.Soc., 70, 2191 (1966).

[38] J.M.Tedder and J.C.Walton, Trans.Faraday Soc., 62, 1859 (1966).

[39] A.P.Stefani, L-Y.Y.Chuang and H.E.Todd, J.Amer.Chem.Soc., 92, 4168 (1970).

[40] A.P.Stefani and H.E.Todd, J.Amer.Chem.Soc., 93, 2982 (1971).

[41] M.Szwarc, The Transition State, Chem.Soc.Spec.Publ., 16, 91 (1962).

[42] R.J.Cvetanović, J.Phys.Chem., 74, 2730 (1970).

[43] M.D.Scheer and R.Klein, J.Phys.Chem., 74, 2732 (1970).

[44] J.A.Kerr and D.M.Timlin, Int.J.Chem.Kinet., 3, 427 (1971).

[45] J.A.Kerr and A.R.Trotman-Dickenson, "Strengths of Chemical Bonds", Handbook of Chemistry and Physics, Chemical Rubber Company, 52nd edition, p.F-177, 1971.

[46] S.W.Benson, F.R.Cruickshank, D.M.Golden, G.R.Haugen, H.E.O'Neal, A.S.Rodgers, R.Shaw and R.Walsh, Chem.Rev., 69, 279 (1969).

[47] J.D.Cox and G.Pilcher, Thermochemistry of Organic and Organometallic Compounds, Academic Press, London, 1970.

2. ABSOLUTE RATE DATA

REACTION: HYDROGEN ATOM + ETHYLENE $\dot{H}(I) + CH_2{=}CH_2(II) \longrightarrow CH_3\dot{C}H_2(III)$

	(I)	(II)	(III)		
$\Delta H^\circ_{f(300)}$	52.1 (218.0)	12.5 (52.3)	26.0 (108.8)	ΔH°	-38.6 (-161.6)
$S^\circ_{(300)}$	27.4 (114.7)	52.4 (219.4)	59.8 (250.4)	ΔS°	-20.0 (-83.7)
$C^\circ_{p(300)}$	5.0 (20.9)	10.3 (43.1)	11.1 (46.5)	$\Delta C^\circ_{p(300)}$	-4.2 (-17.6)
$C^\circ_{p(800)}$	5.0 (20.9)	20.0 (83.7)	22.8 (95.4)	$\Delta C^\circ_{p(800)}$	-2.2 (-9.2)
				$<\Delta C^\circ_p>$	-3.2 (-13.4)

log A	E	log k	Conditions	Reference and Experimental Method
-	-	8.74	298 K 7 torr	[1] H.W.Melville and J.C.Robb, Proc.Roy.Soc.(London) Ser.A., 196, 494 (1949). Static system. H from mercury photosensitised decomposition of H_2. [H] determined by 'blueing' of MoO_3. Reaction rate monitored by change in [H].
8.93	0.81 (3.39)	8.44(300 K) 8.56(463 K)	300-463 K ?	[2] B.de B.Darwent and R.Roberts, Disc.Faraday Soc., 14, 55 (1953). Recalculated on the basis of log k = 9.89 - 17.1/2.3RT for the reaction $H + H_2S \longrightarrow H_2 + HS$ [M.J.Kurylo, N.C.Peterson and W.Braun, J.Chem.Phys., 54, 943 (1971)]. Static system. H or D from photolysis of H_2S or D_2S (λ = 228.8 or 255.0 nm). Rate constants measured relative to reaction $H + H_2S \longrightarrow H_2 + HS$ and $H + C_2H_4 \longrightarrow H_2 + C_2H_3$. Reaction rates measured by rates of formation of H_2, HD and D_2 (mass spectrometer).
-	-	8.70	298 K 7 torr	[3] P.E.M.Allen, H.W.Melville and J.C.Robb, Proc.Roy.Soc.(London) Ser.A., 218, 311 (1953). As for reference [1] with corrections for surface removal of H atoms.
-	-	8.02	304 K 5-150 torr	[4] A.B.Callear and J.C.Robb, Trans.Faraday Soc., 51, 638 (1955). Static system. H from mercury photosensitised decomposition of H_2. [H] from measurement of heat of recombination of H atoms on a filament. Reaction rate monitored by change in [H].

ΔH^o_f, ΔH^o and E in kcal mol^{-1} (kJ mol^{-1}); S^o, C^o_p, ΔS^o and ΔC^o_p in cal mol^{-1} K^{-1} (J mol^{-1} K^{-1}); A and k in l mol^{-1} s^{-1}

log A	E	log k	Conditions	Reference and Experimental Method
-	-	7.86	297 K 435 torr	[5] K.R.Jennings and R.J.Cvetanović, J.Chem.Phys., 35, 1233 (1961). Recalculated on the basis of log k = 10.90 - 7.7/2.3RT for the reaction $H + n\text{-}C_4H_{10} \longrightarrow H_2 + C_4H_9$. [R.R.Baldwin et al., Tenth Symposium (International) on Combustion, Combustion Institute, Pittsburgh, 1965. p.423]. Static system. H atoms from mercury photosensitised decomposition of n-butane. Reaction rate from decrease in rate of formation of H_2 (low temperature distillation; gas burette). Rate constants relative to reaction $H + n\text{-}C_4H_{10} \longrightarrow H_2 + C_4H_9$. This paper gives relative rates for a series of olefins.
10.75	3.4 (14.2)	8.28(300 K)	329-514 K 600 torr	[6] K.Yang, J.Amer.Chem.Soc., 84, 719 (1962); J.Amer.Chem.Soc., 84, 3795 (1962). Recalculated on the basis of log k = 10.81 - 7.83/2.3RT for the reaction $H + C_3H_8 \longrightarrow H_2 + C_3H_7$ [R.R.Baldwin, Trans.Faraday Soc., 60, 527 (1964). Static system. H atoms from γ-radiolysis of propane. Reaction rate measured by decrease in rate of formation of H_2 (gas chromatography) in presence of olefin. Rate constants relative to reaction $H + C_3H_8 \longrightarrow H_2 + C_3H_7$.
-	-	9.18	293 K 200 torr	[7] A.B.Callear and W.P.D.Pereira, Trans.Faraday Soc., 59, 2774 (1963). Recalculated on the basis of log $(k/\ell^2 mol^{-2} s^{-1})$ = 10.65 for the reaction $H + O_2 + H_2 \longrightarrow HO_2 + H_2$ [M.A.A.Clyne and B.A.Thrush, Proc.Roy.Soc.(London) Ser.A., 275, 559 (1963)]. Static system. H from mercury photosensitised decomposition of H_2 in presence of O_2 and C_2H_4. Reaction rate monitored by rate of consumption of C_2H_4. Rate constants relative to reaction $H + O_2 + H_2 \longrightarrow HO_2 + H_2$.
-	-	8.72	297 K 6.6 torr	[8] V.P.Strunin, G.K.Lavrovskaya, A.F.Dodonov and V.L.Tal'roze, Kinetika i Kataliz, 7, 693 (1966). Flow system. H from high frequency discharge of H_2-He mixture. [H] measured by mass spectrometry. Rate constants calculated from theory of diffusion cloud method [V.P.Strunin et al., Kinetika i Kataliz, 7, 385 (1966)].
10.77	3.18 (13.32)	9.54	813 K 0-100 torr	[9] R.R.Baldwin, R.F.Simmons and R.W.Walker, Trans.Faraday Soc., 62, 2486 (1966). Reaction observed in inhibition by C_2H_4 of second explosion limit of $H_2 + O_2$ reaction. Reaction rate followed by pressure change and product analysis (low temperature distillation). Rate constants relative to reaction $H + O_2 \longrightarrow HO + O$.
-	-	8.28	298 K 5 torr	[10] J.V.Michael and R.E.Weston Jr., J.Chem.Phys., 45, 3632 (1966). Flow system. H from microwave discharge on H_2-He mixture. [H] determined by resonance fluorescence and absorption by H of Lyman α-radiation at 121.6 nm. Rate of reaction monitored by changes in [H].

log A	E	log k	Conditions	Reference and Experimental Method
-	-	8.78	298 K 0-75 torr	[11] W.Braun and M.Lenzi, Disc.Faraday Soc., 44, 252 (1967). Static system. H from flash photolysis of unsaturated hydrocarbons. [H] as in reference [10] above.
-	-	7.94	298 K 2 torr	[12] J.M.Brown and B.A.Thrush, Trans.Faraday Soc., 63, 630 (1967); and J.M.Brown, P.B.Coates and B.A.Thrush, Chem.Comm., 843 (1966). Flow system. H from dissociation of purified H_2 with purified Ar in a 50 W microwave discharge. [H] determined by e.s.r. Reaction rate monitored by change in [H].
-	-	8.08	298 K 2 torr	[13] J.M.Brown, B.A.Thrush and A.F.Tuck, Proc.Roy.Soc.(London) Ser.A., 302, 311 (1968). As for reference [12].
-	-	8.63	298 K 5 torr	[14] J.R.Barker and J.V.Michael, J.Chem.Phys., 51, 850 (1969). As for reference [10].
			298 K	[15] J.V.Michael and D.T.Osborne, Chem.Phys.Letters, 3, 402 (1969).
-	-	(i) 8.08	0.5 torr	Static system. H from mercury sensitised decomposition of H_2. [H] determined
-	-	(ii) 8.58	25 torr	as in reference [10] above.
			297 K	[16] A.A.Westenberg and N.De Haas, J.Chem.Phys., 50, 707 (1969).
-	-	(i) 8.34	0.55-2.61 torr	Flow system. H from microwave discharge of H_2-He mixture. [H] determined by
-	-	(ii) 8.15	0.39-2.00 torr	e.s.r. Reaction monitored by change in [H]. Rate constants derived from an
			525 K	extrapolation of 1/P plot vs 1/k to 0; (i) and (iii) M = He.
-	-	(iii) 8.48	0.55-2.61 torr	(ii) M = Ar.
-	-	8.73	298 K 800 torr	[17] G.R.Woolley and R.J.Cvetanović, J.Chem.Phys., 50, 4697 (1969). Recalculated on the basis of log k = 9.89 - 17.1/2.3RT for the reaction $H + H_2S \longrightarrow H_2 + HS$ [M.J.Kurylo, N.C.Peterson and W.Braun, J.Chem.Phys., 54, 943 (1971)]. Static system. H from photolysis of H_2S (λ = 249.0 nm). 'Hot' hydrogen atoms were thermalised by addition of a large excess of carbon dioxide. Reaction rates determined from rate of formation of H_2 in presence and absence of olefin. Rate constants relative to reaction $H + H_2S \longrightarrow H_2 + HS$. This paper gives relative rates for a series of olefins.
-	-	[8.48]	298 K 620-720 torr	[18] R.J.Cvetanović and L.C.Doyle, J.Chem.Phys., 50, 4705 (1969). Static system. H from mercury photosensitised decomposition of H_2. Relative rates obtained from comparison of rates of formation of n-butane ($2C_2H_5 \rightarrow C_4H_{10}$) obtained in presence of C_2H_4 alone or C_2H_4 plus a second olefin. This paper gives relative rates for a series of olefins.

ΔH_f^o, ΔH^o and E in kcal mol^{-1} (kJ mol^{-1}); S^o, C_p^o, ΔS^o and ΔC_p^o in cal mol^{-1} K^{-1} (J mol^{-1} K^{-1}); A and k in l mol^{-1} s^{-1}

log A	E	log k	Conditions	Reference and Experimental Method
10.43	3.30 (13.80)	8.34(345 K)	285-405 K 2 torr	[19] J.H.Knox and D.G.Dalgleish, Int.J.Chem.Kinet., 1, 69 (1969). Flow system. H from radiofrequency discharge of H_2 in Ar. [H] determined by titration with NO_2. Reaction rate monitored by change in [H].
-	-	8.74	298 K 750-1500 torr	[20] J.A.Eyre, T.Hikida and L.M.Dorfman, J.Chem.Phys., 53, 1281 (1970). Static system. H from pulse radiolysis of H_2. [H] determined as in reference [10] above. High pressure limiting rate constant obtained.
-	-	[7.60]	298 K 250-300 torr	[21] R.D.Penzhorn and H.L.Sandoval, J.Phys.Chem., 74, 2065 (1970). Static system. H from photolysis of HBr. Reaction rate determined by comparison of hydrogen yield obtained by photolysis of HBr alone and with added olefin. Rate constant relative to reaction $H + HBr \xrightarrow{2} H_2 + Br$ for which we have taken $\log k_2 = 10.8 - 3.0/2.3RT$, calculated from $\log k_{-2} = 11.43 - 19.7/2.3RT$ [G.Fettis and J.H.Knox, Prog.Reaction Kinetics, 2, 3 (1964)] and the thermodynamic data. The resultant rate constant for $H + C_2H_4$ is obviously too low, even allowing for a possible slight pressure effect, and casts doubt on the calculated value of k_2 for which there appears to be no established experimental value [A.F.Trotman-Dickenson and G.S.Milne, NSRDS-NBS 9, 1967; V.N.Kondratiev, Handbook of Rate Constants of Gas Reactions, Nauka Publishing House, Moscow, 1970].
-	-	8.92	298 K 5-500 torr	[22] M.J.Kurylo, N.C.Peterson and W.Braun, J.Chem.Phys., 53, 2776 (1970). See reference [11]. H from flash photolysis of C_2H_4 (stoichiometric corrections avoided at low [H]). High pressure limiting rate constant calculated by extrapolation of plot of 1/P *vs* $1/k_1$. Extensive data on pressure effect with He as third body.
-	-	8.43	290 K 8-16 torr	[23] M.P.Halstead, D.A.Leathard, R.M.Marshall and J.H.Purnell, Proc.Roy.Soc.(London) Ser.A., 316, 575 (1970). Flow system. H from microwave discharge of H_2-Ar mixture. [H] determined by titration with nitrosyl chloride. Rate constants obtained from computer simulation of rate data from product analyses (gas chromatography) fitted to proposed reaction scheme. This is the only flow discharge system in which full product analyses were carried out.

log A	E	log k	Conditions	Reference and Experimental Method
-	-	8.37	298 K 1-25 torr	[24] J.R.Barker, D.G.Keil, J.V.Michael and D.T.Osborne, J.Chem.Phys., 52, 2079 (1970). Three separate experimental techniques: (i) Flow system. See reference [14]. (ii) Static system. H from mercury photosensitised decomposition of H_2. See reference [15]. (iii) Flow system. H from microwave discharge in H_2. [H] determined by titration with NO_2. Rate monitored by time-of-flight mass spectrometry.
-	-	8.38	298 K 10-15 torr	[25] J.A.Cowfer, D.G.Keil, J.V.Michael and C.Yeh, J.Phys.Chem., 75, 1584 (1971). Same method as for reference [24(ii)]. Rate constants measured relative to rate of reaction $H + CH\equiv CH \xrightarrow{3} C_2H_3$ and calculated from $k_1/k_3 = 8.6$, taking $\log k_3 = 7.43$ at 298 K.

PREFERRED RATE PARAMETERS

$\log k = 10.97 - 2.8/2.3RT$

$= 8.92$ at 298 K

COMMENTS

The preferred Arrhenius parameters were derived partly from data for the reverse reaction coupled with the thermodynamic data and partly from the preferred room-temperature rate constant [22]. Several studies of the reaction $C_2H_5 \xrightarrow{-1} H + C_2H_4$ have been made. Benson and O'Neal [NSRDS-NBS 21, 1970] recommend $\log (k_{-1}/s^{-1}) = 13.5 - 40.7/2.3RT$, which, in conjunction with the entropy change, $\Delta S°_{(300)} = -20.0$ cal mol^{-1} K^{-1} (standard state 1 atm), leads to the above preferred value, $\log A_1 = 10.97$. From this A-factor and the preferred rate constant at 298 K, the activation energy is calculated to be 2.8 kcal mol^{-1}. It should be noted that there is a discrepancy between the above value of A_{-1} and the most recent experimental value, $\log (A_{-1}/s^{-1}) = 14.4$ [L.F.Loucks and K.J.Laidler, Canad.J.Chem., 45, 2795 (1967)]. At the same time, the preferred value of $E_1 = 2.8$ kcal mol^{-1} is slightly higher than the value 1.4 kcal mol^{-1}, calculated from $E_{-1} = 40.7$ kcal mol^{-1} and the $\Delta H°$ for the reaction, but this latter calculated value of E_1 is subject to considerable errors (± 3-4 kcal mol^{-1}).

The selection of a preferred rate constant at 298 K presents some difficulty as seen from the many attempts to measure this rate constant. The initial addition reaction gives rise to a vibrationally excited ethyl radical which can undergo the reverse decomposition reaction unless collisionally deactivated:

$$H + C_2H_4 \longrightarrow C_2H_5^*$$
$$C_2H_5^* \longrightarrow H + C_2H_4$$
$$C_2H_5^* + M \longrightarrow C_2H_5 + M$$

Since the C_2H_5 radical is a small entity, most of the studies of the addition reaction have been made in pressure regions where the

ΔH_f^o, ΔH^o and E in kcal mol^{-1} (kJ mol^{-1}); S^o, C_p^o, ΔS^o and ΔC_p^o in cal mol^{-1} K^{-1} (J mol^{-1} K^{-1}); A and k in l mol^{-1} s^{-1}

bimolecular rate constant for the addition reaction (k_1) is pressure-dependent. The most comprehensive studies of the effect of pressure on k_1 have been carried out by Braun and co-workers, [11,22], who have derived a value of the rate constant at infinite He pressure, $\log k_1 = 8.92 \pm 0.08$ at 298 K, from plots of $1/k_1$ against $1/P$, where P is the total pressure. These plots show marked curvature in contrast to previous erroneous 'linear' extrapolations. The variation of k_1 with He pressure was found to be represented by the equation, $1/k_1 = 1/(8.19 \times 10^8) - (9.03 \times 10^{-9})/P^{0.51}$, where k_1 is in units of $\ell\ mol^{-1}s^{-1}$ and P is in torr.
Several of the other room-temperature rate constants obtained under high pressure conditions [6,17,20], are essentially in agreement with the extrapolated value of Braun et al. [22], notably the data of Dorfman and co-workers [20] over the pressure range 750-1500 torr. In this latter study the rate constant at room temperature, ($\log k_1 = 8.74$), was found to be independent (±10%) of pressure over this range. Braun, [private communication], has estimated from a plot of $1/k$ vs $1/P^{0.51}$ that the rate constants of Dorfman et al. would change by less than 8% over the pressure range 750-1500 torr. Both the studies of Braun [22] and Dorfman [20] were carried out under conditions of very low [H], to avoid stoichiometric corrections which have plagued previous studies.
It should be noted that the rate constant at room temperature recommended here has considerable error limits, amounting to about ±15%.

REACTION: HYDROGEN ATOM + PROPENE

$$\dot{H}(I) + CH_3CH{=}CH_2(II) \xrightarrow{1} CH_3\dot{C}HCH_3(III)$$
$$\xrightarrow{2} CH_3CH_2\dot{C}H_2(IV)$$

	(I)	(II)	(III)	(IV)
$\Delta H^o_{f(300)}$	52.1 (218.0)	4.9 (20.5)	17.3 (72.4)	20.8 (87.0)
$S^o_{(300)}$	27.4 (114.7)	63.8 (267.0)	67.2 (281.0)	68.8 (288.0)
$C^o_{p(300)}$	5.0 (20.9)	15.3 (64.0)	17.5 (73.3)	17.1 (71.5)
$C^o_{p(800)}$	5.0 (20.9)	30.7 (128.4)	33.2 (138.8)	33.7 (141.0)

	Path 1	Path 2
ΔH^o	-39.7 (-166.2)	-36.2 (-151.0)
ΔS^o	-24.0 (-100.0)	-22.4 (-93.8)
$\Delta C^o_{p(300)}$	-2.8 (-11.7)	-3.2 (-13.4)
$\Delta C^o_{p(800)}$	-2.5 (-10.5)	-2.0 (-8.4)
$<\Delta C^o_p>$	-2.7 (-11.3)	-2.6 (-10.9)

Path	log A	E	log k	Conditions	Reference and Experimental Method
1 + 2	-	-	8.12	298 K 7 torr	[1] H.W.Melville and J.C.Robb, Proc.Roy.Soc.(London) Ser.A., 196, 494 (1949). For details see $H + CH_2{=}CH_2$.
1 + 2	9.9	1.71 (7.2)	8.94	300-478 K 100 torr	[2] B.de B.Darwent and R.Roberts, Disc.Faraday Soc., 14, 55 (1953). For details see $H + CH_2{=}CH_2$.

Path	log A	E	log k	Conditions	Reference and Experimental Method
1 + 2	-	-	8.21	298 K 7 torr	[3] P.E.M.Allen, H.W.Melville and J.C.Robb, Proc.Roy.Soc.(London) Ser.A., 218, 311 (1953). For details see $H + CH_2{=}CH_2$.
1 + 2	-	-	8.48	304 K	[4] A.B.Callear and J.C.Robb, Trans.Faraday Soc., 51, 638 (1955). For details see $H + CH_2{=}CH_2$.
1 + 2	-	-	9.17	297 K 435 torr	[5] K.R.Jennings and R.J.Cvetanović, J.Chem.Phys., 35, 1233 (1961). For details see $H + CH_2{=}CH_2$. Recalculated from relative rate constants, $k_{C_3H_6}/k_{C_2H_4} = 1.78$, and preferred value for $H + C_2H_4$ (log k = 8.92 at 298 K).
1 + 2	10.66	3.0 (12.6)	9.06(410 K)	330-490 K 600 torr	[6] K.Yang, J.Amer.Chem.Soc., 84, 719 (1962); 84, 3795 (1962). For details see $H + CH_2{=}CH_2$.
1 + 2	-	-	9.09	298 K 600 torr	[17] G.R.Woolley and R.J.Cvetanović, J.Chem.Phys., 50, 4697 (1969). For details see $H + CH_2{=}CH_2$. Recalculated from relative rate constants, $k_{C_3H_6}/k_{C_2H_4} = 1.49$, and preferred value for $H + C_2H_4$ (log k = 8.92 at 298 K).
1 + 2	-	-	9.11	298 K 600 torr	[18] R.J.Cvetanović and L.C.Doyle, J.Chem.Phys., 50, 4705 (1969). For details see $H + CH_2{=}CH_2$. Recalculated from relative rate constants, $k_{C_3H_6}/k_{C_2H_4} = 1.56$, and preferred value for $H + C_2H_4$ (log k = 8.92 at 298 K).
1 + 2	-	-	8.99	298 K 5-500 torr	[26] M.J.Kurylo, N.C.Peterson and W.Braun, J.Chem.Phys., 54, 4662 (1971). See reference [22] for $H + C_2H_4$. H from flash photolysis of propene in presence of He. Extensive data on pressure and temperature effects on addition reaction. Rate constant from extrapolation of plot of 1/P vs. $1/k_1$.
1	9.79	1.21	8.98(298 K)	177-473 K 50 torr	
1 + 2	-	-	(i) 8.66 (ii) 8.76	298 K 1-5 torr 10-15 torr	[25] J.A.Cowfer, D.G.Keil, J.V.Michael and C.Yeh, J.Phys.Chem., 75, 1584 (1971) (i) Same method as for reference [24.(iii)] of $H + CH_2{=}CH_2$ system. (ii) Same method as for reference [25] of $H + CH_2{=}CH_2$ system. Rate constants measured relative to rate of reaction $H + CH{\equiv}CH \xrightarrow{3} C_2H_3$, and calculated from $k_1/k_3 = 20.8$, taking $\log k_3 = 7.43$ at 298 K.
1 + 2	-	-	(i) 8.66 (ii) 8.79	298 K 1-2.5 torr 0.6-2.2 torr	[27] E.E.Daby, H.Niki and B.Weinstock, J.Phys.Chem., 75, 1601 (1971). Flow system. H or D atoms from microwave discharge of H_2 or D_2. [H] and [Olefin] measured by time-of-flight mass spectrometer. Rate constants calculated from $k_1 = \ln [(O\ell)t_1/(O\ell)t_2]/[H](t_2 - t_1)$ where $(O\ell)t$

ΔH_f^o, ΔH^o and E in kcal mol^{-1} (kJ mol^{-1}); S^o C_p^o, ΔS^o and ΔC_p^o in cal mol^{-1} K^{-1} (J mol^{-1} K^{-1}); A and k in l mol^{-1} s^{-1}

Path	log A	E	log k	Conditions	Reference and Experimental Method
					is the concentration of olefin at time t and [H] is the average H concentration during time interval ($t_2 - t_1$). (i) Data for H addition. (ii) Data for D addition.

PREFERRED RATE PARAMETERS

$\log k_1 = [9.86] - 1.2/2.3RT$
$= 8.97$ at 298 K

$\log k_2 = [9.86 - 2.9/2.3RT]$
$= 7.74$ at 298 K

COMMENTS

Path 1: The preferred parameters are essentially the results of reference [26]. The rate constant at infinite pressure at 298 K was corrected to allow for the 5.7% of non-terminal addition of H to C_3H_6 [W.E.Falconer and W.A.Sunder, Int.J.Chem.Kinet., 3, 395 (1971)]. The measured A-factor has been adjusted to fit the k_1 at infinite pressure at 298 K. The Arrhenius parameters reported in reference [26] were determined at a constant total pressure of 50 torr which at 298 K gave rate constants to within about 10% of the high pressure limiting value. The derived Arrhenius parameters should consequently be taken as minimum values of their high-pressure limits. There are no reliable data on reaction (-1) [Benson and O'Neal, NSRDS-NBS 21, 1970; H.M.Frey and R.Walsh, Chem.Rev., 69, 103 (1969)]. The above preferred parameters for reaction (1) and the thermodynamic data yield $\log (k_{-1}/s^{-1}) = 13.3 - 41.5/2.3RT$ at $T_m = 800$ K. The calculated value of A_{-1} is lower than expected and indicates that A_1 has been under-estimated.

Path 2: The k_2 at infinite pressure is calculated from $k_1 + k_2$ from reference [26] and the percentage (5.7%) of non-terminal addition [W.E.Falconer and W.A.Sunder, Int.J.Chem.Kinet., 3, 395 (1971)]. E_2 is calculated from k_2, assuming $A_1 = A_2$. Benson and O'Neal [NSRDS-NBS 21, 1970] suggest that for the reverse reaction $\log (k_{-2}/s^{-1}) = 13.8 - 38/2.3RT$, and with the thermodynamic data this yields $\log k_2 = 10.7 - 2/2.3RT$, again indicating that the value of A_2 derived above is under-estimated.

As for the $H + C_2H_4$ reaction, the addition of H to C_3H_6 gives rise to a vibrationally excited radical. Braun and co-workers, [26], have shown that the pressure effect of He on the bimolecular rate constant (k_1) takes the same form as for $H + C_2H_4$. $1/k_1 = 1/(9.71 \times 10^8) + (9.92 \times 10^{-10})/P^{0.5}$ where k_1 is in $\ell\ mol^{-1}\ s^{-1}$ and P in torr.

The values of k_1 calculated from the results of Cvetanović and co-workers, [5,17,18] from the ratio $k_{C_3H_6}/k_{C_2H_4}$, together with our preferred value of $k_{C_2H_4}$, are significantly higher than those of Braun et al., [26]. The discrepancy in the $k_{C_3H_6}/k_{C_2H_4}$ ratios has been ascribed, [26], to the lack of an extrapolation to infinite pressure in references [5,17,18]. (See relative rate constants for H atom reactions, p.357).

REACTION: HYDROGEN ATOM + BUTENE-1

$\dot{H}(I) + CH_2{=}CHCH_2CH_3(II) \xrightarrow{1} CH_3\dot{C}HCH_2CH_3(III)$

$\xrightarrow{2} \dot{C}H_2CH_2CH_2CH_3(IV)$

	(I)	(II)	(III)	(IV)		Path 1	Path 2
$\Delta H^\circ_{f(300)}$	52.1 (218.0)	0.0 (0.0)	12.3 (51.5)	15.8 (66.2)	ΔH°	-39.8 (-166.6)	-36.3 (-151.9)
$S^\circ_{(300)}$	27.4 (114.7)	73.6 (308.0)	76.6 (320.4)	76.1 (318.4)	ΔS°	-24.4 (-102.1)	-24.9 (-104.3)
$C^\circ_{p(300)}$	5.0 (20.9)	20.6 (86.2)	23.0 (96.2)	23.2 (97.1)	$\Delta C^\circ_{p(300)}$	-2.6 (-10.9)	-2.4 (-10.5)
$C^\circ_{p(800)}$	5.0 (20.9)	42.0 (175.8)	45.4 (190.0)	45.6 (190.9)	$\Delta C^\circ_{p(800)}$	-1.6 (-6.7)	-1.4 (-5.9)
					$\langle\Delta C^\circ_p\rangle$	-2.1 (-8.8)	-1.9 (-8.0)

Path	log A	E	log k	Conditions	Reference and Experimental Method
1 + 2	-	-	9.02(297 K)	297 K 435 torr	[5] K.R.Jennings and R.J.Cvetanović, J.Chem.Phys., 35, 1233 (1961). For details see $H + CH_2{=}CH_2$. Recalculated from relative rate constants, $k_{C_4H_8\text{-}1}/k_{C_3H_6} = 1.08$ and preferred value for $H + C_3H_6$ (log k = 8.99 at 298 K).
1 + 2	-	-	8.99	298 K 600 torr	[17] G.R.Woolley and R.J.Cvetanović, J.Chem.Phys., 50, 4697 (1969). For details see $H + CH_2{=}CH_2$. Recalculated from relative rate constants, $k_{C_4H_8\text{-}1}/k_{C_3H_6} = 1.0$ and preferred value for $H + C_3H_6$ (log k = 8.99 at 298 K).
1 + 2	-	-	(i) 8.83 (ii) 8.89	298 K 2.9 torr 10-15 torr	[25] J.A.Cowfer, D.G.Keil, J.V.Michael and C.Yeh, J.Phys.Chem., 75. 1584 (1971). (i) Same method as for reference [24.(iii)] of $H + CH_2{=}CH_2$ system. (ii) Same method as for reference [25] of $H + CH_2{=}CH_2$ system. Rate constants measured relative to rate of reaction $H + CH{\equiv}CH \xrightarrow{3} C_2H_3$, and calculated from $k_1/k_3 = 29.6$, taking log $k_3 = 7.43$ at 298 K.
1 + 2	-	-	(i) 8.92 (ii) 8.92	298 K 0.4-1.2 torr 0.6-2.6 torr	[27] E.E.Daby, H.Niki and B.Weinstock, J.Phys.Chem., 75, 1601 (1971). For details see $H + CH_3CH{=}CH_2$. (i) Data for H addition. (ii) Data for D addition.

PREFERRED RATE PARAMETERS

log k_{1+2} = 8.94 at 298 K
log k_1 = 8.91 at 298 K
log k_2 = 7.70 at 298 K

ΔH°_f, ΔH° and E in kcal mol^{-1} (kJ mol^{-1}); S°, C°_p, ΔS° and ΔC°_p in cal mol^{-1} K^{-1} (J mol^{-1} K^{-1}); A and k in l mol^{-1} s^{-1}

COMMENTS

The preferred value of k_{1+2} is the mean of the rate constants from references [5,17,25(i),27(i)]. The separate values of k_1 and k_2 were calculated on the basis of 5.7% non-terminal addition of H to butene-1 [W.E.Falconer and W.A.Sunder, Int.J.Chem.Kinet., 3, 395 (1971)].

REACTION: HYDROGEN ATOM + *cis*-BUTENE-2 $\dot{H}(I) + cis\text{-}CH_3CH{=}CHCH_3(II) \longrightarrow CH_3\dot{C}HCH_2CH_3(III)$

	(I)	(II)	(III)		
$\Delta H^\circ_{f(300)}$	52.1 (218.0)	-1.9 (-7.9)	12.3 (51.5)	ΔH°	-37.9 (-158.5)
$S^\circ_{(300)}$	27.4 (114.6)	73.6 (308.2)	76.7 (320.8)	ΔS°	-24.3 (-101.7)
$C^\circ_{p(300)}$	5.0 (20.9)	19.4 (81.2)	23.0 (96.3)	ΔC°_p	-1.4 (-5.9)

log A	E	log k	Conditions	Reference and Experimental Method
-	-	8.80	298 K 7 torr	[28] H.W.Melville and J.C.Robb, Proc.Roy.Soc.(London) Ser.A., 202, 181 (1950). For details see H + $CH_2{=}CH_2$.
-	-	8.90	298 K 7 torr	[3] P.E.M.Allen, H.W.Melville and J.C.Robb, Proc.Roy.Soc.(London) Ser.A., 218, 311 (1953). For details see H + $CH_2{=}CH_2$.
-	-	8.66	297 K 435 torr	[5] K.R.Jennings and R.J.Cvetanović, J.Chem.Phys., 35, 1233 (1961). For details see H + $CH_2{=}CH_2$. Recalculated from relative rate constants, $k_{cis\text{-}C_4H_8\text{-}2}/k_{C_3H_6} = 0.47$, and preferred value for H + C_3H_6 (log $k_{C_3H_6}$ = 8.99 at 298 K).
11.12	4.3 (18.0)	7.97(298 K)	298-475 K 600 torr	[6] K.Yang, J.Amer.Chem.Soc., 84, 3795 (1962). For details see H + $CH_2{=}CH_2$.
-	-	8.69	300 K 2.5 torr	[29] E.E.Daby and H.Niki, J.Chem.Phys., 51, 1255 (1969). Flow system. H from flash photolysis of H_2. [H] determined by NO_2 titration. Reaction rate from disappearance of olefin measured by mass spectrometry.
-	-	(i) 8.59 (ii) 8.40	298 K 3.2 torr 10-15 torr	[25] J.A.Cowfer, D.G.Keil, J.V.Michael and C.Yeh, J.Phys.Chem., 75, 1584 (1971). (i) See reference [24.(iii)] of H + $CH_2{=}CH_2$ system. (ii) See reference [25] of H + $CH_2{=}CH_2$ system. Rate constants measured relative to rate of reaction H + CH≡CH $\xrightarrow{3}$ C_2H_3, and calculated from k_1/k_3 = 9.1, taking log k_3 = 7.43 at 298 K.

log A	E	log k	Conditions	Reference and Experimental Method
-	-	(i) 8.68 (ii) 8.61	298 K 1.1-2.8 torr 0.6-2.6 torr	[27] E.E.Daby, H.Niki and B.Weinstock, J.Phys.Chem., 75, 1601 (1971). For details see $H + CH_3CH{=}CH_2$. (i) Data for H addition. (ii) Data for D addition.

PREFERRED RATE PARAMETERS

log k = 8.66 at 298 K

COMMENTS

The preferred value of k_1 is the average of the rate constants reported in references [5,25(i),27(i),29]. The Arrhenius parameters of reference [6] appear to be over-estimates, on the basis of the relative rate constants for H + alkenes. There are no kinetic data on the reverse radical decomposition.

REACTION: HYDROGEN ATOM + TRANS-BUTENE-2 $\dot{H}$(I) + trans-$CH_3CH{=}CHCH_3$(II) ⟶ $CH_3\dot{C}HCH_2CH_3$(III)

	(I)	(II)	(III)		
$\Delta H^o_{f(300)}$	52.1 (218.0)	-2.9 (-12.1)	12.3 (51.5)	ΔH^o	-36.9 (-154.4)
$S^o_{(300)}$	27.4 (114.6)	72.4 (302.8)	76.7 (320.8)	ΔS^o	-23.1 (-96.7)
$C^o_{p(300)}$	5.0 (20.9)	20.7 (86.6)	23.0 (96.3)	ΔC^o_p	-2.7 (-11.3)

log A	E	log k	Conditions	Reference and Experimental Method
-	-	8.76	298 K 7 torr	[28] H.W.Melville and J.C.Robb, Proc.Roy.Soc.(London) Ser.A., 202, 181 (1950). For details see $H + CH_2{=}CH_2$.
-	-	8.80	298 K 7 torr	[3] P.E.M.Allen, H.W.Melville and J.C.Robb, Proc.Roy.Soc.(London) Ser.A., 218, 311 (1953). For details see $H + CH_2{=}CH_2$.
-	-	8.76	297 K 435 torr	[5] K.R.Jennings and R.J.Cvetanović, J.Chem.Phys., 35, 1233 (1961). For details see $H + CH_2{=}CH_2$. Recalculated from relative rate constants, $k_{trans\text{-}C_4H_8\text{-}2}/k_{C_3H_6} = 0.59$, and preferred value for $H + C_3H_6$ ($\log k_{C_3H_6} = 8.99$ at 298 K).

ΔH^o_f, ΔH^o and E in kcal mol^{-1} (kJ mol^{-1}); S^o, C^o_p, ΔS^o and ΔC^o_p in cal mol^{-1} K^{-1} (J mol^{-1} K^{-1}); A and k in l mol^{-1} s^{-1}

log A	E	log k	Conditions	Reference and Experimental Method
11.14	4.3 (18.0)	8.00	298-555 K 600 torr	[6] K.Yang, J.Amer.Chem.Soc., 84, 3795 (1962). For details see H + CH_2=CH_2.
-	-	8.78	298 K 5-50 torr	[11] W.Braun and M.Lenzi, Disc.Faraday Soc., 44, 252 (1967). For details see H + CH_2=CH_2.
-	-	8.73	300 K 2.5 torr	[29] E.E.Daby and H.Niki, J.Chem.Phys., 51, 1255 (1969). For details see H + cis-butene-2.
-	-	(i) 8.64 (ii) 8.52	298 K 1.8 torr 10-15 torr	[25] J.A.Cowfer, D.G.Keil, J.V.Michael and C.Yeh, J.Phys.Chem., 75, 1584 (1971). (i) See reference [24(iii)] of H + CH_2=CH_2 system. (ii) See reference [25] of H + CH_2=CH_2 system. Rate constants measured relative to rate of reaction, H + CH≡CH $\xrightarrow{3}$ C_2H_3, and calculated from k_1/k_3 = 11.9, taking log k_3 = 7.43 at 298 K.
-	-	(i) 8.73 (ii) 8.66	298 K 0.4-1.9 torr 0.7-2.2 torr	[27] E.E.Daby, H.Niki and B.Weinstock, J.Phys.Chem., 75, 1601 (1971). Details as for H + CH_3CH=CH_2. (i) Data for H addition. (ii) Data for D addition.

PREFERRED RATE PARAMETERS

log k = 8.75 at 298 K

COMMENTS

The preferred rate constant is the average of the rate constants reported in references [5,11,25(i),27(i),29].

REACTION: HYDROGEN ATOM + 2-METHYLPROPENE (ISOBUTENE)

$$\dot{H}(I) + (CH_3)_2C{=}CH_2(II) \xrightarrow{1} (CH_3)_3\dot{C}(III)$$
$$\xrightarrow{2} (CH_3)_2\dot{C}HCH_2(IV)$$

	(I)	(II)	(III)	(IV)		Path 1	Path 2
$\Delta H^o_{f(300)}$	52.1 (218.0)	-3.6 (-14.9)	6.8 (28.5)	13.7 (57.3)	ΔH^o	-41.7 (-174.4)	-34.8 (-145.6)
$S^o_{(300)}$	27.4 (114.7)	71.3 (298.4)	73.9 (309.0)	75.6 (316.4)	ΔS^o	-24.8 (-103.8)	-23.1 (-96.7)
$C^o_{p(300)}$	5.0 (20.9)	21.6 (90.4)	22.6 (94.6)	22.9 (95.8)	$\Delta C^o_{p(300)}$	-4.0 (-16.7)	-3.7 (-15.5)
$C^o_{p(800)}$	5.0 (20.9)	41.9 (177.0)	43.0 (180.0)		$\Delta C^o_{p(800)}$	-3.9 (-16.3)	

Path	log A	E	log k	Conditions	Reference and Experimental Method
1 + 2	-	-	8.63	298 K 7 torr	[1] H.W.Melville and J.C.Robb, Proc.Roy.Soc.(London) Ser.A., 196, 494 (1949). For details see H + $CH_2{=}CH_2$.
1 + 2	-	-	8.76	298 K 7 torr	[3] P.E.M.Allen, H.W.Melville and J.C.Robb, Proc.Roy.Soc.(London) Ser.A., 218, 311 (1953). For details see H + $CH_2{=}CH_2$.
1 + 2	-	-	9.38	297 K 435 torr	[5] K.R.Jennings and R.J.Cvetanović, J.Chem.Phys., 35, 1233 (1961). For details see H + $CH_2{=}CH_2$. Recalculated from relative rate constants, $k_{iso\text{-}C_4H_8}/k_{C_3H_6}$ = 2.47, and preferred value for H + C_3H_6 (log $k_{C_3H_6}$ = 8.99 at 298 K).
1 + 2	10.58	1.6 (6.7)	9.82(462 K)	425-500 K 600 torr	[6] K.Yang, J.Amer.Chem.Soc., 84, 3795 (1962). For details see H + $CH_2{=}CH_2$.
1 + 2	10.53	1.36 (5.95)	9.67(345 K) 9.53(298 K)	285-405 K 2 torr	[19] D.G.Dalgleish and J.H.Knox, Chem.Comm., 917 (1966); Int.J.Chem.Kinet., 1, 69 (1969). For details see H + $CH_2{=}CH_2$.
1 + 2	-	-	9.36	298 K 5-50 torr	[11] W.Braun and M.Lenzi, Disc.Faraday Soc., 44, 252 (1967). For details see H + $CH_2{=}CH_2$.
1 + 2	-	-	9.40	298 K 600 torr	[17] G.R.Woolley and R.J.Cvetanović, J.Chem.Phys., 50, 4697 (1969). For details see H + $CH_2{=}CH_2$. Recalculated from relative rate constants, $k_{iso\text{-}C_4H_8}/k_{C_3H_6}$ = 2.56, and preferred value for H + C_3H_6,(log $k_{C_3H_6}$ = 8.99 at 298 K).

ΔH^o_f, ΔH^o and E in kcal mol^{-1} (kJ mol^{-1}); S^o, C^o_p, ΔS^o and ΔC^o_p in cal mol^{-1} K^{-1} (J mol^{-1} K^{-1}); A and k in l mol^{-1} s^{-1}

Path	log A	E	log k	Conditions	Reference and Experimental Method
1 + 2	-	-	9.40	298 K 600 torr	[18] R.J.Cvetanović and L.C.Doyle, J.Chem.Phys., 50, 4705 (1969). For details see H + $CH_2{=}CH_2$. Recalculated from relative rate constants, $k_{iso\text{-}C_4H_8}/k_{C_3H_6} = 2.56$, and preferred value for H + C_3H_6 ($\log k_{C_3H_6} = 8.99$ at 298 K).
1 + 2	-	-	9.32	298 K 10-15 torr	[25] J.A.Cowfer, D.G.Keil, J.V.Michael and C.Yeh, J.Phys.Chem., 75, 1584 (1971). Details as for H + C_2H_4 system, reference [24(ii)]. Rate constants measured relative to rate of reaction H + CH≡CH $\xrightarrow{3}$ C_2H_3, and calculated from $k_1/k_3 = 74.5$, taking $\log k_3 = 7.43$ at 298 K.
1 + 2	-	-	(i) ~8.6 (ii) 9.31	298 K ? 1-2.2 torr	[27] E.E.Daby, H.Niki and B.Weinstock, J.Phys.Chem., 75, 1601 (1971). For details see H + $CH_3CH{=}CH_2$. (i) Data for H addition. (ii) Data for D addition.

PREFERRED RATE PARAMETERS

$\log k_1 = 10.49 - 1.5/2.3RT$

$= 9.39$ at 298 K

$\log k_2 = 8.1$ at 298 K

COMMENTS

The preferred rate constant at 298 K is the average value of the rate constants of references [5,11,17,18,25]. The preferred value of $E_1 = 1.5$ kcal mol^{-1} is the average of the results of references [6] and [19] and the preferred value of A_1 was then calculated from k_1 and E_1. From the preferred Arrhenius parameters, together with the thermodynamic data, it can be calculated that $\log(k_{-1}/s^{-1}) = 14.09 - 44.6/2.3RT$ at $T_m = 800$ K. The only experimental value is $\log(k_{-1}/s^{-1}) = 15.5 - 43.6/2.3RT$, [S.W.Benson and H.E.O'Neal, NSRDS-NBS 21, 1970], for which the A factor is undoubtedly too high. The preferred value of k_2 is calculated from k_1 on the basis of 0.5% non-terminal addition of H to isobutene [W.E.Falconer and W.A.Sunder, Int.J.Chem.Kinet., 3, 395 (1971)].

REACTION: HYDROGEN ATOM + PENTENE-1

$$\dot{H}(I) + CH_2{=}CHCH_2CH_2CH_3(II) \xrightarrow{1} CH_3\dot{C}HCH_2CH_2CH_3(III)$$
$$\xrightarrow{2} \dot{C}H_2CH_2CH_2CH_2CH_3(IV)$$

	(I)	(II)	(III)	(IV)		Path 1	Path 2
$\Delta H^o_{f(300)}$	52.1 (218.0)	-4.9 (-20.5)	7.4 (30.9)	10.9 (45.6)	ΔH^o	-39.8 (-166.5)	-36.3 (-151.8)
$S^o_{(300)}$	27.4 (114.7)	83.0 (347.0)	86.0 (359.6)	87.7 (366.9)	ΔS^o	-24.4 (-102.1)	-22.7 (-95.0)
$C^o_{p(300)}$	5.0 (20.9)	26.1 (109.1)	28.5 (119.2)	28.7 (120.0)	$\Delta C^o_{p(300)}$	-2.6 (-10.9)	-2.4 (-10.0)
$C^o_{p(800)}$	5.0 (20.9)	53.0 (221.6)	56.4 (236.0)	56.7 (237.3)	$\Delta C^o_{p(800)}$	-1.6 (-6.7)	-1.3 (-5.4)
					$<\Delta C^o_p>$	-2.1 (-8.8)	-1.9 (-8.0)

Path	log A	E	log k	Conditions	Reference and Experimental Method
1 + 2	-	-	8.94	298 K 600 torr	[18] R.J.Cvetanović and L.C.Doyle, J.Chem.Phys., 50, 4705 (1969). For details see $H + CH_2{=}CH_2$. Recalculated from relative rate constants, $k_{C_5H_{10}-1}/k_{C_3H_6} = 0.89$, and preferred value for $H + C_3H_6$ (log k = 8.99 at 298 K).
1 + 2	-	-	8.89	298 K 10-15 torr	[25] J.A.Cowfer, D.G.Keil, J.V.Michael and C.Yeh, J.Phys.Chem., 75, 1584 (1971). For details see $H + C_2H_4$. Rate constants measured relative to rate of reaction, $H + CH{\equiv}CH \xrightarrow{3} C_2H_3$, and calculated from $k_1/k_3 = 28.6$, taking $\log k_3 = 7.43$ at 298 K.

PREFERRED RATE PARAMETERS

$\log k_{1+2} = 8.92$ at 298 K

COMMENTS

The agreement between the two determinations of k_{1+2} is excellent.

ΔH^o_f, ΔH^o and E in kcal mol^{-1} (kJ mol^{-1}); S^o, C^o_p, ΔS^o and ΔC^o_p in cal mol^{-1} K^{-1} (J mol^{-1} K^{-1}); A and k in l mol^{-1} s^{-1}

REACTION: HYDROGEN ATOM + *cis*-PENTENE-2

$\dot{H}(I) + cis\text{-}CH_3CH{=}CHCH_2CH_3(II) \xrightarrow{1} CH_3CH_2\dot{C}HCH_2CH_3(III)$

$\xrightarrow{2} CH_3\dot{C}HCH_2CH_2CH_3(IV)$

	(I)	(II)	(III)	(IV)		Path 1	Path 2
$\Delta H^\circ_{f(300)}$	52.1 (218.0)	-6.6 (-27.6)	7.4 (31.0)	7.4 (31.0)	ΔH°	-38.1 (-159.5)	-38.1 (-159.5)
$S^\circ_{(300)}$	27.4 (114.7)	82.3 (344.6)	86.0 (360.1)	86.0 (360.1)	ΔS°	-23.7 (-99.2)	-23.7 (-99.2)
$C^\circ_{p(300)}$	5.0 (20.9)	25.8 (108.0)	28.5 (119.3)	28.5 (119.3)	ΔC°_p	-2.3 (-9.6)	-2.3 (-9.6)

Path	log A	E	log k	Conditions	Reference and Experimental Method
1 + 2	-	-	8.54	298 K 7 torr	[1] H.W.Melville and J.C.Robb, Proc.Roy.Soc.(London) Ser.A., 196, 494 (1949). For details see $H + CH_2{=}CH_2$.
1 + 2	-	-	8.76	298 K 7 torr	[3] P.E.M.Allen, H.W.Melville and J.C.Robb, Proc.Roy.Soc.(London) Ser.A., 218, 311 (1953). For details see $H + CH_2{=}CH_2$.
1 + 2	-	-	8.58	298 K 600 torr	[18] R.J.Cvetanović and L.C.Doyle, J.Chem.Phys., 50, 4705 (1969). For details see $H + CH_2{=}CH_2$. Recalculated from relative rate constants, $k_{cis\text{-}C_5H_{10}\text{-}2}/k_{C_3H_6} = 0.39$, and preferred value for $H + C_3H_6$ (log k = 8.99 at 298 K).

PREFERRED RATE PARAMETERS

$\log k_{1+2} = 8.58$ at 298 K

REACTION: HYDROGEN ATOM + TRANS-PENTENE-2

$\dot{H}(I)$ + trans-$CH_3CH{=}CHCH_2CH_3(II) \xrightarrow{1} CH_3CH_2\dot{C}HCH_2CH_3(III)$

$\xrightarrow{2} CH_3\dot{C}HCH_2CH_2CH_3(IV)$

	(I)	(II)	(III)	(IV)
$\Delta H^\circ_{f(300)}$	52.1 (218.0)	-7.7 (-32.2)	7.4 (31.0)	7.4 (31.0)
$S^\circ_{(300)}$	27.4 (114.7)	82.3 (344.6)	86.0 (360.1)	86.0 (360.1)
$C^\circ_{p(300)}$	5.0 (20.9)	25.8 (108.0)	28.5 (119.3)	28.5 (119.3)
$C^\circ_{p(800)}$	5.0 (20.9)	52.6 (220.4)	56.4 (236.3)	56.4 (236.3)

	Path 1	Path 2
ΔH°	-37.0 (-154.9)	-37.0 (-154.9)
ΔS°	-23.7 (-99.2)	-23.7 (-99.2)
$\Delta C^\circ_{p(300)}$	-2.3 (-9.6)	-2.3 (-9.6)
$\Delta C^\circ_{p(800)}$	-1.2 (-5.0)	-1.2 (-5.0)
$\langle\Delta C^\circ_p\rangle$	-1.8 (-7.5)	-1.8 (-7.5)

Path	log A	E	log k	Conditions	Reference and Experimental Method
1 + 2	-	-	8.61	298 K 600 torr	[18] R.J.Cvetanović and L.C.Doyle, J.Chem.Phys., 50, 4705 (1969). For details see $H + CH_2{=}CH_2$. Recalculated from relative rate constants, $k_{trans\text{-}C_5H_{10}\text{-}2}/k_{C_3H_6} = 0.44$, and preferred value for $H + C_3H_6$ (log k = 8.99 at 298 K).

PREFERRED RATE PARAMETERS

Reasonable as given above

ΔH°_f, ΔH° and E in kcal mol^{-1} (kJ mol^{-1}); S°, C°_p, ΔS° and ΔC°_p in cal mol^{-1} K^{-1} (J mol^{-1} K^{-1}); A and k in l mol^{-1} s^{-1}

REACTION: HYDROGEN ATOM + 2-METHYLBUTENE-1

$\dot{H}(I) + CH_2{=}C(CH_3)CH_2CH_3(II) \xrightarrow{1} CH_3\dot{C}(CH_3)CH_2CH_3(III)$

$\xrightarrow{2} \dot{C}H_2CH(CH_3)CH_2CH_3(IV)$

	(I)	(II)	(III)	(IV)		Path 1	Path 2
$\Delta H^\circ_{f(300)}$	52.1 (218.1)	-8.3 (-34.8)	1.8 (7.5)	8.8 (36.8)	ΔH°	-42.0 (-175.9)	-35.0 (-146.5)
$S^\circ_{(300)}$	27.4 (114.7)	81.1 (339.6)	83.3 (348.8)	85.0 (355.9)	ΔS°	-25.2 (-105.5)	-23.5 (-98.4)
$C^\circ_{p(300)}$	5.0 (20.9)	26.7 (111.8)	28.1 (117.7)	28.4 (118.9)	ΔC°_p	-3.6 (-15.1)	-3.3 (-13.8)

Path	log A	E	log k	Conditions	Reference and Experimental Method
				298 K	[27] E.E.Daby, H.Niki and B.Weinstock, J.Phys.Chem., 75, 1601 (1971).
1 + 2	-	-	(i) 8.96	0.7-1.3 torr	For details see H + $CH_3CH{=}CH_2$.
			(ii) 9.31	1.2-2.6 torr	(i) Data for H addition. (ii) Data for D addition.

PREFERRED RATE PARAMETERS

Reasonable as given above

REACTION: HYDROGEN ATOM + 3-METHYLBUTENE-1

$\dot{H}(I) + CH_2{=}CHCH(CH_3)_2(II) \xrightarrow{1} CH_3\dot{C}HCH(CH_3)_2(III)$

$\xrightarrow{2} \dot{C}H_2CH_2CH(CH_3)_2(IV)$

	(I)	(II)	(III)	(IV)		Path 1	Path 2
$\Delta H^\circ_{f(300)}$	52.1 (218.1)	-6.8 (-28.5)	5.3 (22.2)	8.8 (36.8)	ΔH°	-40.0 (167.5)	-36.5 (-152.8)
$S^\circ_{(300)}$	27.4 (114.7)	[80 (335)]	83.3 (348.8)	85.0 (355.9)	ΔS°	[-24 (-100)]	[-22 (-92)]
$C^\circ_{p(300)}$	5.0 (20.9)	[26.8 (112.2)]	28.3 (118.5)	28.4 (118.9)	ΔC°_p	[-3.5 (-14.7)]	[-3.4 (-14.2)]

Path	log A	E	log k	Conditions	Reference and Experimental Method
				298 K	[27] E.E.Daby, H.Niki and B.Weinstock, J.Phys.Chem., 75, 1601 (1971).
1 + 2	-	-	(i) 8.87	0.6-2.7 torr	For details see H + $CH_3CH{=}CH_2$.
			(ii) 8.88	0.6-2.6 torr	(i) Data for H addition. (ii) Data for D addition.

PREFERRED RATE PARAMETERS

Reasonable as given above

REACTION: HYDROGEN ATOM + 2-METHYLBUTENE-2 (TRIMETHYLETHYLENE)

$\dot{H}(I) + (CH_3)_2C{=}CHCH_3(II) \xrightarrow{1} (CH_3)_2CH\dot{C}HCH_3(III)$

$\xrightarrow{2} (CH_3)_2\dot{C}CH_2CH_3(IV)$

	(I)	(II)	(III)	(IV)		Path 1	Path 2
$\Delta H^o_{f(300)}$	52.1 (218.0)	-11.3 (-47.3)	5.3 (22.2)	1.8 (7.6)	ΔH^o	-35.5 (-148.5)	-39.0 (-163.2)
$S^o_{(300)}$	27.4 (114.7)	79.9 (334.4)	83.3 (348.5)	84.6 (353.8)	ΔS^o	-24.0 (-100.4)	-22.7 (-95.4)
$C^o_{p(300)}$	5.0 (20.9)	26.8 (112.1)	28.3 (118.4)	28.1 (117.6)	ΔC^o_p	-3.5 (-14.7)	-3.7 (-15.5)

Path	log A	E	log k	Conditions	Reference and Experimental Method
1 + 2	-	-	8.75	298 K 8 torr	[28] H.W.Melville and J.C.Robb, Proc.Roy.Soc.(London) Ser.A., 202, 181 (1950). For details see $H + CH_2{=}CH_2$.
1 + 2	-	-	8.87	298 K 8 torr	[3] P.E.M.Allen, H.W.Melville and J.C.Robb, Proc.Roy.Soc.(London) Ser.A., 218, 311 (1953). For details see $H + CH_2{=}CH_2$.
1 + 2	-	-	9.00	298 K 600 torr	[18] R.J.Cvetanović and L.C.Doyle, J.Chem.Phys., 50, 4705 (1969). For details see $H + CH_2{=}CH_2$. Recalculated from relative rate constants, $k_{(CH_3)_2C=CHCH_3}/k_{C_3H_6} = 1.03$, and the preferred value for $H + C_3H_6$ (log k = 8.99 at 298 K).
1 + 2	-	-	(i) 8.91 (ii) 8.97	298 K 0.7-1.3 torr 0.6-2.6 torr	[27] E.E.Daby, H.Niki and B.Weinstock, J.Phys.Chem., 75, 1601 (1971). For details see $H + CH_3CH{=}CH_2$. (i) Data for H addition. (ii) Data for D addition.

PREFERRED RATE PARAMETERS

$\log k_{1+2} = 8.96$ at 298 K

COMMENTS

The preferred rate constant is the average of the rate constants of references [18] and [27(i)], which are in good agreement.

ΔH^o_f, ΔH^o and E in kcal mol^{-1} (kJ mol^{-1}); S^o, C^o_p, ΔS^o and ΔC^o_p in cal mol^{-1} K^{-1} (J mol^{-1} K^{-1}); A and k in l mol^{-1} s^{-1}

REACTION: HYDROGEN ATOM + 2,3-DIMETHYLBUTENE-2 (TETRAMETHYLETHYLENE) $\dot{H}(I) + (CH_3)_2C{=}C(CH_3)_2(II) \longrightarrow (CH_3)_2CH\dot{C}(CH_3)_2(III)$

	(I)	(II)	(III)		
$\Delta H^\circ_{f(300)}$	52.1 (218.0)	-19.6 (-81.9)	-5.2 (-21.7)	ΔH°	-37.7 (-157.7)
$S^\circ_{(300)}$	27.4 (114.7)	87.4 (365.3)	90.0 (376.0)	ΔS°	-24.8 (-103.8)
$C^\circ_{p(300)}$	5.0 (20.9)	33.0 (137.9)	34.3 (143.2)	ΔC°_p	-3.7 (-15.5)

log A	E	log k	Conditions	Reference and Experimental Method
-	-	8.62	298 K 7 torr	[28] H.W.Melville and J.C.Robb, Proc.Roy.Soc.(London) Ser.A., 202, 181 (1950). For details see $H + CH_2{=}CH_2$.
-	-	8.68	298 K 7 torr	[3] P.E.M.Allen, H.W.Melville and J.C.Robb, Proc.Roy.Soc.(London) Ser.A., 218, 311 (1953). For details see $H + CH_2{=}CH_2$.
-	-	8.90	297 K 435 torr	[5] K.R.Jennings and R.J.Cvetanović, J.Chem.Phys., 35, 1233 (1961). For details see $H + CH_2{=}CH_2$. Recalculated from relative rate constants, $k_{(CH_3)_2C=C(CH_3)_2}/k_{C_3H_6} = 0.81$, and the preferred value for $H + C_3H_6$ (log k = 8.99 at 298 K).
-	-	8.51	300 K 600 torr	[6] K.Yang, J.Amer.Chem.Soc., 84, 3795 (1962). For details see $H + CH_2{=}CH_2$.
-	-	8.93	298 K 600 torr	[18] R.J.Cvetanović and L.C.Doyle, J.Chem.Phys., 50, 4705 (1969). For details see $H + CH_2{=}CH_2$. Recalculated from relative rate constants, $k_{(CH_3)_2C=C(CH_3)_2}/k_{C_3H_6} = 0.87$, and the preferred value for $H + C_3H_6$ (log k = 8.99 at 298 K).
-	-	(i) 8.85 (ii) 8.93	298 K 0.7-1.3 torr 0.6-2.6 torr	[27] E.E.Daby, H.Niki and B.Weinstock, J.Phys.Chem., 75, 1601 (1971). For details see $H + CH_3CH{=}CH_2$. (i) Data for H addition. (ii) Data for D addition.

PREFERRED RATE PARAMETERS

log k = 8.89 at 298 K

COMMENTS

The preferred rate constant is the average of the rate constants reported in references [5,18,27(i)].

REACTION: HYDROGEN ATOM + 2,3,3-TRIMETHYLBUTENE-1

$$\dot{H}(I) + CH_2{=}C(CH_3)C(CH_3)_2CH_3(II) \xrightarrow{1} \dot{C}H_2CH(CH_3)C(CH_3)_2CH_3(III)$$
$$\xrightarrow{2} (CH_3)_2\dot{C}C(CH_3)_2CH_3(IV)$$

	(I)	(II)	(III)	(IV)		Path 1	Path 2
$\Delta H^o_{f(300)}$	52.1 (218.0)	-22.0 (-92.1)	-5.9 (-24.7)	-11.9 (-49.8)	ΔH^o	-36.0 (-150.7)	-42.0 (-175.9)
$S^o_{(300)}$	27.4 (114.7)	93.0 (389.4)	96.9 (405.8)	95.2 (398.5)	ΔS^o	-23.5 (-98.4)	-25.2 (-105.5)
$C^o_{p(300)}$	5.0 (20.9)	38.0 (159.1)	39.7 (166.2)	39.4 (165.0)	ΔC^o_p	-3.3 (-13.8)	-3.6 (-15.1)

Path	log A	E	log k	Conditions	Reference and Experimental Method
1 + 2	-	-	8.78	298 K 7 torr	[1] H.W.Melville and J.C.Robb, Proc.Roy.Soc.(London) Ser.A., 196, 494 (1949). For details see $H + CH_2{=}CH_2$.
1 + 2	-	-	8.84	298 K 7 torr	[3] P.E.M.Allen, H.W.Melville and J.C.Robb, Proc.Roy.Soc.(London) Ser.A., 218, 311 (1953). For details see $H + CH_2{=}CH_2$.

PREFERRED RATE PARAMETERS

$\log k_1 = [9.20]$ at 298 K

COMMENTS

Tentative value based on H + isobutene.

ΔH^o_f, ΔH^o and E in kcal mol^{-1} (kJ mol^{-1}); S^o, C^o_p, ΔS^o and ΔC^o_p in cal mol^{-1} K^{-1} (J mol^{-1} K^{-1}); A and k in l mol^{-1} s^{-1}

REACTION: HYDROGEN ATOM + VINYL FLUORIDE

$$\dot{H}(I) + CH_2{=}CHF(II) \xrightarrow{1} CH_2F\dot{C}H_2(III)$$
$$\xrightarrow{2} CH_3\dot{C}HF(IV)$$

	(I)	(II)	(III)	(IV)		Path 1	Path 2
$\Delta H^\circ_{f(300)}$	52.1 (218.0)	-31.6 (-132.3)	[-15 (-63)]	[-13 (-54)]	ΔH°	[-36 (-151)]	[-34 (-142)]
$S^\circ_{(300)}$	27.4 (114.7)	60.4 (252.9)	66.6 (278.9)	66.8 (279.7)	ΔS°	-21.2 (-88.8)	-21.0 (-87.9)
$C^\circ_{p(300)}$	5.0 (20.9)	11.9 (49.8)	14.1 (59.0)	14.0 (58.6)	ΔC°_p	-2.8 (-11.7)	-2.9 (-12.1)

Path	log A	E	log k	Conditions	Reference and Experimental Method
1 + 2	-	-	[8.81]	298 K 150-260 torr	[21] R.D.Penzhorn and H.L.Sandoval, J.Phys.Chem., 74, 2065 (1970). For details see $H + CH_2{=}CH_2$. Recalculated from relative rate constants, $k_{C_2H_3F}/k_{C_2H_4} = 0.786$, and preferred value for $H + C_2H_4$ (log k = 8.92 at 298 K).

PREFERRED RATE PARAMETERS

Tentatively as given above

COMMENTS

A much lower rate constant is obtained from the ratio $k_{C_2H_3F}/k_{HBr}$ reported in reference [21], and taking the calculated value for $\log k_{HBr} = 10.8 - 3.0/2.3RT$ (see $H + C_2H_4$, reference [22]). At the same time it is possible that $k_{C_2H_4}$ and/or $k_{C_2H_3F}$ derived from [21] may be slightly pressure dependent.

REACTION: HYDROGEN ATOM + 1,1-DIFLUOROETHYLENE

$$\dot{H}(I) + CF_2{=}CH_2(II) \xrightarrow{1} CF_2H\dot{C}H_2(III)$$
$$\xrightarrow{2} \dot{C}F_2CH_3(IV)$$

	(I)	(II)	(III)	(IV)		Path 1	Path 2
$\Delta H^\circ_{f(300)}$	52.1 (218.0)	[-71.5 (-299.4)]	[-73 (-306)]	[-68 (-285)]	ΔH°	[-54 (-226)]	[-49 (-205)]
$S^\circ_{(300)}$	27.4 (114.7)	64.9 (271.7)	70.9 (296.9)	68.7 (287.6)	ΔS°	-21.4 (-89.6)	-23.6 (-98.8)
$C^\circ_{p(300)}$	5.0 (20.9)	14.8 (62.0)	15.9 (66.6)	16.0 (67.0)	ΔC°_p	-3.9 (-16.3)	-3.8 (-15.9)

Path	log A	E	log k	Conditions	Reference and Experimental Method
1 + 2	-	-	[9.07]	298 K 260 torr	[21] R.D.Penzhorn and H.L.Sandoval, J.Phys.Chem., 74, 2065 (1970). For details see $H + C_2H_4$ and $H + CH_2{=}CHF$. Recalculated from relative rate constants, $k_{C_2H_2F_2}/k_{C_2H_4} = 1.452$, and preferred value for $H + C_2H_4$ (log k = 8.92 at 298 K).

PREFERRED RATE PARAMETERS

Tentatively as given above

COMMENTS

See $H + CH_2{=}CHF$.

ΔH°_f, ΔH° and E in kcal mol^{-1} (kJ mol^{-1}); S°, C°_p, ΔS° and ΔC°_p in cal mol^{-1} K^{-1} (J mol^{-1} K^{-1}); A and k in l mol^{-1} s^{-1}

<u>REACTION</u>: HYDROGEN ATOM + <u>CIS</u>-DIFLUOROETHYLENE $\dot{H}(I)$ + <u>cis</u>-CHF=CHF(II) ⟶ $CH_2F\dot{C}HF(III)$

	(I)	(II)	(III)		
$\Delta H^\circ_{f(300)}$	52.1 (218.0)	-75.2 (-314.9)	[-56 (-234)]	ΔH°	[-33 (-138)]
$S^\circ_{(300)}$	27.4 (114.7)	65.6 (274.7)	72.8 (304.8)	ΔS°	-20.2 (-84.6)
$C^\circ_{p(300)}$	5.0 (20.9)	13.6 (56.9)	15.0 (62.8)	ΔC°_p	-3.6 (-15.1)

log A	E	log k	Conditions	Reference and Experimental Method
-	-	[8.75]	298 K 273 torr	[21] R.D.Penzhorn and H.L.Sandoval, <u>J.Phys.Chem.</u>, <u>74</u>, 2056 (1970). For details see H + C_2H_4 and H + CH_2=CHF. Recalculated from relative rate constants, $k_{\underline{cis}-C_2H_2F_2}/k_{C_2H_4}$ = 0.699, and preferred value for H + C_2H_4 (log k = 8.92 at 298 K).

<u>PREFERRED RATE PARAMETERS</u>

Tentatively as given above

<u>COMMENTS</u>

See H + CH_2=CHF.

REACTION: HYDROGEN ATOM + *TRANS*-DIFLUOROETHYLENE $\dot{H}(I) + \textit{trans}\text{-}CHF{=}CHF(II) \longrightarrow CH_2F\dot{C}HF(III)$

	(I)	(II)	(III)		
$\Delta H^\circ_{f(300)}$	52.1 (218.0)	-75.2 (-314.9)	[-56 (-234)]	ΔH°	[-33 (-138)]
$S^\circ_{(300)}$	27.4 (114.7)	65.6 (274.7)	72.8 (304.8)	ΔS°	-20.2 (-84.6)
$C^\circ_{p(300)}$	5.0 (20.9)	13.6 (56.9)	15.0 (62.8)	ΔC°_p	-3.6 (-15.1)

log A	E	log k	Conditions	Reference and Experimental Method
-	-	[8.98]	298 K 255-547 torr	[21] R.D.Penzhorn and H.L.Sandoval, *J.Phys.Chem.*, *74*, 2065 (1970). For details see $H + CH_2{=}CH_2$ and $H + CH_2{=}CHF$. Recalculated from relative rate constants, $k_{\textit{trans}\text{-}C_2H_2F_2}/k_{C_2H_4} = 1.168$, and preferred value for $H + CH_2{=}CH_2$ (log k = 8.92 at 298 K).

PREFERRED RATE PARAMETERS

Tentatively as given above

COMMENTS

See $H + CH_2{=}CHF$.

ΔH^o_f, ΔH^o and E in kcal mol^{-1} (kJ mol^{-1}); S^o, C^o_p, ΔS^o and ΔC^o_p in cal mol^{-1} K^{-1} (J mol^{-1} K^{-1}); A and k in l mol^{-1} s^{-1}

REACTION: HYDROGEN ATOM + TRIFLUOROETHYLENE

$$\dot{H}(I) + CHF{=}CF_2(II) \xrightarrow{1} CH_2F\dot{C}F_2(III)$$
$$\xrightarrow{2} \dot{C}HFCHF_2(IV)$$

	(I)	(II)	(III)	(IV)		Path 1	Path 2
$\Delta H^\circ_{f(300)}$	52.1 (218.0)	-117.3 (-491.1)	[-103 (-431)]	[-106 (-444)]	ΔH°	[-38 (-159)]	[-41 (-172)]
$S^\circ_{(300)}$	27.4 (114.7)	70.1 (293.5)	75.9 (317.8)	75.9 (317.8)	ΔS°	-21.6 (-90.4)	-21.6 (-90.4)
$C^\circ_{p(300)}$	5.0 (20.9)	16.5 (69.1)	16.8 (70.3)	16.8 (70.3)	ΔC°_p	-4.7 (-19.7)	-4.7 (-19.7)

Path	log A	E	log k	Conditions	Reference and Experimental Method
1 + 2	-	-	[9.13]	298 K 150-260 torr	[21] R.D.Penzhorn and H.L.Sandoval, J.Phys.Chem., 74, 2065 (1970). For details see H + $CH_2{=}CH_2$ and H + $CH_2{=}CHF$. Recalculated from relative rate constants, $k_{C_2HF_3}/k_{C_2H_4}$ = 1.650, and preferred value for H + $CH_2{=}CH_2$ (log k = 8.92 at 298 K).

PREFERRED RATE PARAMETERS

Tentatively as given above

COMMENTS

See H + $CH_2{=}CHF$.

REACTION: HYDROGEN ATOM + TETRAFLUOROETHYLENE $\dot{H}(I) + CF_2{=}CF_2(II) \longrightarrow \dot{C}F_2CF_2H(III)$

	(I)	(II)	(III)		
$\Delta H^\circ_{f(300)}$	52.1 (218.0)	-157.9 (-661.1)	[-154 (-645)]	ΔH°	[-48 (-201)]
$S^\circ_{(300)}$	27.4 (114.7)	74.6 (312.4)	79.6 (333.3)	ΔS°	-22.4 (-93.8)
$C^\circ_{p(300)}$	5.0 (20.9)	19.4 (81.2)	19.6 (82.1)	ΔC°_p	-4.8 (-20.1)

log A	E	log k	Conditions	Reference and Experimental Method
-	-	7.78	298 K 8 torr	[3] P.E.M.Allen, H.W.Melville and J.C.Robb, Proc.Roy.Soc.(London) Ser.A., 218, 311 (1953). For details see $H + C_2H_4$.
-	-	[9.14]	298 K 150-310 torr	[21] R.D.Penzhorn and H.L.Sandoval, J.Phys.Chem., 74, 2065 (1970). For details see $H + C_2H_4$ and $H + CH_2{=}CHF$. Recalculated from relative rate constants, $k_{C_2F_4}/k_{C_2H_4} = 1.690$, and preferred value for $H + C_2H_4$ (log k = 8.92 at 298 K).

PREFERRED RATE PARAMETERS

Tentatively log k = 9.14 at 298 K

COMMENTS

See $H + CH_2{=}CHF$.

ΔH^o_f, ΔH^o and E in kcal mol^{-1} (kJ mol^{-1}); S^o, C^o_p, ΔS^o and ΔC^o_p in cal mol^{-1} K^{-1} (J mol^{-1} K^{-1}); A and k in l mol^{-1} s^{-1}

REACTION: HYDROGEN ATOM + TETRACHLOROETHYLENE

$\dot{H}(I) + CCl_2{=}CCl_2(II) \longrightarrow CHCl_2\dot{C}Cl_2(III)$

	(I)	(II)	(III)		
$\Delta H^\circ_{f(300)}$	52.1 (218.0)	-2.7 (-11.3)	[5 (21)]	ΔH°	[-44 (-184)]
$S^\circ_{(300)}$	27.4 (114.7)	84.2 (352.5)	88.8 (371.8)	ΔS°	-22.8 (-95.5)
$C^\circ_{p(300)}$	5.0 (20.9)	22.8 (95.5)	23.9 (100.0)	ΔC°_p	-3.9 (-16.3)

log A	E	log k	Conditions	Reference and Experimental Method
-	-	Probably small	298 K 8 torr	[3] P.E.M.Allen, H.W.Melville and J.C.Robb, Proc.Roy.Soc.(London) Ser.A., 218, 311 (1953). For details see H + C_2H_4. No numerical result obtained.

PREFERRED RATE PARAMETERS

No estimate possible

REACTION: HYDROGEN ATOM + BUTADIENE-1,3

$\dot{H}(I) + CH_2{=}CHCH{=}CH_2(II) \xrightarrow{1} CH_3\dot{C}HCH{=}CH_2(III)$

$\xrightarrow{2} \dot{C}H_2CH_2CH{=}CH_2(IV)$

	(I)	(II)	(III)	(IV)		Path 1	Path 2
$\Delta H^\circ_{f(300)}$	52.1 (218.0)	26.1 (109.2)	30.3 (126.8)	45.7 (191.2)	ΔH°	-47.9 (-200.5)	-32.5 (-136.0)
$S^\circ_{(300)}$	27.4 (114.7)	68.0 (284.5)	73.0 (305.4)	76.2 (319.0)	ΔS°	-22.4 (-93.8)	-19.2 (-80.4)
$C^\circ_{p(300)}$	5.0 (20.9)	19.1 (80.0)	20.0 (83.7)	20.8 (87.1)	ΔC°_p	-4.1 (-17.1)	-3.3 (-13.8)

Path	log A	E	log k	Conditions	Reference and Experimental Method
1 + 2	-	-	No result	298 K 9 torr	[28] H.W.Melville and J.C.Robb, Proc.Roy.Soc.(London) Ser.A., 202, 181 (1950). For details see H + $CH_2{=}CH_2$.

Path	log A	E	log k	Conditions	Reference and Experimental Method
1 + 2	-	-	9.68	297 K 435 torr	[5] K.R.Jennings and R.J.Cvetanović, J.Chem.Phys., 35, 1233 (1961). For details see $H + CH_2{=}CH_2$. Recalculated from relative rate constants, $k_{C_4H_6}/k_{C_3H_6} = 4.85$, and preferred value for $H + C_3H_6$ (log k = 8.99 at 298 K).
1 + 2	10.61	1.3 (5.4)	9.66	304 K 600 torr	[6] K.Yang, J.Amer.Chem.Soc., 84, 3795 (1962). For details see $H + CH_2{=}CH_2$.
1 + 2	-	-	9.69	298 K 600 torr	[17] G.R.Woolley and R.J.Cvetanović, J.Chem.Phys., 50, 4697 (1969). For details see $H + CH_2{=}CH_2$. Recalculated from relative rate constants, $k_{C_4H_6}/k_{C_3H_6} = 5.01$, and preferred value for $H + C_3H_6$ (log k = 8.99 at 298 K).
1 + 2	-	-	9.72	298 K 600 torr	[18] R.J.Cvetanović and L.C.Doyle, J.Chem.Phys., 50, 4705 (1969). For details see $H + CH_2{=}CH_2$. Recalculated from relative rate constants, $k_{C_4H_6}/k_{C_3H_6} = 5.32$, and preferred value for $H + C_3H_6$ (log k = 8.99 at 298 K).
1 + 2	-	-	(i) 9.70 (ii) 9.50	298 K 1.3 torr 1.6-2.6 torr	[27] E.E.Daby, H.Niki and B.Weinstock, J.Phys.Chem., 75, 1601 (1971). For details see $H + CH_2{=}CH_2$. (i) Data for H addition. (ii) Data for D addition.

PREFERRED RATE PARAMETERS

$\log k_{1+2} = 10.61 - 1.3/2.3RT$

$= 9.69$ at 298 K

COMMENTS

The preferred rate constant at 298 K is the average of references [5,6,17,18,27] which are in excellent agreement. The Arrhenius parameters of reference [6] also seem reasonable. Path 1 will be the dominant route for addition.

ΔH_f^o, ΔH^o and E in kcal mol^{-1} (kJ mol^{-1}); S^o, C_p^o, ΔS^o and ΔC_p^o in cal mol^{-1} K^{-1} (J mol^{-1} K^{-1}); A and k in l mol^{-1} s^{-1}

REACTION: HYDROGEN ATOM + CYCLOHEXENE $\dot{H}(I) + C_6H_{10}(II) \longrightarrow \dot{C}_6H_{11}(III)$

	(I)	(II)	(III)		
$\Delta H^\circ_{f(300)}$	52.1 (218.0)	-0.8 (-3.3)	12.7 (53.2)	ΔH°	-38.6 (-161.6)
$S^\circ_{(300)}$	27.4 (114.7)	74.3 (311.1)	76.6 (320.7)	ΔS°	-25.1 (-105.1)
$C^\circ_{p(300)}$	5.0 (20.9)	25.3 (105.9)	26.2 (109.7)	ΔC°_p	-4.1 (-17.2)

log A	E	log k	Conditions	Reference and Experimental Method
-	-	8.76	298 K 9 torr	[28] H.W.Melville and J.C.Robb, Proc.Roy.Soc.(London) Ser.A., 202, 181 (1950). For details see H + C_2H_4.
-	-	8.90	298 K 9 torr	[3] P.E.M.Allen, H.W.Melville and J.C.Robb, Proc.Roy.Soc.(London) Ser.A., 218, 311 (1953). For details see H + C_2H_4.

PREFERRED RATE PARAMETERS

log k = [8.60] at 298 K

COMMENTS

Tentative, based on k for H + cis-butene-2.

REACTION: HYDROGEN ATOM + BENZENE $\dot{H}(I) + C_6H_6(II) \longrightarrow \dot{C}_6H_7(III)$

	(I)	(II)	(III)		
$\Delta H^\circ_{f(300)}$	52.1 (218.0)	19.8 (82.9)	43.0 (180.0)	ΔH°	-29.0 (-121.3)
$S^\circ_{(300)}$	27.4 (114.7)	64.5 (270.1)	69.0 (288.9)	ΔS°	-23.0 (-96.3)
$C^\circ_{p(300)}$	5.0 (20.9)	19.7 (82.5)	24.1 (100.9)	ΔC°_p	-0.6 (-2.5)

log A	E	log k	Conditions	Reference and Experimental Method
-	-	8.77	298 K 7 torr	[28] H.W.Melville and J.C.Robb, Proc.Roy.Soc.(London) Ser.A., 202, 181 (1950). For details see $H + CH_2{=}CH_2$.
-	-	8.03	298 K 7 torr	[3] P.E.M.Allen, H.W.Melville and J.C.Robb, Proc.Roy.Soc.(London) Ser.A., 218, 311 (1953). For details see $H + CH_2{=}CH_2$.
-	-	7.74	298 K 435 torr	[5] K.R.Jennings and R.J.Cvetanović, J.Chem.Phys., 35, 1233 (1961). For details see $H + CH_2{=}CH_2$. Recalculated from relative rate constants, $k_{C_6H_6}/k_{C_3H_6} = 0.056$, and preferred value for $H + C_3H_6$ (log k = 8.99 at 298 K).
10.42	4.3 (18.0)	7.28(300 K)	300-480 K 600 torr	[6] K.Yang, J.Amer.Chem.Soc., 84, 3795 (1962). For details see $H + CH_2{=}CH_2$.
9.9	3.3 (13.8)	7.57(298 K)	300-357 K 800 torr	[30] M.C.Sauer,Jr. and B.Ward, J.Phys.Chem., 71, 3971 (1967). Static system. H from pulse radiolysis of H_2. Reaction rate monitored by rate of change in optical absorption due to adduct radical (λ = 302 nm) Rate constants derived from kinetic analysis of optical absorption decay curves.

PREFERRED RATE PARAMETERS

$\log k = 10.6 - 4/2.3RT$

$= 7.7$ at 298 K

COMMENTS

The preferred rate constant at 298 K is based on references [5,30], and the preferred E is within the range of references [6] and [30] leading to a higher preferred A-factor than the experimental values. For the reverse reaction, E_{-1} has been estimated to be 31 kcal mol^{-1} [D.G.L.James and R.D.Suart, Trans.Faraday Soc., 64, 2752 (1968)], which, with the thermodynamic data, leads to E_1 = 2.6 kcal mol^{-1}. The preferred value of A_1, together with the ΔS°, gives $\log (A_{-1}/s^{-1}) \simeq 14$, which is reasonable.

ΔH^o_f, ΔH^o and E in kcal mol^{-1} (kJ mol^{-1}); S^o, C^o_p, ΔS^o and ΔC^o_p in cal mol^{-1} K^{-1} (J mol^{-1} K^{-1}); A and k in l mol^{-1} s^{-1}

REACTION: HYDROGEN ATOM + TOLUENE $\dot{H}(I) + CH_3C_6H_5(II) \longrightarrow CH_3\dot{C}_6H_6(III)$

	(I)	(II)	(III)		
$\Delta H^\circ_{f(300)}$	52.1 (218.0)	12.0 (50.2)	26.8 (112.2)	ΔH°	-37.3 (-156.2)
$S^\circ_{(300)}$	27.4 (114.7)	78.2 (327.4)	[69 (289)]	ΔS°	[-37 (-155)]
$C^\circ_{p(300)}$	5.0 (20.9)	25.0 (104.7)	[29.3 (122.7)]	$\Delta C^\circ_{p(300)}$	[-0.7 (-2.9)]
$C^\circ_{p(800)}$	5.0 (20.9)	55.7 (233.2)	[62.3 (260.9)]	$\Delta C^\circ_{p(800)}$	[+1.6 (+6.7)]
				$\langle \Delta C^\circ_p \rangle$	+0.4 (+1.7)

log A	E	log k	Conditions	Reference and Experimental Method
-	-	7.38	298 K 7 torr	[3] P.E.M.Allen, H.W.Melville and J.C.Robb, Proc.Roy.Soc.(London) Ser.A., 218. 311 (1953). For details see $H + CH_2{=}CH_2$.
-	-	8.0(298 K)	298-373 K 8000 torr	[30] M.C.Sauer,Jr. and B.Ward, J.Phys.Chem., 71, 3971 (1967). For details see $H + C_6H_6$.

PREFERRED RATE PARAMETERS

log k = 8.0 at 298 K

COMMENTS

The preferred rate constant seems rather high in relation to $k_{benzene}$. The results of reference [30] include any contribution from the reaction, $H + C_6H_5CH_3 \longrightarrow H_2 + C_6H_5CH_2$, for which the rate constant is estimated to be log k = 10.5 - 6/2.3RT [S.W.Benson and G.R.Haugen, J.Phys.Chem., 71, 4404 (1967)], leading to log k = 6.1 at 298 K, indicating a negligible contribution to $k_{toluene}$.

REACTION: HYDROGEN ATOM + ACETYLENE $\dot{H}(I) + CH{\equiv}CH(II) \longrightarrow CH_2{=}\dot{C}H(III)$

	(I)	(II)	(III)		
$\Delta H^o_{f(300)}$	52.1 (218.0)	54.2 (226.6)	[69 (288)]	ΔH^o	[-37 (-155)]
$S^o_{(300)}$	27.4 (114.7)	48.0 (200.7)	56.3 (235.7)	ΔS^o	-19.1 (-79.9)
$C^o_{p(300)}$	5.0 (20.9)	10.5 (43.9)	9.7 (40.6)	ΔC^o_p	-5.8 (-24.3)

log A	E	log k	Conditions	Reference and Experimental Method
10.57	3.9 (16.3)	7.71	298 K 15 torr	[31] E.L.Tollefson and D.J.LeRoy, J.Chem.Phys., 16, 1057 (1948). Flow system. H from thermal dissociation of H_2 on a hot filament. [H] by thermal conductivity measurements of tungsten filament. Reaction rate from decrease in [H] in presence of C_2H_2. The method could not differentiate between the two possible sets of reactions $H + C_2H_2 \rightarrow C_2H_3$, $H + C_2H_3 \rightarrow C_2H_2 + H_2$ and $H + C_2H_2 \rightarrow C_2H + H_2$, $H + C_2H \rightarrow C_2H_2$. The reported rate constant refers to the overall second-order reaction.
8.04	1.5 (6.3)	6.91	290 K 315 torr	[32] J.R.Dingle and D.J.LeRoy, J.Chem.Phys., 18, 1632 (1950). As above for reference [31].
-	-	7.38	313 K 1.35-3.00 torr	[33] G.G.Volpi and F.Zocchi, J.Chem.Phys., 44, 4010 (1966). Flow system. H from microwave discharge in H_2 - Ar mixture [H] determined by mass spectrometry. Rate constant determined by change in [H]. Linear plot of k^{-1} against P^{-1} obtained, from which extrapolated second-order k was derived.
-	-	8.56	300 K 10 torr	[10] J.V.Michael and R.E.Weston,Jr., J.Chem.Phys., 45, 3632 (1966). Flow system. H from microwave discharge in H_2 - He mixture. [H] determined by resonance fluorescence and absorption by H of Lyman α radiation at 121.6 nm. Reaction rate monitored by changes in [H]. Extrapolated second-order k obtained from plot of k against P^{-1}, but no pressure dependence found below 2 torr.

ΔH^o_f, ΔH^o and E in kcal mol^{-1} (kJ mol^{-1}); S^o, C^o_p, ΔS^o and ΔC^o_p in cal mol^{-1} K^{-1} (J mol^{-1} K^{-1}); A and k in l mol^{-1} s^{-1}

log A	E	log k	Conditions	Reference and Experimental Method
-	-	>7.37	298 K 0.51-1.67 torr	[34] J.V.Michael and H.Niki, J.Chem.Phys., 46, 4969 (1967). Flow system. H from microwave discharge in H_2. [H] determined by titration with NO_2. Reaction rate measured from decrease in [H]. $[C_2H_2]$ determined by time-of-flight mass spectrometry. For H + C_2D_2 reaction, second-order k obtained from plot of k against P^{-1}, but extrapolations not possible for H + C_2H_2 and D + C_2D_2 data.
-	-	7.86 (H + C_2D_2)	298 K 0.50-1.56 torr	
-	-	>7.20 (D + C_2D_2)	298 K 0.68-1.63 torr	

PREFERRED RATE PARAMETERS

Data insufficient to make a meaningful selection of a preferred rate constant

COMMENTS

The mechanism appears to be

$$H + C_2H_2 \longrightarrow C_2H_3^*$$
$$C_2H_3^* \longrightarrow C_2H_2 + H$$
$$C_2H_3^* + M \longrightarrow C_2H_3 + M$$
$$C_2H_3 + H \longrightarrow C_2H_2 + H_2$$

so that the apparent second-order rate constant for H + C_2H_2 should be pressure dependent, which is well established, references [10,33,34].

Unfortunately, there is insufficient data on the effect of pressure, particularly at high pressures, to make a meaningful extrapolation to obtain the high-pressure second-order rate constant. An experimental approach of the type applied to the H + C_2H_4 reaction, reference [22], is obviously needed. All of the above rate constants would appear to be considerably below the high-pressure limiting rate constant on the basis of the H + C_2H_4 results. The above Arrhenius parameters are also suspect.

REACTION: HYDROGEN ATOM + PROPYNE

$$\dot{H}(I) + CH_3C{\equiv}CH(II) \xrightarrow{1} CH_3CH{=}\dot{C}H(III)$$
$$\xrightarrow{2} CH_3\dot{C}{=}CH_2(IV)$$

	(I)	(II)	(III)	(IV)		Path 1	Path 2
$\Delta H^\circ_{f(300)}$	52.1 (218.0)	44.4 (185.9)	60.7 (254.1)	57.7 (241.4)	ΔH°	-35.8 (-149.9)	-38.8 (-162.4)
$S^\circ_{(300)}$	27.4 (114.7)	61.5 (251.5)	67.3 (281.7)	67.3 (281.7)	ΔS°	-21.6 (-90.4)	-21.6 (-90.4)
$C^\circ_{p(300)}$	5.0 (20.9)	14.6 (61.1)	15.2 (63.6)	15.2 (63.6)	ΔC°_p	-4.4 (-18.4)	-4.4 (-18.4)

Path	log A	E	log k	Conditions	Reference and Experimental Method
1 + 2	-	-	8.38	298 K 2 torr	[12] J.M.Brown and B.A.Thrush, Trans.Faraday Soc., 63, 630 (1967). For details see $H + C_2H_4$.

PREFERRED RATE PARAMETERS

log k_{1+2} > 8.4 at 298 K

COMMENTS

The above value is considered to be a lower limit on the grounds that the reaction is likely to be in the pressure-dependent region at 2 torr, on the basis of the data for $H + C_3H_6$, reference [26].

ΔH°_f, ΔH° and E in kcal mol^{-1} (kJ mol^{-1}); S°, C°_p, ΔS° and ΔC°_p in cal mol^{-1} K^{-1} (J mol^{-1} K^{-1}); A and k in l mol^{-1} s^{-1}

REACTION: NITROGEN ATOM + ETHYLENE $N(I) + CH_2{=}CH_2(II) \longrightarrow$?

	(I)	(II)	
$\Delta H^\circ_{f(300)}$	113.0 (473.1)	12.5 (52.3)	ΔH°
$S^\circ_{(300)}$	36.6 (153.2)	52.4 (219.4)	ΔS°
$C^\circ_{p(300)}$	5.0 (20.9)	10.3 (43.1)	ΔC°_p

log A	E	log k	Conditions	Reference and Experimental Method
7.76	0	7.76(473-600 K)	300-600 K 2.8 torr	[35] J.T.Herron, J.Chem.Phys., 33, 1273 (1960). Flow system. N atoms from microwave discharge through N_2. Changes in the concentrations of stable materials were observed by a mass spectrometer. Relative [N] were also determined by mass spectrometry, and absolute [N] from titration with NO. Rate constants calculated from -d[N]/dt on the basis of a first-order disappearance of N according to scheme: $N + C_2H_4 \xrightarrow{1} [NC_2H_4]$ $N + [NC_2H_4] \xrightarrow{2} N_2 + C_2H_4$ $[NC_2H_4] \xrightarrow{3}$ products The adduct radical $[NC_2H_4]$ was not detected, and the products of reaction (3) were not identified or specified.
7.36	0.4 (1.7)	7.18(483 K)	293-672 K 3 torr	[36] E.M.Levy and C.A.Winkler, Can.J.Chem., 40, 686 (1962). Flow system. N atoms from a condensed discharge in N_2. The reaction was stopped at various times by means of a movable catalytic probe. The rate constant was calculated from the overall stoichiometry on the assumption that the hydrogen cyanide produced in the reaction was proportional to the [N] consumed: $N + C_2H_4 \longrightarrow HCN$ + products A summary of earlier work is given in this paper.
8.39	1.9 (8.0)	7.37(407 K)	291-523 K 4.8 torr	[37] L.I.Avramenko and V.M.Krasnen'kov, Izv.Akad.Nauk SSSR, Ser.Khim., 600 (1964). Flow system. N atoms from discharge through N_2. Reaction rate determined from the rate of formation of HCN, assumed to arise from the initial reaction, $N + CH_2{=}CH_2 \longrightarrow HCN + CH_3$. Rate constants calculated from an assumed "sequence of formation" of products, [L.I.Avramenko and R.V.Kolesnikova, Dokl.AN SSSR, 42, 349 (1953); Izv.Otd.Khim.N., (1955), 386].

log A	E	log k	Conditions	Reference and Experimental Method
-	-	< 6.85	320 K 1-3 torr	[38] J.T.Herron, J.Phys.Chem., 69, 2736 (1965). Experimental details as for reference [35] above. Relative partial pressures of H determined by mass spectrometry. Mechanism involving H atoms postulated: $N + C_2H_4 \xrightarrow{1} HCN + CH_3$ $N + CH_3 \xrightarrow{2} HCN + 2H$ $H + C_2H_4 \xrightarrow{3} C_2H_5$ $H + C_2H_5 \xrightarrow{4} 2CH_3$ $N + C_2H_5 \xrightarrow{5} NH + C_2H_4$ $N + NH \xrightarrow{6} N_2 + H$ Under conditions of low reactants flow with [N] in large excess and with low percentage conversion, the rate constant is given by $k_{C_2H_4} = \ln(R_0/R)/\int_0^t [N]dt$ where R_0 and R are the partial pressures of the reactants at times zero and t, and [N] is the concentration of N atoms. The rate of decay of N was sufficiently small that the integral in the above equation could be replaced by $[N]_{av}t$, where $[N]_{av}$ is the average N concentration. This work supersedes data of reference [35].
-	-	∿ 7.0	340 K 2 torr	[39] J.T.Herron, J.Phys.Chem., 70, 2803 (1966). As for references [35] and [38]. N_2 was further purified in these experiments.
-	-	7.01	298 K 2.6 torr	[40] K.D.Foster, P.Kebarle and H.B.Dunford, Can.J.Chem., 44, 2691 (1966). Flow system. N atoms from microwave discharge through N_2. [N] measured directly by mass spectrometer. Reactions with $CH_2{=}CH_2$, $CD_2{=}CD_2$ and $CD_2{=}CDH$ were studied. Above rate constant is for reaction with $CD_2{=}CDH$ and was calculated as for the data of reference [38].
7.20	0.7 (2.9)	6.90(509 K)	318-701 K 3 torr	[41] G.Paraskevopoulos and C.A.Winkler, J.Phys.Chem., 71, 947 (1967). As for reference [36] above. [N] was determined by titration with NO. Stoichiometry was established by analysis of the products to be 1.53 ± 0.7 moles of HCN/mole of C_2H_4 reacted.
-	-	7.18	313 K 2.8 torr	[42] N.Madhavan and W.E.Jones, J.Phys.Chem., 72, 1812 (1968). Similar to system of reference [36]. Rate constants were calculated by the diffusion flame technique. Experiments with C_2H_4 were performed to check system.

ΔH_f^o, ΔH^o and E in kcal mol^{-1} (kJ mol^{-1}); S^o, C_p^o, ΔS^o and ΔC_p^o in cal mol^{-1} K^{-1} (J mol^{-1} K^{-1}); A and k in l mol^{-1} s^{-1}

log A	E	log k	Conditions	Reference and Experimental Method
7.4	0.76 (3.18)	7.0(435 K)	320-550 K 2.5 torr	[43] J.T.Herron and R.E.Huie, J.Phys.Chem., 72, 2538 (1968). Experimental details as for references [35,38,39]. Temperature measurements were made with CD_2=CDH as reactant to prevent interference from background peaks in mass spectrometer arising from out-gassing of ion-source.

PREFERRED RATE PARAMETERS

log k = 7.30 - 0.7/2.3RT

COMMENTS

There is general agreement on the second-order rate constants and Arrhenius parameters for the reaction of N(^{4}S) atoms with ethylene. On thermochemical grounds, the reaction must proceed through the formation of an adduct radical, but there is no reliable evidence on the structure of the adduct, and its subsequent reactions are also speculative. The work of Herron, [38,39,43], has shown that the subsequent reactions involve H atom reactions. It has also been pointed out that the reactions of atomic nitrogen (^{4}S) with alkenes and alkynes, to yield products in their ground electronic states, are spin forbidden, and account for the low A-factors.

A comparison of the rate constants for N atom addition to the series of alkenes reveals that $k_{C_2H_4}$ is distinctly smaller than for the other members of the series, but the others have remarkably similar rate constants (log k ≃ 7.5 at 435 K). This suggests that Arrhenius parameters should not be too different for the series of alkenes, and marked differences are probably due to internal compensation of the Arrhenius parameters.

REACTION: NITROGEN ATOM + PROPENE $N(I) + CH_2{=}CHCH_3(II) \longrightarrow ?$

	(I)	(II)	
$\Delta H^\circ_{f(300)}$	113.0 (473.1)	4.9 (20.5)	ΔH°
$S^\circ_{(300)}$	36.6 (153.2)	63.8 (267.1)	ΔS°
$C^\circ_{p(300)}$	5.0 (20.9)	15.3 . (64.0)	ΔC°_p

log A	E	log k	Conditions	Reference and Experimental Method
-	-	7.28	340 K 2 torr	[39] J.T.Herron, J.Phys.Chem., 70, 2803 (1966). For details see $N + CH_2{=}CH_2$. Identical results were obtained with $CD_3CD{=}CD_2$. The following general mechanism was proposed: N + olefin $\longrightarrow$ products + H? N + products $\longrightarrow$ products' + H H + olefin $\longrightarrow$ R H + R $\longrightarrow$ RH N + R $\longrightarrow$ olefin + NH R + R $\longrightarrow$ R_2 (or RH + R-H) N + NH $\longrightarrow$ N_2 + H Rate constants were calculated from the usual approximation for one reactant greatly in excess ([N] >> [olefin]): $k = \ell n([O\ell]_0/[O\ell]_t)/[N]t$
8.18	1.65 (6.90)	7.35(435 K)	338-697 K 3 torr	[41] G.Paraskevopoulos and C.A.Winkler, J.Phys.Chem., 71, 947 (1967). For details see $N + CH_2{=}CH_2$. The following reactions were proposed: $N + C_3H_6 \longrightarrow HCN + C_2H_4 + H$ $\longrightarrow CH_3CN + CH_3$ $N + C_2H_4 \longrightarrow HCN + CH_3$ $N + CH_3 \longrightarrow HCN + 2H$ $N + CH_3CN \longrightarrow 2HCN + H$ The ratio of rate constants for N addition to ethylene and propene was obtained from a treatment of the C_3H_6 reaction as a sequence of two competitive consecutive reactions. Thus, $k_{C_3H_6}/k_{C_2H_4}$ = 0.46 at 338 K, and 0.20 at 697 K, which compare favourably with the ratios calculated from the absolute rate constants, $k_{C_3H_6}/k_{C_2H_4}$ = 0.40 at 338 K, and 0.19 at 697 K (data of this reference).

ΔH°_f, ΔH° and E in kcal mol^{-1} (kJ mol^{-1}); S°, C°_p, ΔS° and ΔC°_p in cal mol^{-1} K^{-1} (J mol^{-1} K^{-1}); A and k in l mol^{-1} s^{-1}

log A	E	log k	Conditions	Reference and Experimental Method
-	-	7.40	313 K 1.75 torr	[42] N.Madhavan and W.E.Jones, J.Phys.Chem., 72, 1812 (1968). For details see N + CH_2=CH_2.
7.95	1.02 (4.27)	7.44(435 K)	320-550 K 3 torr	[43] J.T.Herron and R.E.Huie, J.Phys.Chem., 72, 2538 (1968). For details see N + CH_2=CH_2. The reactant was CD_3CD=CD_2.

PREFERRED RATE PARAMETERS

log k = 8.07 - 1.3/2.3RT

COMMENTS

See comments for N + CH_2=CH_2 system.

REACTION: NITROGEN ATOM + BUTENE-1

N(I) + CH_2=$CHCH_2CH_3$(II) ⟶ ?

	(I)	(II)	
$\Delta H^\circ_{f(300)}$	113.0 (473.1)	0.0 (0.0)	ΔH°
$S^\circ_{(300)}$	36.6 (153.2)	73.6 (308.2)	ΔS°
$C^\circ_{p(300)}$	5.0 (20.9)	20.6 (86.3)	ΔC°_p

log A	E	log k	Conditions	Reference and Experimental Method
-	-	7.30	340 K 3 torr	[39] J.T.Herron, J.Phys.Chem., 70, 2803 (1966). For details see N + CH_2=CH_2.
8.19	1.31 (5.48)	7.53(435 K)	320-550 K 2.5 torr	[43] J.T.Herron and R.E.Huie, J.Phys.Chem., 72, 2538 (1968). For details see N + CH_2=CH_2.

PREFERRED RATE PARAMETERS

As given in reference [43]

COMMENTS

See comments for N + CH_2=CH_2 system.

REACTION: NITROGEN ATOM + *cis*-BUTENE-2 — N(I) + *cis*-CH_3CH=$CHCH_3$(II) ⟶ ?

	(I)	(II)	
$\Delta H^\circ_{f(300)}$	113.0 (473.1)	-2.0 (-8.4)	ΔH°
$S^\circ_{(300)}$	36.6 (153.2)	73.6 (308.2)	ΔS°
$C^\circ_{p(300)}$	5.0 (20.9)	19.4 (81.2)	ΔC°_p

log A	E	log k	Conditions	Reference and Experimental Method
8.37	1.98 (8.29)	7.38(435 K)	320-550 K 2.5 torr	[43] J.T.Herron and R.E.Huie, J.Phys.Chem., 72, 2538 (1968). For details see N + CH_2=CH_2.

PREFERRED RATE PARAMETERS

As given above

COMMENTS

See comments for N + CH_2=CH_2 system.

ΔH^o_f, ΔH^o and E in kcal mol^{-1} (kJ mol^{-1}); S^o, C^o_p, ΔS^o and ΔC^o_p in cal mol^{-1} K^{-1} (J mol^{-1} K^{-1}); A and k in l mol^{-1} s^{-1}

REACTION: NITROGEN ATOM + TRANS-BUTENE-2

$N(I)$ + trans-$CH_3CH{=}CHCH_3(II)$ $\longrightarrow$?

	(I)	(II)	
$\Delta H^\circ_{f(300)}$	113.0 (473.1)	-3.0 (-12.6)	ΔH°
$S^\circ_{(300)}$	36.6 (153.2)	72.4 (303.0)	ΔS°
$C^\circ_{p(300)}$	5.0 (20.9)	20.7 (86.6)	ΔC°_p

log A	E	log k	Conditions	Reference and Experimental Method
-	-	7.23	340 K 1-3 torr	[39] J.T.Herron, J.Phys.Chem., 70, 2803 (1966). For details see $N + CH_2{=}CH_2$.
8.53	2.1 (8.8)	7.48(435 K)	320-550 K 2.5 torr	[43] J.T.Herron and R.E.Huie, J.Phys.Chem., 72, 2538 (1968). For details see $N + CH_2{=}CH_2$.

PREFERRED RATE PARAMETERS

As given in reference [43]

COMMENTS

See comments for $N + CH_2{=}CH_2$ system.

REACTION: NITROGEN ATOM + 2-METHYLPROPENE (ISOBUTENE) $N(I) + (CH_3)_2C{=}CH_2(II) \longrightarrow ?$

	(I)	(II)	
$\Delta H^\circ_{f(300)}$	113.0 (473.1)	-3.6 (-14.9)	ΔH°
$S^\circ_{(300)}$	36.6 (153.2)	71.3 (298.4)	ΔS°
$C^\circ_{p(300)}$	5.0 (20.9)	21.6 (90.4)	ΔC°_p

log A	E	log k	Conditions	Reference and Experimental Method
-	-	7.62	340 K 1-3 torr	[39] J.T.Herron, J.Phys.Chem., 70, 2803 (1966). For details see $N + CH_2{=}CH_2$
7.89	0.55 (2.30)	7.61(435 K)	320-550 K 2.5 torr	[43] J.T.Herron and R.E.Huie, J.Phys.Chem., 72, 2538 (1968). For details see $N + CH_2{=}CH_2$

PREFERRED RATE PARAMETERS

As given in reference [43]

COMMENTS

See comments for $N + CH_2{=}CH_2$ system.

ΔH^o_f, ΔH^o and E in kcal mol^{-1} (kJ mol^{-1}); S^o, C^o_p, ΔS^o and ΔC^o_p in cal mol^{-1} K^{-1} (J mol^{-1} K^{-1}); A and k in l mol^{-1} s^{-1}

REACTION: NITROGEN ATOM + 2-METHYLBUTENE-2 $N(I) + (CH_3)_2C{=}CHCH_3(II) \longrightarrow ?$

	(I)	(II)	
$\Delta H^\circ_{f(300)}$	113.0 (473.1)	-11.3 (-47.3)	ΔH°
$S^\circ_{(300)}$	36.6 (153.2)	79.9 (334.4)	ΔS°
$C^\circ_{p(300)}$	5.0 (20.9)	26.8 (112.1)	ΔC°_p

log A	E	log k	Conditions	Reference and Experimental Method
7.97	0.86 (3.60)	7.54(435 K)	320-550 K 2.5 torr	[43] J.T.Herron and R.E.Huie, J.Phys.Chem., 72, 2538 (1968). For details see $N + CH_2{=}CH_2$

PREFERRED RATE PARAMETERS

As given above

COMMENTS

See comments for $N + CH_2{=}CH_2$ system.

REACTION: NITROGEN ATOM + 2,3-DIMETHYLBUTENE-2 $N(I) + (CH_3)_2C{=}C(CH_3)_2(II) \longrightarrow ?$

	(I)	(II)	
$\Delta H^\circ_{f(300)}$	113.0 (473.1)	-19.6 (-81.9)	ΔH°
$S^\circ_{(300)}$	36.6 (153.2)	87.4 (365.3)	ΔS°
$C^\circ_{p(300)}$	5.0 (20.9)	33.0 (137.9)	ΔC°_p

log A	E	log k	Conditions	Reference and Experimental Method
-	-	7.38	340 K 1-3 torr	[39] J.T.Herron, J.Phys.Chem., 70, 2803 (1966). For details see $N + CH_2{=}CH_2$.
8.23	1.37 (5.74)	7.54(435 K)	320-550 K 2.5 torr	[43] J.T.Herron and R.E.Huie, J.Phys.Chem., 72, 2538 (1968). For details see $N + CH_2{=}CH_2$.

PREFFRRED RATE PARAMETERS

As given in reference [43]

COMMENTS

See comments for N + $CH_2=CH_2$ system.

REACTION: NITROGEN ATOM + TRANS-PERFLUOROBUTENE-2 $N(I) + trans\text{-}CF_3CF=CFCF_3(II) \longrightarrow ?$

	(I)	(II)	
$\Delta H^\circ_{f(300)}$	113.0 (473.1)		ΔH°
$S^\circ_{(300)}$	36.6 (153.2)		ΔS°
$C^\circ_{p(300)}$	5.0 (20.9)		ΔC°_p

log A	E	log k	Conditions	Reference and Experimental Method
-	-	(i) 7.52 (ii) 7.72	313 K 1.75 torr	[42] N.Madhavan and W.E.Jones, J.Phys.Chem., 72, 1812 (1968). For details see N + $CH_2=CH_2$. (i) N from condensed discharge. (ii) N from microwave discharge. Reactant contained 98% trans, 2% cis.

PREFERRED RATE PARAMETERS

log k = 7.6 at 313 K

COMMENTS

See comments for N + $CH_2=CH_2$ system.

ΔH^o_f, ΔH^o and E in kcal mol^{-1} (kJ mol^{-1}); S^o, C^o_p, ΔS^o and ΔC^o_p in cal mol^{-1} K^{-1} (J mol^{-1} K^{-1}); A and k in l mol^{-1} s^{-1}

REACTION: NITROGEN ATOM + BUTADIENE-1,3 $N(I) + CH_2{=}CHCH{=}CH_2(II) \longrightarrow ?$

	(I)	(II)	
$\Delta H^\circ_{f(300)}$	113.0 (473.1)	26.1 (109.2)	ΔH°
$S^\circ_{(300)}$	36.6 (153.2)	68.0 (284.5)	ΔS°
$C^\circ_{p(300)}$	5.0 (20.9)	19.1 (79.9)	ΔC°_p

log A	E	log k	Conditions	Reference and Experimental Method
-	-	7.54	340 K 1-3 torr	[39] J.T.Herron, J.Phys.Chem., 70, 2803 (1966). For details see $N + CH_2{=}CH_2$.

PREFERRED RATE PARAMETERS

As given above

COMMENTS

See comments for $N + CH_2{=}CH_2$ system

REACTION: NITROGEN ATOM + ACETYLENE $N(I) + CH{\equiv}CH(II) \longrightarrow C_2H_2N(III)$

	(I)	(II)	
$\Delta H^\circ_{f(300)}$	113.0 (473.1)	54.2 (226.9)	ΔH°
$S^\circ_{(300)}$	36.6 (153.2)	48.0 (201.0)	ΔS°
$C^\circ_{p(300)}$	5.0 (20.9)	10.5 (44.0)	ΔC°_p

log A	E	log k	Conditions	Reference and Experimental Method
7.34	2.7 (11.3)	6.0(440 K)	330-550 K ? torr	[44] L.I.Avramenko and V.M.Krasnen'kov, Izv.Akad.Nauk SSSR, Ser.Khim., 882 (1964). For details see $N + CH_2{=}CH_2$.

log A	E	log k	Conditions	Reference and Experimental Method
-	-	<6.3(440 K)	330-550 K 2.5 torr	[43] J.T.Herron and R.E.Huie, J.Phys.Chem., 72, 2538 (1968). For details see $N + CH_2{=}CH_2$.
8.82	2.0 (8.3)	7.83(440 K)	303-469 K 1-11 torr	[45] C.A.Arrington, O.O.Bernardini and G.B.Kistiakowsky, Proc.Roy.Soc.(London) Ser.A., 310, 161 (1969). Flow system. N from microwave discharge through N_2. [N] measured by titration with N_2O. Reactants and products were analyzed by mass spectrometry. Rate constants were calculated from the expression $-d[N]/dt = k[N][C_2H_2]$. With initial $[C_2H_2]/[N] > 1$, a comparatively slow reaction occurred, giving cyanogen, hydrogen and cyanoacetylene as products. The mechanism was proposed to be $C_2H_2 + N \longrightarrow C_2H_2N^*$ $C_2H_2N^* \longrightarrow C_2H_2 + N$ $C_2H_2N^* + N_2 \longrightarrow C_2H_2N + N_2$ $C_2H_2N \longrightarrow$ products and the pressure dependence of the apparent second-order rate constant was in accordance with this mechanism.

PREFERRED RATE PARAMETERS

log k < 6.3 at 440 K

COMMENTS

There is a large discrepancy between the results of references [43] and [45]. We favour the data of reference [43] on the grounds that the experimental method gives results for other N atoms reactions which have been independently confirmed, whereas the data of reference [45] are unsupported. In addition, it has been shown by Herron, [J.Phys.Chem., 70, 2803 (1966)], that it is necessary to carry out additional purification procedures even with "prepurified" N_2 to avoid gross errors in the determination of rate constants. In reference [45], "prepurified" N_2 was used without further purification.

See comments for $N + CH_2{=}CH_2$ system.

ΔH_f^o, ΔH^o and E in kcal mol^{-1} (kJ mol^{-1}); S^o, C_p^o, ΔS^o and ΔC_p^o in cal mol^{-1} K^{-1} (J mol^{-1} K^{-1}); A and k in l mol^{-1} s^{-1}

REACTION: NITROGEN ATOM + PROPYNE $N(I) + CH_3C{\equiv}CH(II) \longrightarrow ?$

	(I)	(II)	
$\Delta H^\circ_{f(300)}$	113.0 (473.1)	44.4 (185.9)	ΔH°
$S^\circ_{(300)}$	36.6 (153.2)	59.3 (247.3)	ΔS°
$C^\circ_{p(300)}$	5.0 (20.9)	14.6 (61.1)	ΔC°_p

log A	E	log k	Conditions	Reference and Experimental Method
7.84	1.48 (6.20)	7.10(435 K)	320-550 K 1-3 torr	[43] J.T.Herron and R.E.Huie, J.Phys.Chem., 72, 2538 (1968). For details see $N + CH_2{=}CH_2$.
8.8	1.63 (6.66)	7.77(384 K)	283-485 K 2.5 torr	[46] W.C.Solomon, K.H.Hofmann and J.H.Warnatz, Int.J.Chem.Kinet., 2, 457 (1970). Flow system. N from discharge through N_2. Reactants, in a stream of He carrier gas, were passed through a nozzle-beam sampling system into the ionization chamber of a mass spectrometer. [N] were determined by mass spectrometry (calibrated for N by NO titration). Overall rate constants for the disappearance of C_3H_4 in excess N were calculated from $\ell n([C_3H_4]_1/[C_3H_4]_2) = k_{C_3H_4}[N](t_2 - t_1)$ where $[C_3H_4]_1$ is the concentration at time t_1, and [N] is the average between t_1 and t_2.

PREFERRED RATE PARAMETERS

log k = 7.84 - 1.48/2.3RT

COMMENTS

There is a fairly large discrepancy in the rate constants determined in references [43] and [46]. It was pointed out in reference [46] that the rate constant, which was the experimental overall second-order rate constant for the disappearance of C_3H_4, should not be identified with the rate constant of the bimolecular reaction of N with C_3H_4 since the measured k was independent of both $[N]_i$ and $[C_3H_4]_i$. For this reason the data of reference [43] are preferred.

A comparison of the rate constants for the reactions of N atoms with the series of alkynes shows that at 435 K the values are all close to log k = 7.4, apart from $k_{propyne}$. The slight differences in the Arrhenius parameters for the series of reactions may be due to some internal compensation.

REACTION: NITROGEN ATOM + BUTYNE-1 $N(I) + CH{\equiv}CCH_2CH_3(II) \longrightarrow ?$

	(I)	(II)	
$\Delta H^\circ_{f(300)}$	113.0 (473.1)	39.7 (166.2)	ΔH°
$S^\circ_{(300)}$	36.6 (153.2)	69.6 (291.4)	ΔS°
$C^\circ_{p(300)}$	5.0 (20.9)	19.5 (81.6)	ΔC°_p

log A	E	log k	Conditions	Reference and Experimental Method
8.54	2.24 (9.38)	7.41(435 K)	320-550 K 2.5 torr	[43] J.T.Herron and R.E.Huie, J.Phys.Chem., 72, 2538 (1968). For details see $N + CH_2{=}CH_2$.

PREFERRED RATE PARAMETERS

As given above

COMMENTS

See comments for $N + CH_2{=}CH_2$ system and for $N + CH_3C{\equiv}CH$ system.

REACTION: NITROGEN ATOM + BUTYNE-2 $N(I) + CH_3C{\equiv}CCH_3(II) \longrightarrow ?$

	(I)	(II)	
$\Delta H^\circ_{f(300)}$	113.0 (473.1)	34.9 (146.1)	ΔH°
$S^\circ_{(300)}$	36.6 (153.2)	62.8 (262.9)	ΔS°
$C^\circ_{p(300)}$	5.0 (20.9)	18.6 (77.9)	ΔC°_p

log A	E	log k	Conditions	Reference and Experimental Method
8.27	1.84 (7.70)	7.35(435 K)	320-550 K 2.5 torr	[43] J.T.Herron and R.E.Huie, J.Phys.Chem., 72, 2538 (1968). For details see $N + CH_2CH_2$.

ΔH^o_f, ΔH^o and E in kcal mol^{-1} (kJ mol^{-1}); S^o, C^o_p, ΔS^o and ΔC^o_p in cal mol^{-1} K^{-1} (J mol^{-1} K^{-1}); A and k in l mol^{-1} s^{-1}

PREFERRED RATE PARAMETERS

As given above

COMMENTS

See comments for N + $CH_2=CH_2$ system and for N + $CH_3C\equiv CH$ system.

Reaction: Nitrogen Atom + Pentyne-1 N(I) + $HC\equiv CCH_2CH_2CH_2CH_3$(II) → ?

	(I)	(II)	
$\Delta H^\circ_{f(300)}$	113.0 (473.1)	34.7 (145.3)	ΔH°
$S^\circ_{(300)}$	36.6 (153.2)	79.0 (330.8)	ΔS°
$C^\circ_{p(300)}$	5.0 (20.9)	25.0 (104.7)	ΔC°_p

log A	E	log k	Conditions	Reference and Experimental Method
8.47	2.08 (8.71)	7.43(435 K)	320-550 K 2.5 torr	[43] J.T.Herron and R.E.Huie, J.Phys.Chem., 72, 2538 (1968). For details see N + $CH_2=CH_2$.

PREFERRED RATE PARAMETERS

As given above

COMMENTS

See comments for N + $CH_2=CH_2$ system and for N + $CH_3C\equiv CH$ system.

REACTION: NITROGEN ATOM + HEXYNE-1

$N(I) + HC{\equiv}CCH_2CH_2CH_2CH_3(II) \longrightarrow ?$

	(I)	(II)	
$\Delta H^\circ_{f(300)}$	113.0 (473.1)	29.8 (124.8)	ΔH°
$S^\circ_{(300)}$	36.6 (153.2)	88.4 (370.1)	ΔS°
$C^\circ_{p(300)}$	5.0 (20.9)	30.5 (127.7)	ΔC°_p

log A	E	log k	Conditions	Reference and Experimental Method
8.66	2.45 (10.26)	7.43(435 K)	320-550 K 2.5 torr	[43] J.T.Herron and R.E.Huie, J.Phys.Chem., 72, 2538 (1968). For details see $N + CH_2{=}CH_2$.

PREFERRED RATE PARAMETERS

As given above

COMMENTS

See comments for $N + CH_2{=}CH_2$ system and for $N + CH_3C{\equiv}CH$ system.

REACTION: NITROGEN ATOM + HEXYNE-3

$N(I) + CH_3CH_2C{\equiv}CCH_2CH_3(II) \longrightarrow ?$

	(I)	(II)	
$\Delta H^\circ_{f(300)}$	113.0 (473.1)	25.5 (106.8)	ΔH°
$S^\circ_{(300)}$	36.6 (153.2)	89.7 (375.6)	ΔS°
$C^\circ_{p(300)}$	5.0 (20.9)	28.5 (119.3)	ΔC°_p

log A	E	log k	Conditions	Reference and Experimental Method
8.53	2.19 (9.17)	7.43(435 K)	320-550 K 2.5 torr	[43] J.T.Herron and R.E.Huie, J.Phys.Chem., 72, 2538 (1968). For details see $N + CH_2{=}CH_2$.

PREFERRED RATE PARAMETERS

As given above

COMMENTS

See comments for $N + CH_2{=}CH_2$ system and for $N + CH_3C{\equiv}CH$ system.

ΔH^o_f, ΔH^o and E in kcal mol^{-1} (kJ mol^{-1}); S^o, C^o_p, ΔS^o and ΔC^o_p in cal mol^{-1} K^{-1} (J mol^{-1} K^{-1}); A and k in l mol^{-1} s^{-1}

REACTION: SULPHUR ATOM + ETHYLENE $S(I) + CH_2{=}CH_2(II) \longrightarrow \overline{CH_2CH_2S}(III)$

	(I)	(II)	(III)		
$\Delta H^\circ_{f(300)}$	66.3 (277.6)	12.5 (52.3)	19.9 (83.3)	ΔH°	-58.9 (-246.6)
$S^\circ_{(300)}$	40.1 (167.9)	52.4 (219.4)	62.4 (261.3)	ΔS°	-30.1 (-126.0)
$C^\circ_{p(300)}$	5.7 (23.9)	10.3 (43.1)	12.9 (54.0)	ΔC°_p	-3.1 (-13.0)

log A	E	log k	Conditions	Reference and Experimental Method
-	-	-	298 K 250 torr	[47] O.P.Strausz and H.E.Gunning, J.Amer.Chem.Soc., 84, 4080 (1962). Static system. $S(^1D)$ atoms produced from photolysis (λ 255-229 nm) of COS. In the presence of C_2H_4 the sole product $\overline{CH_2CH_2S}$ was formed, and in the presence of C_3H_6 the only product was $CH_3\overline{CHCH_2S}$. The rate constant ratio was determined to be $k_{C_3H_6}/k_1 = 3.6$ from the analysis of the episulfide products (mass spectrometry), from photolysis of COS in a mixture with C_2H_4 and C_3H_6. It was noted that the participation of ground state $S(^3P)$ atoms in this system could not be ruled out and that the ratio of rate constants probably referred to the reactions of a mixture of $S(^1D)$ and $S(^3P)$
-	-	-	298 K 100-1250 torr	[48] H.A.Wiebe, A.R.Knight, O.P.Strausz and H.E.Gunning, J.Amer.Chem.Soc., 87, 1443 (1965). Similar experimental set up to reference [47] above, but with gas chromatographic analyses of the products. With C_2H_4 the product vinyl mercaptan (45%) was found in addition to the episulfide, and with C_3H_6 the products propenethiol (10-15%) and allyl mercaptan (10-15%) as well as the episulfide. Experiments with added CO_2 to bring about the collisional relaxation $S(^1D) \longrightarrow S(^3P)$ showed that the mercaptans were formed exclusively from (^1D) atoms, while the episulfides may be formed by either (^1D) or (^3P) S atoms.
-	-	-	298 K ? torr	[49] E.M.Lown, E.L.Dedio, O.P.Strausz and H.E.Gunning, J.Amer.Chem.Soc., 89, 1056 (1967). Same experimental details as for references [47] and [48] above. Relative rate constants were obtained for $S(^3P)$ reactions with pairs of olefins from analyses of the episulfide products (gas chromatography) from runs with added CO_2 to affect $S(^1D) \longrightarrow S(^3P)$. Results were confirmed by producing $S(^3P)$ from the $Hg(^3P_1)$ photosensitization of COS.

log A	E	log k	Conditions	Reference and Experimental Method
-	-	8.86	300 K 60-250 torr	[50] R.J.Donovan, D.Husain, R.W.Fair, O.P.Strausz and H.E.Gunning, Trans.Faraday Soc., 66, 1635 (1970). Static system. $S(^3P)$ produced from flash photolysis of CS_2 moderated with 10^4 excess of Ar. [S] was determined by vacuum ultra-violet absorption spectroscopy involving the $S(4s^3S_1 \leftarrow 3p^3P_2)$ transition at 180.7 nm. Absolute rate constants were calculated from -d[S]/dt in the $p(C_2H_4)$ range 1.3-5.0 torr x 10^2.
(i) [10.0] (ii) 10.06 (iii)10.03 (iv) 10.02	[1.5] 1.5 1.5 1.5 (6.3)	[8.9](298 K) 8.96 8.93 8.92	298-450 K ? torr	[51] O.P.Strausz, W.B.O'Callaghan, E.M.Lown and H.E.Gunning, J.Amer.Chem.Soc., 93, 559 (1971). Static system. $S(^3P)$ produced from photolysis of COS in excess CO_2 or by Hg photosensitization in presence of pairs of olefins. Relative rate constants derived from analyses of episulfide products (gas chromatography), and hence relative Arrhenius parameters. The above Arrhenius parameters for $CH_2=CH_2$ were obtained from the rate constant of reference [52] and an assumed A-factor of log A_1 = 10.0. (i) Addition to $CH_2=CH_2$. (ii) Addition to $CD_2=CD_2$. (iii) Addition to $CD_2=CH_2$. (iv) Addition to CHD=CHD. Arrhenius parameters for the deuteroethylenes obtained from relative values and assumed parameters for $CH_2=CH_2$.
-	-	8.95	298 K ∿ 200 torr	[52] J.Connor, A.Van Roodselaar, R.W.Fair and O.P.Strausz, J.Amer.Chem.Soc., 93, 560 (1971). Static system. $S(^3P)$ from flash photolysis of COS in presence of olefin and large excess of CO_2. [S] from kinetic spectroscopy. Absolute rate constants obtained.

PREFERRED RATE PARAMETERS

log k = 8.91 at 298 K

COMMENTS

The preferred rate constant is the mean of the two absolute determinations of references [50] and [52]. The assumed Arrhenius parameters for $CH_2=CH_2$ and the derived parameters for the deuteroethylenes [51] require confirmation by the technique for measuring absolute rate constants before they are accepted. The work of Strausz and Gunning [47,48,49] has established that $S(^3P)$ atoms react with alkenes to form episulfides. For a review of S atom reactions, see H.E.Gunning and O.P.Strausz, Advan.Photochem., 1, 220 (1963).

ΔH_f^o, ΔH^o and E in kcal mol^{-1} (kJ mol^{-1}); S^o, C_p^o, ΔS^o and ΔC_p^o in cal mol^{-1} K^{-1} (J mol^{-1} K^{-1}); A and k in l mol^{-1} s^{-1}

REACTION: SULPHUR ATOM + PROPENE $S(I) + CH_3CH{=}CH_2(II) \longrightarrow CH_3\overline{CHCH_2S}(III)$

	(I)	(II)	(III)		
$\Delta H^\circ_{f(300)}$	66.3 (277.6)	4.9 (20.5)	12.9 (54.0)	ΔH°	-58.3 (-244.1)
$S^\circ_{(300)}$	40.1 (167.9)	63.8 (267.1)	69.4 (290.6)	ΔS°	-34.5 (-144.5)
$C^\circ_{p(300)}$	5.7 (23.9)	15.3 (64.1)	18.6 (77.9)	ΔC°_p	-2.4 (-10.0)

log A	E	log k	Conditions	Reference and Experimental Method
-	-	9.49	298 K 250 torr	[47] O.P.Strausz and H.E.Gunning, J.Amer.Chem.Soc., 84, 4080 (1962). For details see $S + CH_2{=}CH_2$. Recalculated from relative rate constants $k_1/k_{C_2H_4} = 3.6$ and preferred value for $S + CH_2{=}CH_2$ (log k = 8.91 at 298 K). In this first quantitative study of S atom additions it was not possible to eliminate $S(^1D)$ reactions entirely.
-	-	9.75	298 K 1330 torr	[49] E.M.Lown, E.L.Dedio, O.P.Strausz and H.E.Gunning, J.Amer.Chem.Soc., 89, 1056 (1967). For details see $S + CH_2{=}CH_2$. Recalculated from the relative rate constants $k_1/k_{C_2H_4} = 6.9$ and preferred value for $S + CH_2{=}CH_2$ (log k = 8.91 at 298 K).
10.0	0.36 (1.51)	9.74(298 K)	298-450 K ? torr	[51] O.P.Strausz, W.B.O'Callaghan, E.M.Lown and H.E.Gunning, J.Amer.Chem.Soc., 93, 559 (1971). For details see $S + CH_2{=}CH_2$. Arrhenius parameters calculated from $E_{C_2H_4} - E_1 = 1.14$ and $A_1/A_{C_2H_4} = 1.0$ and assumed parameter for $S + CH_2{=}CH_2$, $E_{C_2H_4} = 1.5$ and $\log A_{C_2H_4} = 10.0$.
-	-	9.78	298 K ∿ 200 torr	[52] J.Connor, A. Van Roodselaar, R.W.Fair and O.P.Strausz, J.Amer.Chem.Soc., 93, 560 (1971). For details see $S + CH_2{=}CH_2$.

PREFERRED RATE PARAMETERS

log k = 9.76 at 298 K

COMMENTS

The preferred rate constant is the mean of the results of references [49,51,52]. The Arrhenius parameters require confirmation by the technique for measuring absolute rate constants before they are accepted. See comments for $S + CH_2{=}CH_2$.

REACTION: SULPHUR ATOM + BUTENE-1 $S(I) + CH_3CH_2CH{=}CH_2(II) \longrightarrow CH_3CH_2\overline{CHCH_2S}(III)$

	(I)	(II)	(III)		
$\Delta H^\circ_{f(300)}$	66.3 (277.6)	0.0 (0.0)	7.9 (33.1)	ΔH°	-58.4 (-244.5)
$S^\circ_{(300)}$	40.1 (167.9)	73.6 (308.2)	78.8 (339.1)	ΔS°	-34.9 (-136.9)
$C^\circ_{p(300)}$	5.7 (23.9)	20.6 (86.3)	24.1 (100.9)	ΔC°_p	-2.2 (-9.2)

log A	E	log k	Conditions	Reference and Experimental Method
-	-	-	298 K 5-515 torr	[53] K.S.Sidhu, E.M.Lown, O.P.Strausz and H.E.Gunning, J.Amer.Chem.Soc., 88, 254 (1966). Further extensions of experiments described in $S + CH_2{=}CH_2$ system [47,48,49]. No quantitative kinetic data.
-	-	9.91	298 K 1330 torr	[49] E.M.Lown, E.L.Dedio, O.P.Strausz and H.E.Gunning, J.Amer.Chem.Soc., 89, 1056 (1967). For details see $S + CH_2{=}CH_2$. Recalculated from relative rate constants $k_1/k_{C_2H_4} = 10$ and preferred value for $S + CH_2{=}CH_2$ (log k = 8.91 at 298 K).
9.88	-0.22 (-0.92)	10.04(298 K)	298-450 K ? torr	[51] O.P.Strausz, W.B.O'Callaghan, E.M.Lown and H.E.Gunning, J.Amer.Chem.Soc., 93, 559 (1971). For details see $S + CH_2{=}CH_2$. Arrhenius parameters calculated from $E_{C_2H_4} - E_1 = 1.72$ and $A_1/A_{C_2H_4} = 0.75$ and assumed parameters for $S + CH_2{=}CH_2$, $E_{C_2H_4} = 1.5$ and $\log A_{C_2H_4} = 10.0$.
-	-	9.95	298 K ~ 200 torr	[52] J.Connor, A.Van Roodselaar, R.W.Fair and O.P.Strausz, J.Amer.Chem.Soc., 93, 560 (1971). For details see $S + CH_2{=}CH_2$.

PREFERRED RATE PARAMETERS

log k = 9.97 at 298 K

COMMENTS

The preferred rate constant is the mean of the above results. See comments for $S + CH_2{=}CH_2$. The Arrhenius parameters require confirmation by the technique for absolute rate constants before they are accepted.

ΔH°_f, ΔH° and E in kcal mol^{-1} (kJ mol^{-1}); S°, C°_p, ΔS° and ΔC°_p in cal mol^{-1} K^{-1} (J mol^{-1} K^{-1}); A and k in l mol^{-1} s^{-1}

REACTION: SULPHUR ATOM + cis-BUTENE-2 $S(I) + cis\text{-}CH_3CH{=}CHCH_3(II) \longrightarrow CH_3CH\overline{CH(CH_3)S}(III)$

	(I)	(II)	(III)		
$\Delta H^\circ_{f(300)}$	66.3 (277.6)	-1.9 (-7.9)	5.8 (24.3)	ΔH°	-58.6 (-245.4)
$S^\circ_{(300)}$	40.1 (167.9)	73.6 (308.2)	76.4 (319.9)	ΔS°	-37.3 (-156.2)
$C^\circ_{p(300)}$	5.7 (23.9)	19.4 (81.2)	24.2 (101.3)	ΔC°_p	-0.9 (-3.8)

log A	E	log k	Conditions	Reference and Experimental Method
-	-	-	298 K 5-515 torr	[53] K.S.Sidhu, E.M.Lown, O.P.Strausz and H.E.Gunning, J.Amer.Chem.Soc., 88, 254 (1966). Further extensions of experiments described in $S + CH_2{=}CH_2$ system [47,48,49]. The episulfide product retains the geometry of the parent alkene not only with $S(^1D)$ but also with $S(^3P)$ addition.
-	-	10.11	298 K 1330 torr	[49] E.M.Lown, E.L.Dedio, O.P.Strausz and H.E.Gunning, J.Amer.Chem.Soc., 89, 1056 (1967). For details see $S + CH_2{=}CH_2$. Recalculated from relative rate constants $k_1/k_{C_2H_4}$ = 16 and preferred value for $S + CH_2{=}CH_2$ (log k = 8.91 at 298 K).
9.72	-0.59 (-2.47)	10.15(298 K)	298-450 K ? torr	[51] O.P.Strausz, W.B.O'Callaghan, E.M.Lown and H.E.Gunning, J.Amer.Chem.Soc., 93, 559 (1971). For details see $S + CH_2{=}CH_2$. Arrhenius parameters calculated from $E_{C_2H_4} - E_1$ = 2.09 and $A_1/A_{C_2H_4}$ = 0.53 and assumed parameters for $S + CH_2{=}CH_2$, $E_{C_2H_4}$ = 1.5 and log $A_{C_2H_4}$ = 10.0.

PREFERRED RATE PARAMETERS

log k = 10.13 at 298 K

COMMENTS

The preferred rate constant is the mean of the above results. The Arrhenius parameters require confirmation by the technique for absolute rate constants before they are accepted. See comments for $S + CH_2{=}CH_2$.

REACTION: SULPHUR ATOM + TRANS-BUTENE-2

$S(I) + trans\text{-}CH_3CH{=}CHCH_3(II) \longrightarrow CH_3\overline{CHCH(CH_3)S}(III)$

	(I)	(II)	(III)		
$\Delta H^\circ_{f(300)}$	66.3 (277.6)	-2.9 (-12.1)	5.8 (24.3)	ΔH°	-57.6 (-241.2)
$S^\circ_{(300)}$	40.1 (167.9)	72.4 (302.8)	76.4 (319.9)	ΔS°	-36.1 (-151.2)
$C^\circ_{p(300)}$	5.7 (23.9)	20.7 (86.6)	24.2 (101.3)	ΔC°_p	-2.2 (-9.2)

log A	E	log k	Conditions	Reference and Experimental Method
-	-	-	298 K 5-515 torr	[53] K.S.Sidhu, E.M.Lown, O.P.Strausz and H.E.Gunning, J.Amer.Chem.Soc., 88, 254 (1966). Further extensions of experiments described in $S + CH_2{=}CH_2$ system [47,48,49]. The episulfide product retains the geometry of the parent alkene not only with $S(^1D)$ but also with $S(^3P)$ addition.
-	-	10.21	298 K 1300 torr	[49] E.M.Lown, E.L.Dedio, O.P.Strausz and H.E.Gunning, J.Amer.Chem.Soc., 89, 1056 (1967). For details see $S + CH_2{=}CH_2$. Recalculated from relative rate constants $k_1/k_{C_2H_4} = 20$ and preferred value for $S + CH_2{=}CH_2$ (log k = 8.91 at 298 K).
9.81	-0.51 (-2.14)	10.18(298 K)	298-450 K ? torr	[51] O.P.Strausz, W.B.O'Callaghan, E.M.Lown and H.E.Gunning, J.Amer.Chem.Soc., 93, 559 (1971). For details see $S + CH_2{=}CH_2$. Arrhenius parameters calculated from $E_{C_2H_4} - E_1 = 2.01$ and $A_1/A_{C_2H_4} = 0.65$ and assumed parameters for $S + CH_2{=}CH_2$, $E_{C_2H_4} = 1.5$ and $\log A_{C_2H_4} = 10.0$.
-	-	10.08	298 K ∿ 200 torr	[52] J.Connor, A.Van Roodselaar, R.W.Fair and O.P.Strausz, J.Amer.Chem.Soc., 93, 560 (1971). For details see $S + CH_2{=}CH_2$.

PREFERRED RATE PARAMETERS

log k = 10.16 at 298 K

COMMENTS

The preferred rate constant is the mean of the above results, although there is some discrepancy between the data of references [49,51] and [52]. The Arrhenius parameters require confirmation by the technique for absolute rate constants before they are accepted. See comments for $S + CH_2{=}CH_2$.

ΔH°_f, ΔH° and E in kcal mol^{-1} (kJ mol^{-1}); S°, C°_p, ΔS° and ΔC°_p in cal mol^{-1} K^{-1} (J mol^{-1} K^{-1}); A and k in l mol^{-1} s^{-1}

REACTION: SULPHUR ATOM + 2-METHYLPROPENE (ISOBUTENE) S(I) + $(CH_3)_2C{=}CH_2$(II) ⟶ $(CH_3)_2\overline{CCH_2S}$(III)

	(I)	(II)	(III)		
$\Delta H^\circ_{f(300)}$	66.3 (277.6)	-3.6 (-15.1)	4.9 (20.5)	ΔH°	-57.8 (-242.0)
$S^\circ_{(300)}$	40.1 (167.9)	71.3 (298.5)	74.5 (311.9)	ΔS°	-36.9 (-154.5)
$C^\circ_{p(300)}$	5.7 (23.9)	21.6 (90.4)	24.5 (102.6)	ΔC°_p	-2.8 (-11.7)

log A	E	log k	Conditions	Reference and Experimental Method
-	-	-	298 K 5-515 torr	[53] K.S.Sidhu, E.M.Lown, O.P.Strausz and H.E.Gunning, J.Amer.Chem.Soc., 88, 254 (1966). Further extensions of experiments described in S + $CH_2{=}CH_2$ system [47,48,49]. No quantitative kinetic data.
-	-	10.61	298 K 1330 torr	[49] E.M.Lown, E.L.Dedio, O.P.Strausz and H.E.Gunning, J.Amer.Chem.Soc., 89, 1056 (1967). For details see S + $CH_2{=}CH_2$. Recalculated from relative rate constants $k_1/k_{C_2H_4} = 50$ and preferred value for S + $CH_2{=}CH_2$ (log k = 8.91 at 298 K).
9.99	-0.86 (-3.60)	10.62(298 K)	298-450 K ? torr	[51] O.P.Strausz, W.B.O'Callaghan, E.M.Lown and H.E.Gunning, J.Amer.Chem.Soc., 93, 559 (1971). For details see S + $CH_2{=}CH_2$. Arrhenius parameters calculated from $E_{C_2H_4} - E_1 = 2.36$ and $A_1/A_{C_2H_4} = 0.97$ and assumed parameters for S + $CH_2{=}CH_2$, $E_{C_2H_4} = 1.5$ and $\log A_{C_2H_4} = 10.0$.
-	-	10.56	298 K ∿ 200 torr	[52] J.Connor, A.Van Roodselaar, R.W.Fair and O.P.Strausz, J.Amer.Chem.Soc., 93, 560 (1971). For details see S + $CH_2{=}CH_2$.

PREFERRED RATE PARAMETERS

log k = 10.60 at 298 K

COMMENTS

The preferred rate constant is the mean of the above results. The Arrhenius parameters require confirmation by the technique for absolute rate constants before they are accepted. See comments for S + $CH_2{=}CH_2$.

REACTION: SULPHUR ATOM + PENTENE-1

$S(I) + CH_3(CH_2)_2CH{=}CH_2(II) \longrightarrow CH_3(CH_2)_2\overline{CHCH_2S}(III)$

	(I)	(II)	(III)		
$\Delta H^\circ_{f(300)}$	66.3 (277.6)	-4.9 (-20.5)	3.0 (12.6)	ΔH°	-58.4 (-244.5)
$S^\circ_{(300)}$	40.1 (167.9)	83.0 (347.5)	88.2 (369.3)	ΔS°	-34.9 (-146.1)
$C^\circ_{p(300)}$	5.7 (23.9)	26.1 (109.3)	29.6 (123.9)	ΔC°_p	-2.2 (-9.2)

log A	E	log k	Conditions	Reference and Experimental Method
-	-	9.91	298 K 1330 torr	[49] E.M.Lown, E.L.Dedio, O.P.Strausz and H.E.Gunning, J.Amer.Chem.Soc., 89, 1056 (1967). For details see $S + CH_2{=}CH_2$. Recalculated from relative rate constants $k_1/k_{C_2H_4}$ = 10 and preferred value for $S + CH_2{=}CH_2$ (log k = 8.91 at 298 K).

PREFERRED RATE PARAMETERS

As given above

COMMENTS

See comments for $S + CH_2{=}CH_2$ system.

ΔH^o_f, ΔH^o and E in kcal mol^{-1} (kJ mol^{-1}); S^o, C^o_p, ΔS^o and ΔC^o_p in cal mol^{-1} K^{-1} (J mol^{-1} K^{-1}); A and k in l mol^{-1} s^{-1}

REACTION: SULPHUR ATOM + 2-METHYLBUTENE-1 $S(I) + CH_3CH_2C(CH_3){=}CH_2(II) \longrightarrow CH_3CH_2\overline{C(CH_3)CH_2S}(III)$

	(I)	(II)	(III)		
$\Delta H^\circ_{f(300)}$	66.3 (277.6)	-8.3 (-34.8)	-0.1 (-0.4)	ΔH°	-58.1 (-243.3)
$S^\circ_{(300)}$	40.1 (167.9)	81.1 (339.6)	83.9 (351.3)	ΔS°	-37.3 (-156.2)
$C^\circ_{p(300)}$	5.7 (23.9)	26.7 (111.7)	30.0 (125.6)	ΔC°_p	-2.4 (-10.0)

log A	E	log k	Conditions	Reference and Experimental Method
-	-	10.65	298 K 1330 torr	[49] E.M.Lown, E.L.Dedio, O.P.Strausz and H.E.Gunning, J.Amer.Chem.Soc., 89, 1056 (1967). For details see $S + CH_2{=}CH_2$. Recalculated from relative rate constants $k_1/k_{C_2H_4} = 56$ and preferred value for $S + CH_2{=}CH_2$ (log k = 8.91 at 298 K).
9.89	-1.33 (-5.57)	10.87(298 K)	298-450 K ? torr	[51] O.P.Strausz, W.B.O'Callaghan, E.M.Lown and H.E.Gunning, J.Amer.Chem.Soc., 93, 559 (1971). For details see $S + CH_2{=}CH_2$. Arrhenius parameters calculated from $E_{C_2H_4} - E_1 = 2.83$, $A_1/A_{C_2H_4} = 0.78$ and assumed parameters for $S + CH_2{=}CH_2$, $E_{C_2H_4} = 1.5$ and $\log A_1 = 10.0$.

PREFERRED RATE PARAMETERS

log k = 10.87 at 298 K

COMMENTS

There is a fairly large discrepancy in the two reported rate constants for this reaction. We have selected the data of reference [51] on the grounds that they are more recent results of the same authors. The Arrhenius parameters require confirmation by the technique for absolute rate constants before they are accepted. See comments for $S + CH_2{=}CH_2$.

REACTION: SULPHUR ATOM + 2-METHYLBUTENE-2 (TRIMETHYLETHYLENE) $S(I) + (CH_3)_2C{=}CH(CH_3)(II) \longrightarrow (CH_3)_2\overline{CCH(CH_3)S}(III)$

	(I)	(II)	(III)		
$\Delta H^o_{f(300)}$	66.3 (277.6)	-11.3 (-47.3)	-2.2 (-9.2)	ΔH^o	-57.2 (-239.5)
$S^o_{(300)}$	40.1 (167.9)	79.9 (334.4)	81.5 (341.2)	ΔS^o	-38.5 (-161.2)
$C^o_{p(300)}$	5.7 (23.9)	26.8 (112.1)	30.1 (126.0)	ΔC^o_p	-2.4 (-10.0)

log A	E	log k	Conditions	Reference and Experimental Method
9.71	-1.51 (-6.32)	10.81(298 K)	298-450 K ? torr	[51] O.P.Strausz, W.B.O'Callaghan, E.M.Lown and H.E.Gunning, J.Amer.Chem.Soc., 93, 559 (1971). For details see $S + CH_2{=}CH_2$. Arrhenius parameters calculated from $E_{C_2H_4} - E_1 = 3.01$ and $A_1/A_{C_2H_4} = 0.51$ and assumed parameters for $S + CH_2{=}CH_2$, $E_{C_2H_4} = 1.5$ and $\log A_{C_2H_4} = 10.0$.

PREFERRED RATE PARAMETERS

log k = 10.81 at 298 K

COMMENTS

The above Arrhenius parameters require confirmation by the technique for absolute rate constants before they are accepted. See comments for $S + CH_2{=}CH_2$.

ΔH^o_f, ΔH^o and E in kcal mol^{-1} (kJ mol^{-1}); S^o, C^o_p, ΔS^o and ΔC^o_p in cal mol^{-1} K^{-1} (J mol^{-1} K^{-1}); A and k in l mol^{-1} s^{-1}

REACTION: SULPHUR ATOM + 2,3-DIMETHYLBUTENE-2 (TETRAMETHYLETHYLENE) $S(I) + (CH_3)_2C{=}C(CH_3)_2(II) \longrightarrow (CH_3)_2\overline{CC(CH_3)_2S}(III)$

	(I)	(II)	(III)		
$\Delta H^\circ_{f(300)}$	66.3 (277.6)	-19.6 (-81.9)	-10.2 (-42.7)	ΔH°	-56.9 (-238.2)
$S^\circ_{(300)}$	40.1 (167.9)	87.4 (365.3)	86.7 (363.0)	ΔS°	-40.8 (-170.8)
$C^\circ_{p(300)}$	5.7 (23.9)	33.0 (137.9)	36.1 (151.2)	ΔC°_p	-2.6 (-10.9)

log A	E	log k	Conditions	Reference and Experimental Method
9.70	-1.86 (-7.79)	11.06(298 K)	298-450 K ? torr	[51] O.P.Strausz, W.B.O'Callaghan, E.M.Lown and H.E.Gunning, J.Amer.Chem.Soc., 93, 559 (1971). For details see $S + CH_2{=}CH_2$. Arrhenius parameters calculated from $E_{C_2H_4} - E_1 = 3.36$ and $A_1/A_{C_2H_4} = 0.50$ and assumed parameters for $S + CH_2{=}CH_2$, $E_{C_2H_4} = 1.5$ and $\log A_{C_2H_4} = 10.0$.
-	-	10.79	298 K ∿ 200 torr	[52] J.Connor, A.Van Roodselaar, R.W.Fair and O.P.Strausz, J.Amer.Chem.Soc., 93, 560 (1971). For details see $S + CH_2{=}CH_2$.

PREFERRED RATE PARAMETERS

log k = 10.93 at 298 K

COMMENTS

The preferred rate constant is the mean of the above two results, which show some discrepancy. The Arrhenius parameters require confirmation by the technique for absolute rate constants before they are accepted. See comments for $S + CH_2{=}CH_2$.

REACTION: SULPHUR ATOM + VINYL FLUORIDE $S(I) + CH_2{=}CHF(II) \longrightarrow \overline{CH_2CHFS}(III)$

	(I)	(II)	(III)	
$\Delta H^o_{f(300)}$	66.3 (277.6)	-31.6 (-132.3)		ΔH^o
$S^o_{(300)}$	40.1 (167.9)	60.4 (252.9)		ΔS^o
$C^o_{p(300)}$	5.7 (23.9)	11.9 (49.8)		ΔC^o_p

log A	E	log k	Conditions	Reference and Experimental Method
10.15	2.23 (9.34)	8.51(298 K)	298-450 K ? torr	[51] O.P.Strausz, W.B.O'Callaghan, E.M.Lown and H.E.Gunning, J.Amer.Chem.Soc., 93, 559 (1971). For details see $S + CH_2{=}CH_2$. Arrhenius parameters calculated from $E_{C_2H_4} - E_1 = -0.73$ and $A_1/A_{C_2H_4} = 1.4$ and assumed parameters for $S + CH_2{=}CH_2$, $E_{C_2H_4} = 1.5$ and $\log A_{C_2H_4} = 10.0$.

PREFERRED RATE PARAMETERS

log k = 8.51 at 298 K

COMMENTS

The above Arrhenius parameters require confirmation by the technique for absolute rate constants before they are accepted. See comments for $S + CH_2{=}CH_2$.

ΔH^o_f, ΔH^o and E in kcal mol^{-1} (kJ mol^{-1}); S^o, C^o_p, ΔS^o and ΔC^o_p in cal mol^{-1} K^{-1} (J mol^{-1} K^{-1}); A and k in l mol^{-1} s^{-1}

REACTION: SULPHUR ATOM + cis-DIFLUOROETHYLENE $S(I) + cis\text{-}CHF{=}CHF(II) \longrightarrow \overline{CHFCHFS}(III)$

	(I)	(II)	(III)	
$\Delta H^\circ_{f(300)}$	66.3 (277.6)	-75.2 (-314.9)		ΔH°
$S^\circ_{(300)}$	40.1 (167.9)	65.6 (274.7)		ΔS°
$C^\circ_{p(300)}$	5.7 (23.9)	13.6 (56.9)		ΔC°_p

log A	E	log k	Conditions	Reference and Experimental Method
10.23	4.21 (17.63)	7.15(298 K)	298-450 K ? torr	[51] O.P.Strausz, W.B.O'Callaghan, E.M.Lown and H.E.Gunning, J.Amer.Chem.Soc., 93, 559 (1971). For details see $S + CH_2{=}CH_2$. Arrhenius parameters calculated from $E_{C_2H_4} - E_1 = -2.71$ and $A_1/A_{C_2H_4} = 1.7$ and assumed parameters for $S + CH_2{=}CH_2$, $E_{C_2H_4} = 1.5$ and $\log A_{C_2H_4} = 10.0$.

PREFERRED RATE PARAMETERS

log k = 7.15 at 298 K

COMMENTS

The above Arrhenius parameters require confirmation by the technique for absolute rate constants before they are accepted. See comments for $S + CH_2{=}CH_2$.

REACTION: SULPHUR ATOM + TRANS-DIFLUOROETHYLENE   S(I) + trans-CHF=CHF(II) ⟶ CHFCHFS(III)

	(I)	(II)	(III)	
$\Delta H^\circ_{f(300)}$	66.3 (277.6)	-75.2 (-314.9)		ΔH°
$S^\circ_{(300)}$	40.1 (167.9)	65.6 (274.7)		ΔS°
$C^\circ_{p(300)}$	5.7 (23.9)	13.6 (56.9)		ΔC°_p

log A	E	log k	Conditions	Reference and Experimental Method
10.53	4.12 (17.3)	7.51(298 K)	298-450 K ? torr	[51] O.P.Strausz, W.B.O'Callaghan, E.M.Lown and H.E.Gunning, J.Amer.Chem.Soc., 93, 559 (1971). For details see $S + CH_2{=}CH_2$. Arrhenius parameters calculated from $E_{C_2H_4} - E_1 = -2.62$ and $A_1/A_{C_2H_4} = 3.4$ and assumed parameters for $S + CH_2{=}CH_2$, $E_{C_2H_4} = 1.5$ and $\log A_{C_2H_4} = 10.0$.

PREFERRED RATE PARAMETERS

log k = 7.51 at 298 K

COMMENTS

The above Arrhenius parameters require confirmation by the technique for absolute rate constants before they are accepted. See comments for $S + CH_2{=}CH_2$.

ΔH^o_f, ΔH^o and E in kcal mol^{-1} (kJ mol^{-1}); S^o, C^o_p, ΔS^o and ΔC^o_p in cal mol^{-1} K^{-1} (J mol^{-1} K^{-1}); A and k in l mol^{-1} s^{-1}

REACTION: SULPHUR ATOM + TETRAFLUOROETHYLENE $S(I) + CF_2{=}CF_2(II) \longrightarrow \overline{CF_2CF_2S}(III)$

	(I)	(II)	(III)	
$\Delta H^\circ_{f(300)}$	66.3 (277.6)	-157.9 (-661.1)		ΔH°
$S^\circ_{(300)}$	40.1 (167.9)	74.6 (312.4)		ΔS°
$C^\circ_{p(300)}$	5.7 (23.9)	19.4 (81.2)		ΔC°_p

log A	E	log k	Conditions	Reference and Experimental Method
10.15	2.9 (12.1)	8.02(298 K)	298-450 K ? torr	[51] O.P.Strausz, W.B.O'Callaghan, E.M.Lown and H.E.Gunning, J.Amer.Chem.Soc., 93, 559 (1971). For details see $S + CH_2{=}CH_2$. Arrhenius parameters calculated from $E_{C_2H_4} - E_1 = -1.4$ and $A_1/A_{C_2H_4} = 1.4$ and assumed parameters for $S + CH_2{=}CH_2$, $E_{C_2H_4} = 1.5$ and $\log A_{C_2H_4} = 10.0$.

PREFERRED RATE PARAMETERS

log k = 8.02 at 298 K

COMMENTS

The above Arrhenius parameters require confirmation by the technique for absolute rate constants before they are accepted. See comments for $S + CH_2{=}CH_2$.

REACTION: SULPHUR ATOM + 2-FLUOROPROPENE

$S(I) + CH_3CF{=}CH_2(II) \longrightarrow CH_3\overline{CFCH_2S}(III)$

	(I)	(II)	(III)	
$\Delta H^\circ_{f(300)}$	66.3 (277.6)			ΔH°
$S^\circ_{(300)}$	40.1 (167.9)			ΔS°
$C^\circ_{p(300)}$	5.7 (23.9)			ΔC°_p

log A	E	log k	Conditions	Reference and Experimental Method
10.08	3.0 (12.6)	7.88(298 K)	298-450 K ? torr	[51] O.P.Strausz, W.B.O'Callaghan, E.M.Lown and H.E.Gunning, J.Amer.Chem.Soc., 93, 559 (1971). For details see $S + CH_2{=}CH_2$. Arrhenius parameters calculated from $E_{C_2H_4} - E_1 = -1.5$ and $A_1/A_{C_2H_4} = 1.2$ and assumed parameters for $S + CH_2{=}CH_2$, $E_{C_2H_4} = 1.5$ and $\log A_{C_2H_4} = 10.0$.

PREFERRED RATE PARAMETERS

$\log k = 7.88$ at 298 K

COMMENTS

The above Arrhenius parameters require confirmation by the technique for absolute rate constants before they are accepted. See comments for $S + CH_2{=}CH_2$.

ΔH°_f, ΔH° and E in kcal mol^{-1} (kJ mol^{-1}); S°, C°_p, ΔS° and ΔC°_p in cal mol^{-1} K^{-1} (J mol^{-1} K^{-1}); A and k in l mol^{-1} s^{-1}

REACTION: SULPHUR ATOM + VINYL CHLORIDE $S(I) + CH_2{=}CHCl(II) \longrightarrow \overline{CH_2CHClS}(III)$

	(I)	(II)	(III)	
$\Delta H^\circ_{f(300)}$	66.3 (277.6)	8.6 (36.0)		ΔH°
$S^\circ_{(300)}$	40.1 (167.9)	63.0 (263.6)		ΔS°
$C^\circ_{p(300)}$	5.7 (23.9)	13.0 (54.4)		ΔC°_p

log A	E	log k	Conditions	Reference and Experimental Method
-	-	9.06	298 K 1330 torr	[49] E.M.Lown, E.L.Dedio, O.P.Strausz and H.E.Gunning, J.Amer.Chem.Soc., 89, 1056 (1967). For details see $S + CH_2{=}CH_2$. Recalculated from relative rate constants $k_1/k_{C_2H_4} = 1.4$ and preferred value for $S + CH_2{=}CH_2$ (log k = 8.91 at 298 K).
10.53	2.02 (8.46)	9.05(298 K)	298-450 K ? torr	[51] O.P.Strausz, W.B.O'Callaghan, E.M.Lown and H.E.Gunning, J.Amer.Chem.Soc., 93, 559 (1971). For details see $S + CH_2{=}CH_2$. Arrhenius parameters calculated from $E_{C_2H_4} - E_1 = -0.52$ and $A_1/A_{C_2H_4} = 3.4$ and assumed parameters for $S + CH_2{=}CH_2$, $E_{C_2H_4} = 1.5$ and $\log A_{C_2H_4} = 10.0$.

PREFERRED RATE PARAMETERS

log k = 10.06 at 298 K

COMMENTS

The Arrhenius parameters require confirmation by the technique for absolute rate constants before they are accepted. See comments for $S + CH_2{=}CH_2$.

REACTION: SULPHUR ATOM + BUTADIENE-1,3

$S(I) + CH_2{=}CHCH{=}CH_2(II) \longrightarrow CH_2{=}CH\overline{CHCH_2S}(III)$

	(I)	(II)	(III)		
$\Delta H^\circ_{f(300)}$	66.3 (277.6)	26.1 (109.2)	37.8 (158.3)	ΔH°	-54.6 (-228.6)
$S^\circ_{(300)}$	40.1 (167.9)	68.0 (284.5)	86.8 (363.4)	ΔS°	-41.3 (-172.9)
$C^\circ_{p(300)}$	5.7 (23.9)	19.1 (80.0)	21.6 (90.4)	ΔC°_p	-3.2 (-13.4)

log A	E	log k	Conditions	Reference and Experimental Method
10.38	-0.54 (-2.26)	10.78(298 K)	298-450 K ? torr	[51] O.P.Strausz, W.B.O'Callaghan, E.M.Lown and H.E.Gunning, J.Amer.Chem.Soc., 93, 559 (1971). For details see $S + CH_2{=}CH_2$. Arrhenius parameters calculated from $E_{C_2H_4} - E_1 = 2.04$ and $A_1/A_{C_2H_4} = 2.4$ and assumed parameters for $S + CH_2{=}CH_2$, $E_{C_2H_4} = 1.5$ and $\log A_{C_2H_4} = 10.0$.

PREFERRED RATE PARAMETERS

log k = 10.78 at 298 K

COMMENTS

The above Arrhenius parameters require confirmation by the technique for absolute rate constants before they are accepted. See comments for $S + CH_2{=}CH_2$.

ΔH°_f, ΔH° and E in kcal mol^{-1} (kJ mol^{-1}); S°, C°_p, ΔS° and ΔC°_p in cal mol^{-1} K^{-1} (J mol^{-1} K^{-1}); A and k in l mol^{-1} s^{-1}

REACTION: SULPHUR ATOM + CYCLOPENTENE

S (I) +  (II) ⟶ 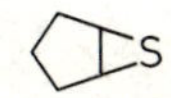 (III)

	(I)	(II)	(III)		
$\Delta H^\circ_{f(300)}$	66.3 (277.6)	8.6 (36.0)	17.2 (72.0)	ΔH°	-57.7 (-241.6)
$S^\circ_{(300)}$	40.1 (167.9)	70.8 (296.4)	75.5 (316.1)	ΔS°	-35.4 (-148.2)
$C^\circ_{p(300)}$	5.7 (23.9)	18.1 (75.8)	20.8 (87.1)	ΔC°_p	-3.0 (-12.6)

log A	E	log k	Conditions	Reference and Experimental Method
-	-	10.16	298 K 1330 torr	[49] E.M.Lown, E.L.Dedio, O.P.Strausz and H.E.Gunning, J.Amer.Chem.Soc., 89, 1056 (1967). For details see $S + CH_2{=}CH_2$. Recalculated from relative rate constants $k_1/k_{C_2H_4} = 18$ and preferred value for $S + CH_2{=}CH_2$ (log k = 8.91 at 298 K).
9.83	-0.65 (-2.72)	10.31(298 K)	298-450 K ? torr	[51] O.P.Strausz, W.B.O'Callaghan, E.M.Lown and H.E.Gunning, J.Amer.Chem.Soc., 93, 559 (1971). For details see $S + CH_2{=}CH_2$. Arrhenius parameters calculated from $E_{C_2H_4} - E_1 = 2.15$ and $A_1/A_{C_2H_4} = 0.67$ and assumed parameters for $S + CH_2{=}CH_2$, $E_{C_2H_4} = 1.5$ and $\log A_{C_2H_4} = 10.0$.

PREFERRED RATE PARAMETERS

log k = 10.24 at 298 K

COMMENTS

The preferred rate constant is the mean of the above results. The Arrhenius parameters require confirmation by the technique for absolute rate constants before they are accepted. See comments for $S + CH_2{=}CH_2$.

REACTION: SULPHUR ATOM + ACETYLENE   S(I) + CH≡CH(II) ⟶ CH=CHS(III)

	(I)	(II)	(III)	
$\Delta H^\circ_{f(300)}$	66.3 (277.6)	54.2 (226.6)		ΔH°
$S^\circ_{(300)}$	40.1 (167.9)	48.0 (200.7)		ΔS°
$C^\circ_{p(300)}$	5.7 (23.9)	10.5 (43.9)		ΔC°_p

log A	E	log k	Conditions	Reference and Experimental Method
10.79	3.5 (14.7)	8.22(298 K)	298-450 K ? torr	[51] O.P.Strausz, W.B.O'Callaghan, E.M.Lown and H.E.Gunning, J.Amer.Chem.Soc., 93, 559 (1971). For details see S + CH_2=CH_2. Arrhenius parameters calculated from $E_{C_2H_4} - E_1 = -2.0$ and $A_1/A_{C_2H_4} = 6.2$ and assumed parameters for S + CH_2=CH_2, $E_{C_2H_4} = 1.5$ and $\log A_{C_2H_4} = 10.0$.

PREFERRED RATE PARAMETERS

log k = 8.22 at 298 K

COMMENTS

The above Arrhenius parameters require confirmation by the technique for absolute rate constants before they are accepted. See comments for S + CH_2=CH_2.

ΔH^o_f, ΔH^o and E in kcal mol^{-1} (kJ mol^{-1}); S^o, C^o_p, ΔS^o and ΔC^o_p in cal mol^{-1} K^{-1} (J mol^{-1} K^{-1}); A and k in l mol^{-1} s^{-1}

REACTION: SULPHUR ATOM + PROPYNE $S(I) + CH_3C{\equiv}CH(II) \longrightarrow CH_3C{=}CHS(III)$

	(I)	(II)	(III)	
$\Delta H^\circ_{f(300)}$	66.3 (277.6)	44.4 (185.9)		ΔH°
$S^\circ_{(300)}$	40.1 (167.9)	61.5 (257.5)		ΔS°
$C^\circ_{p(300)}$	5.7 (23.9)	14.6 (61.1)		ΔC°_p

log A	E	log k	Conditions	Reference and Experimental Method
10.79	2.4 (10.0)	9.03(298 K)	298-450 K ? torr	[51] O.P.Strausz, W.B.O'Callaghan, E.M.Lown and H.E.Gunning, J.Amer.Chem.Soc., 93, 559 (1971). For details see $S + CH_2{=}CH_2$. Arrhenius parameters calculated from $E_{C_2H_4} - E_1 = -0.9$ and $A_1/A_{C_2H_4} = 6.2$ and assumed parameters for $S + CH_2{=}CH_2$, $E_{C_2H_4} = 1.5$ and $\log A_{C_2H_4} = 10.0$.

PREFERRED RATE PARAMETERS

log k = 9.03 at 298 K

COMMENTS

The above Arrhenius parameters require confirmation by the technique for absolute rate constants before they are accepted. See comments for $S + CH_2{=}CH_2$.

REACTION: SULPHUR ATOM + BUTYNE-2 $S(I) + CH_3C{\equiv}CCH_3(II) \longrightarrow CH_3\overline{C{=}C(CH_3)S}(III)$

	(I)	(II)	(III)	
$\Delta H^\circ_{f(300)}$	66.3 (277.6)	34.9 (146.1)		ΔH°
$S^\circ_{(300)}$	40.1 (167.9)	69.1 (289.2)		ΔS°
$C^\circ_{p(300)}$	5.7 (23.9)	18.6 (77.8)		ΔC°_p

log A	E	log k	Conditions	Reference and Experimental Method
10.43	0.20 (0.80)	10.28(298 K)	298-450 K ? torr	[51] O.P.Strausz, W.B.O'Callaghan, E.M.Lown and H.E.Gunning, J.Amer.Chem.Soc., 93, 559 (1971). For details see $S + CH_2{=}CH_2$. Arrhenius parameters calculated from $E_{C_2H_4} - E_1 = 1.3$ amd $A_1/A_{C_2H_4} = 2.7$ and assumed parameters for $S + CH_2{=}CH_2$, $E_{C_2H_4} = 1.5$ and $\log A_{C_2H_4} = 10.0$.

PREFERRED RATE PARAMETERS

log k = 10.28 at 298 K

COMMENTS

The above Arrhenius parameters require confirmation by the technique for absolute rate constants before they are accepted. See comments for $S + CH_2{=}CH_2$.

ΔH^o_f, ΔH^o and E in kcal mol^{-1} (kJ mol^{-1}); S^o, C^o_p, ΔS^o and ΔC^o_p in cal mol^{-1} K^{-1} (J mol^{-1} K^{-1}); A and k in l mol^{-1} s^{-1}

REACTION: CHLORINE ATOM + ETHYLENE $\dot{C\ell}(I) + CH_2{=}CH_2(II) \longrightarrow CH_2C\ell\dot{C}H_2(III)$

	(I)	(II)	(III)		
$\Delta H^\circ_{f(300)}$	28.9 (121.0)	12.5 (52.4)	19.8 (88.7)	ΔH°	-21.6 (-90.3)
$S^\circ_{(300)}$	39.5 (165.4)	52.4 (219.4)	68.2 (285.4)	ΔS°	-23.7 (-99.2)
$C^\circ_{p(300)}$	5.2 (21.9)	10.3 (43.1)	14.8 (61.9)	ΔC°_p	-0.7 (-2.9)

log A	E	log k	Conditions	Reference and Experimental Method
10.2	0	10.2(373 K)	323-423 K 20-100 torr	[54] R.Eckling, P.Goldfinger, G.Huybrechts, G.Martens, L.Meyers and S.Smoes, Chem.Ber., 93, 3014 (1960). Calculated from data of various sources.
-	-	-	298 K 80-120 torr	[55] F.S.Dainton, D.A.Lomax and M.Weston, Trans.Faraday Soc., 58, 308 (1962). Static system. $C\ell$ from photolysis (λ = 365 nm) of $C\ell_2$. Reaction rates measured by rate of decrease of pressure (Bourdon gauge). Empirical rate law: $\text{rate} = k_o I^{\frac{1}{2}}_{abs}[O\ell]^{\frac{1}{2}}[C\ell_2]/(a + b[C\ell_2])^{\frac{1}{2}}$ explained by the following reactions: $\dot{C\ell} + O\ell \xrightarrow{1} C\ell\dot{O}\ell^*$ $C\ell\dot{O}\ell^* \xrightarrow{2} \dot{C\ell} + O\ell$ $C\ell\dot{O}\ell^* + M \xrightarrow{3} C\ell\dot{O}\ell + M$ $C\ell\dot{O}\ell + C\ell_2 \xrightarrow{4} C\ell O\ell C\ell + \dot{C\ell}$ $2\ C\ell\dot{O}\ell \xrightarrow{5}$ dimer or $C\ell O\ell C\ell + O\ell$ $C\ell\dot{O}\ell + \dot{C\ell} \xrightarrow{6} C\ell O\ell C\ell$ or $C\ell_2 + O\ell$ $2\ \dot{C\ell} + M \xrightarrow{7} C\ell_2 + M$ $\dot{C\ell}$ + wall $\xrightarrow{8}$ removal of $\dot{C\ell}$ with termination predominantly by reaction (6). Rate constants for reaction (1) were not obtained.
-	-	9.3	308 K 4-280 torr	[56] A.K.E.Hagopian, J.H.Knox and E.A.Thompson, Bull.Soc.Chim.Belg., 71, 764 (1962). Static system. $C\ell$ from photolysis (W-Filament lamp) of $C\ell_2$. Competitive technique (cf. reference [57]) using reaction of $C\ell$ with C_3H_8. Mechanism for chain propagation:

log A	E	log k	Conditions	Reference and Experimental Method
				$\dot{C}l + Ol \xrightarrow{1} Cl\dot{O}l^*$ $Cl\dot{O}l^* \xrightarrow{2} \dot{C}l + Ol$ $Cl\dot{O}l^* + M_i \xrightarrow{3} Cl\dot{O}l + M_i$ $Cl\dot{O}l + Cl_2 \xrightarrow{4} ClOlCl + \dot{C}l$ $\dot{C}l + C_3H_8 \xrightarrow{9} \dot{C}_3H_7 + HCl$ $\dot{C}_3H_7 + Cl_2 \xrightarrow{10} C_3H_7Cl + \dot{C}l$ leading to the rate expression: $\text{Relative rate} = \frac{[ClOlCl][RH]}{[Ol][C_3H_7Cl]} = \frac{k_1}{k_9}\left\{\frac{\Sigma k_3[M_i]}{k_2 + \Sigma k_3[M_i]}\right\}$ where $[ClOlCl]$ and $[C_3H_7Cl]$ are the final concentrations of products and [RH] and $[Ol]$ are the initial concentrations of reactants. k_1/k_9 is obtained from the high pressure limit of the relative rate expression. (See also Cl + cis-CHCl=CHCl system, reference [57]). The rate of addition of Cl_2 to C_2H_4 could not be measured owing to extensive surface reaction which could not be suppressed.
-	-	10.65	310 K 150-600 torr	[57] J.A.Franklin, P.Goldfinger and G.Huybrechts, Ber.Bunsenges, 72, 173 (1968). Static system. Cl from photolysis (λ = 435.8 nm) of Cl_2. Cl reacted with mixtures of C_2H_4 and C_2H_6 with and without added SF_6 or CO_2. $-d[C_2H_4]/dt$ measured by rate of decrease of total pressure $(-dp(total)/dt)$, $-d[C_2H_6]/dt$ obtained from $-d[Cl_2]/dt + dp(total)/dt$. Stoichiometry confirmed to correspond to $C_2H_4 + Cl_2 \longrightarrow 1,2\text{-}C_2H_4Cl_2$ and $C_2H_6 + Cl_2 \longrightarrow C_2H_5Cl + HCl$ by analysis of products (gas chromatography). Proposed mechanism: $\dot{C}l + C_2H_4 \xrightarrow{1} \dot{C}_2H_4Cl^*$ $\dot{C}_2H_4Cl^* \xrightarrow{2} \dot{C}l + C_2H_4$ $\dot{C}_2H_4Cl^* + M \xrightarrow{3} \dot{C}_2H_4Cl + M$ $\dot{C}_2H_4Cl + Cl_2 \xrightarrow{4} 1,2\text{-}C_2H_4Cl_2 + \dot{C}l$ $\dot{C}l + C_2H_6 \xrightarrow{11} \dot{C}_2H_5 + HCl$ $\dot{C}_2H_5 + Cl_2 \xrightarrow{12} C_2H_5Cl + \dot{C}l$ which leads to a rate law of the form $\text{Rate} = k_o[1 + (k/P)]$

ΔH_f^o, ΔH^o and E in kcal mol^{-1} (kJ mol^{-1}); S^o, C_p^o, ΔS^o and ΔC_p^o in cal mol^{-1} K^{-1} (J mol^{-1} K^{-1}); A and k in l mol^{-1} s^{-1}

$\log A_P$	E	log k	Conditions	Reference and Experimental Method
				as confirmed by the results, where $k_0 = k_{11}/k_1$ and P is the total pressure of the system. Extrapolations of R against 1/P gave $k_{11}/k_1 = 0.38$ for three different sets of pressure conditions. Taking $\log k_{11} = 10.23$ at 310 K [G.C.Fettis and J.H.Knox, Prog.Reaction.Kinetics, 2, 3 (1964).] gives $\log k_1 = 10.65$
[10.63]	-	-	352 K	[58] P.C.Beadle, J.H.Knox, F.Placido and K.C.Waugh, Trans.Faraday Soc., 65, 1571 (1969). A-factor calculated from transition state theory.
-	-	-	300-400 K 300 torr	[59] J.H.Knox and K.C.Waugh, Trans.Faraday Soc., 65, 1585 (1969). As for reference [56] above.

PREFERRED RATE PARAMETERS

log k = 10.65

COMMENTS

The data of reference [57] appear to be reliable. They are based on a mechanism involving chemically activated CH_2CH_2Cl radicals and collisional deactivation of the radicals by all molecular species in the system. This type of mechanism has been established by Knox and Riddick [Trans.Faraday Soc., 62, 1190 (1966)] in a very comprehensive study in which it was shown that chemically activated chloroalkyl radicals do not undergo transfer reactions with Cl_2, as previously believed.

The above preferred rate constant is, however, subject to any uncertainty in the measurement of the rate constant of the reaction $Cl + C_2H_6 \xrightarrow{11} HCl + C_2H_5$. Davis, Braun and Bass [Int.J.Chem.Kinet., 2, 101 (1970)] have recently made the absolute measurement, $\log k_{11} = 10.61$ at 298 K, from flash photolysis and resonance fluorescence studies. This latter value is about a factor of 2 greater than the value on which k_1 is based, and if accepted would give $\log k_1 = 11.03$. The absolute measurements of k_{11} require confirmation before they are accepted in preference to the earlier data for this reaction based on competitive experiments.

There is little reliable experimental evidence on the temperature dependence of k_1, but it seems likely that the above preferred rate constant is temperature independent.

From the thermodynamic data and the rate constant of the reverse reaction listed by Benson and O'Neal [NSRDS-NBS 21 (1970)], $\log(k_{-1}/s^{-1}) = 13 - 23.9/2.3RT$, it can be calculated that $\log k_1 = 9.6 - 2.9/2.3RT$ at 298 K. The agreement with the preferred rate parameters is not very good, but is within the accumulated errors which are quite large in this case.

REACTION: CHLORINE ATOM + PROPENE

$\dot{C}\ell(I) + CH_3CH{=}CH_2(II) \xrightarrow{1} CH_3CHC\ell\dot{C}H_2(III)$

$\xrightarrow{2} CH_3\dot{C}HCH_2C\ell(IV)$

	(I)	(II)	(III)	(IV)		Path 1	Path 2
$\Delta H^\circ_{f(300)}$	28.9 (121.0)	4.9 (20.5)	[12 (50)]	[12 (50)]	ΔH°	[-22 (-92)]	[-22 (-92)]
$S^\circ_{(300)}$	39.5 (165.4)	63.8 (267.0)	77.0 (322.4)	76.8 (321.6)	ΔS°	-26.3 (-110.1)	-26.5 (-111.0)
$C^\circ_{p(300)}$	5.2 (21.8)	15.3 (64.0)	21.2 (88.8)	20.3 (85.0)	ΔC°_p	0.7 (2.9)	-0.2 (-0.8)

Path	log A	E	log k	Conditions	Reference and Experimental Method
1 + 2	-	-	10.25(308 K)	308-568 K 18-162 torr	[56] A.K.E.Hagopian, J.H.Knox and E.A.Thompson, Bull.Soc.Chim.Belg., 71, 764 (1962). For details see $C\ell + CH_2{=}CH_2$.

PREFERRED RATE PARAMETERS

$\log k_{1+2} > 10.25$

COMMENTS

The above rate constant seems low in relation to $\log k_1 = 10.65$ for $H + CH_2{=}CH_2$. In addition, a very low rate constant for the $H + CH_2{=}CH_2$ reaction was reported in reference [56]; consequently, we suggest that the above rate constant should be regarded as a lower limit. The rate constant is probably temperature independent.

ΔH^o_f, ΔH^o and E in kcal mol^{-1} (kJ mol^{-1}); S^o, C^o_p, ΔS^o and ΔC^o_p in cal mol^{-1} K^{-1} (J mol^{-1} K^{-1}); A and k in l mol^{-1} s^{-1}

REACTION: CHLORINE ATOM + 2-METHYLPROPENE (ISOBUTENE) $\dot{C\ell}(I) + (CH_3)_2C{=}CH_2(II) \xrightarrow{1} (CH_3)_2CC\ell\dot{C}H_2(III)$

$\xrightarrow{2} (CH_3)_2\dot{C}CH_2C\ell(IV)$

	(I)	(II)	(III)	(IV)		Path 1	Path 2
$\Delta H^\circ_{f(300)}$	28.9 (121.0)	-3.6 (-15.1)	[2 (8)]	[2 (8)]	ΔH°	[-23 (-96)]	[-23 (-96)]
$S^\circ_{(300)}$	39.5 (165.4)	71.3 (298.5)	84.0 (351.7)	85.7 (358.8)	ΔS°	-26.8 (-112.2)	-25.1 (-105.1)
$C^\circ_{p(300)}$	5.2 (21.8)	21.6 (90.4)	27.7 (116.0)	25.3 (105.9)	ΔC°_p	0.9 (3.8)	-1.5 (-6.3)

Path	log A	E	log k	Conditions	Reference and Experimental Method
1 + 2	-	-	10.4	308 K 2-30 torr	[56] A.K.E.Hagopian, J.H.Knox and E.A.Thompson, Bull.Soc.Chim.Belg., 71, 764 (1962). For details see $C\ell + CH_2{=}CH_2$.

PREFERRED RATE PARAMETERS

$\log k_{1+2} > 10.4$

COMMENTS

See comments for $C\ell + CH_3CH{=}CH_2$ system. The rate constant is probably temperature independent. The enthalpy changes indicate that both paths 1 and 2 should be operative.

REACTION: CHLORINE ATOM + 1,1-DIFLUOROETHYLENE

$$\dot{C\ell}(I) + CF_2{=}CH_2(II) \xrightarrow{1} CF_2C\ell\dot{C}H_2(III)$$
$$\xrightarrow{2} \dot{C}F_2CH_2C\ell(IV)$$

	(I)	(II)	(III)	(IV)		Path 1	Path 2
$\Delta H^\circ_{f(300)}$	28.9 (121.0)	[-71.5 (-299.4)]	[-71 (-297)]	[-68 (-285)]	ΔH°	[-28 (-117)]	[-25 (-105)]
$S^\circ_{(300)}$	39.5 (165.4)	64.9 (271.7)	71.7 (300.2)	78.3 (327.8)	ΔS°	-32.7 (-136.9)	-26.1 (-109.3)
$C^\circ_{p(300)}$	5.2 (21.8)	14.8 (62.0)	19.7 (82.5)	18.8 (78.7)	ΔC°_p	-0.3 (-1.3)	-1.2 (-5.0)

Path	log A	E	log k	Conditions	Reference and Experimental Method
1 + 2	-	-	-	293-323 K 130-270 torr	[60] N.R.Bergamin, E.Castellano and H.J.Schumacher, Zeit.Physik.Chem.Frankfurt, 40, 246 (1964). Static system. $C\ell$ from photolysis (λ = 436 nm) of $C\ell_2$. Reaction rates determined from rate of pressure decrease (quartz spiral manometer) stoichiometry confirmed to be $C\ell_2 + O\ell = C\ell O\ell C\ell$ by pressure change and product analysis. Rate law found to be: $d[C\ell O\ell C\ell]/dt = k[C\ell_2]I_a^{\frac{1}{2}}$ and mechanism proposed: $C\ell_2 + h\nu \xrightarrow{I} 2\,\dot{C\ell}$ $\dot{C\ell} + O\ell \xrightarrow{1+2} C\ell\dot{O}\ell$ $C\ell\dot{O}\ell + C\ell_2 \xrightarrow{3} C\ell O\ell C\ell + \dot{C\ell}$ $2\,C\ell\dot{O}\ell \xrightarrow{4}$ products Steady state treatment gives $d[C\ell O\ell C\ell]/dt = (k_3/k_4^{\frac{1}{2}})[C\ell_2]I_a^{\frac{1}{2}}$ so that reaction (3) is the rate controlling step, and the experimental activation energy (1.9 kcal mol^{-1}) was equated to E_3. Results indicate that reactions (1) and (2) are very fast, probably with $E_{1+2} \simeq 0$ kcal mol^{-1}.

PREFERRED RATE PARAMETERS

No quantitative data available

COMMENTS

The reaction is clearly fast with E close to 0 kcal mol^{-1}.

ΔH^o_f, ΔH^o and E in kcal mol^{-1} (kJ mol^{-1}); S^o, C^o_p, ΔS^o and ΔC^o_p in cal mol^{-1} K^{-1} (J mol^{-1} K^{-1}); A and k in l mol^{-1} s^{-1}

REACTION: CHLORINE ATOM + TETRAFLUOROETHYLENE $\dot{C\ell}(I) + CF_2{=}CF_2(II) \longrightarrow \dot{C}F_2CF_2C\ell(III)$

	(I)	(II)	(III)		
$\Delta H^\circ_{f(300)}$	28.9 (121.0)	-157.9 (-661.1)	[-157 (-657)]	ΔH°	[-28 (-117)]
$S^\circ_{(300)}$	39.5 (165.4)	74.6 (312.4)	81.0 (339.1)	ΔS°	-33.1 (-138.6)
$C^\circ_{p(300)}$	5.2 (21.8)	19.4 (81.2)	23.6 (98.8)	ΔC°_p	-1.0 (-4.2)

log A	E	log k	Conditions	Reference and Experimental Method
-	-	-	303-333 K 150-546 torr	[61] E.Castellano, N.R.Bergamin and H.J.Schumacher, Zeit.Physik.Chem.Frankfurt, 27, 112 (1961). Same method, rate law and mechanism as for $C\ell + CH_2{=}CF_2$ reaction. Overall activation energy (1 kcal mol^{-1}) was equated to E_3 for the rate-controlling step, $C\ell\dot{O}\ell + C\ell_2 \xrightarrow{3} C\ell O\ell C\ell + \dot{C\ell}$

PREFERRED RATE PARAMETERS

No quantitative data available

COMMENTS

The reaction is clearly fast with E close to 0 kcal mol^{-1}.

REACTION: CHLORINE ATOM + TRIFLUOROCHLOROETHYLENE

$$\dot{C}\ell(I) + CF_2{=}CFC\ell(II) \xrightarrow{1} \dot{C}F_2CFC\ell_2(III)$$
$$\xrightarrow{2} CF_2C\ell\dot{C}FC\ell(IV)$$

	(I)	(II)	(III)	(IV)		Path 1	Path 2
$\Delta H^\circ_{f(300)}$	28.9 (121.0)	[-125 (-523)]			ΔH°		
$S^\circ_{(300)}$	39.5 (165.4)	77.1 (322.8)			ΔS°		
$C^\circ_{p(300)}$	5.2 (21.8)	20.0 (83.7)			ΔC°_p		

Path	log A	E	log k	Conditions	Reference and Experimental Method
1 + 2	-	-	-	303-333 K 150-546 torr	[61] E.Castellano, N.R.Bergamin and H.J.Schumacher, Zeit.Physik.Chem.Frankfurt, 27, 112 (1961). Same method, rate law and mechanism as for $C\ell + CH_2{=}CF_2$ reaction. Overall activation energy (2.3 kcal mol^{-1}) was equated to E_3 for the rate-controlling step, $C\ell\dot{O}\ell + C\ell_2 \xrightarrow{3} C\ell OC\ell C\ell + \dot{C}\ell$ It is not possible to distinguish which path is more favourable, from these experiments.

PREFERRED RATE PARAMETERS

No quantitative data available

COMMENTS

The reaction is clearly fast with E close to 0 kcal mol^{-1}.

ΔH°_f, ΔH° and E in kcal mol^{-1} (kJ mol^{-1}); S°, C°_p, ΔS° and ΔC°_p in cal mol^{-1} K^{-1} (J mol^{-1} K^{-1}); A and k in l mol^{-1} s^{-1}

REACTION: CHLORINE ATOM + 1,1-DIFLUORO-2,2-DICHLOROETHYLENE

$$\dot{C\ell}(I) + CF_2{=}CC\ell_2(II) \xrightarrow{1} \dot{C}F_2CC\ell_3(III)$$
$$\xrightarrow{2} CF_2C\ell\dot{C}C\ell_2(IV)$$

	(I)	(II)	(III)	(IV)
$\Delta H^\circ_{f(300)}$	28.9 (121.0)	-79.3 (-332.0)	[-73 (-306)]	[-82 (-343)]
$S^\circ_{(300)}$	39.5 (165.4)	69.4 (290.6)	90.9 (380.6)	85.6 (358.4)
$C^\circ_{p(300)}$	5.2 (21.8)	21.1 (88.3)	26.3 (109.7)	25.8 (108.0)

	Path 1	Path 2
ΔH°	[-23 (-96)]	[-31 (-130)]
ΔS°	-18.0 (-75.4)	-23.3 (-97.6)
ΔC°_p	-0.1 (-0.4)	-0.5 (-2.1)

Path	log A	E	log k	Conditions	Reference and Experimental Method
1 + 2	-	-	-	303-343 K 77-350 torr	[62] C.Vallana, E.Castellano and H.J.Schumacher, Zeit.Physik.Chem.Frankfurt,. 46, 294 (1965). Same method, rate law and mechanism as for $C\ell$ + $CH_2{=}CF_2$ reaction. Overall activation energy (5.4 kcal mol^{-1}) was equated to E_3 for the rate-controlling step, $C\ell\dot{O}\ell + C\ell_2 \xrightarrow{3} C\ell O\ell C\ell + \dot{C\ell}$ It is not possible to distinguish which path is more favourable from these experiments.

PREFERRED RATE PARAMETERS

No quantitative data available

COMMENTS

The reaction is clearly fast with E close to 0 kcal mol^{-1}.

REACTION: CHLORINE ATOM + VINYL CHLORIDE

$\dot{C}\ell(I) + CH_2{=}CHC\ell(II) \xrightarrow{1} CH_2C\ell\dot{C}HC\ell(III)$

$\xrightarrow{2} \dot{C}H_2CHC\ell_2(IV)$

	(I)	(II)	(III)	(IV)
$\Delta H^\circ_{f(300)}$	28.9 (121.0)	8.6 (36.0)	[14 (59)]	[15 (63)]
$S^\circ_{(300)}$	39.5 (165.4)	63.0 (263.6)	78.4 (328.3)	74.9 (313.6)
$C^\circ_{p(300)}$	5.2 (21.8)	13.0 (54.4)	17.5 (73.3)	18.0 (75.3)

	Path 1	Path 2
ΔH°	[-24 (-100)]	[-23 (-96)]
ΔS°	-24.1 (-100.8)	-27.6 (-115.6)
ΔC°_p	-0.7 (-2.9)	-0.2 (-0.8)

Path	log A	E	log k	Conditions	Reference and Experimental Method
1 + 2	10.2	0	10.2	323-423 K 20-100 torr	[54] R.Eckling, P.Goldfinger, G.Huybrechts, G.Martens, L.Meyers and S.Smoes, Chem.Ber., 93, 3014 (1960). Calculated from data of various sources.
1 + 2	9.5	1.0 (4.2)	8.8(313 K)	298-328 K 50 torr	[63] P.B.Ayscough, A.J.Cocker, F.S.Dainton, S.Hirst and M.Weston, Proc.Chem.Soc., 244 (1961). Preliminary communication summarising results but with no experimental details.
1 + 2	-	-	-	298 K 52-200 torr	[55] F.S.Dainton, D.A.Lomax and M.Weston, Trans.Faraday Soc., 58, 308 (1962). See $C\ell + CH_2{=}CH_2$ system.
1 + 2	10.3	1.5 (6.3)	9.3(328 K)	303-353 K 65-350 torr	[64] P.B.Ayscough, A.J.Cocker, F.S.Dainton and S.Hirst, Trans.Faraday Soc., 58, 318 (1962). Static system. $C\ell$ from photolysis (λ = 365.5 nm) of $C\ell_2$ in presence of pairs of chloroethylenes. Relative reaction rates were measured by analysis of addition products (gas chromatography) and checked against pressure decrease during reaction (Bourdon gauge). Proposed reaction scheme:

$$\dot{C}\ell + O\ell \xrightarrow{1,2} C\ell\dot{O}\ell^*$$
$$\dot{C}\ell + O\ell' \xrightarrow{1',2'} C\ell\dot{O}\ell'^*$$
$$C\ell\dot{O}\ell^* \xrightarrow{3} \dot{C}\ell + O\ell$$
$$C\ell\dot{O}\ell'^* \xrightarrow{3'} \dot{C}\ell + O\ell'$$
$$C\ell\dot{O}\ell^* + C\ell_2 \xrightarrow{4} O\ell + \dot{C}\ell_3(\rightarrow C\ell_2 + \dot{C}\ell)$$
$$C\ell\dot{O}\ell'^* + C\ell_2 \xrightarrow{4'} O\ell' + \dot{C}\ell_3(\rightarrow C\ell_2 + \dot{C}\ell)$$
$$C\ell\dot{O}\ell^* + C\ell_2 \xrightarrow{5} C\ell\dot{O}\ell + C\ell_2$$
$$C\ell\dot{O}\ell'^* + C\ell_2 \xrightarrow{5'} C\ell\dot{O}\ell' + C\ell_2$$
$$C\ell\dot{O}\ell + C\ell_2 \xrightarrow{6} C\ell O\ell C\ell + \dot{C}\ell$$
$$C\ell\dot{O}\ell' + C\ell_2 \xrightarrow{6'} C\ell O\ell'C\ell + \dot{C}\ell$$

ΔH^o_f, ΔH^o and E in kcal mol^{-1} (kJ mol^{-1}); S^o, C^o_p, ΔS^o and ΔC^o_p in cal mol^{-1} K^{-1} (J mol^{-1} K^{-1}); A and k in l mol^{-1} s^{-1}

Path	log A	E	log k	Conditions	Reference and Experimental Method
					leading to the rate expression, $$\frac{d[C\ell O\ell C\ell]/dt}{d[C\ell O\ell' C\ell]/dt} = \frac{k_{1+2}[O\ell]}{k_{1'+2'}[O\ell']} \; \frac{1 + (k_{3'} + k_{4'}[C\ell_2])/k_{5'}[C\ell_2]}{1 + (k_3 + k_4[C\ell_2])/k_5[C\ell_2]}$$ Since it was not possible to determine the separate rate constants for the above reactions of the vibrationally excited adduct radicals, k_{1+2} and $k_{1'+2'}$ could not be determined directly. Making reasonable assumptions about the rate constants for reactions 3,3',4,4',5 and 5' indicates that the differences in the rate constants for Cℓ addition to chloroethylenes are not much larger than the experimental errors, and consequently, for CH_2=CHCℓ, cis-CHCℓ=CHCℓ and CHCℓ=C$C\ell_2$, it was assumed that E_{1+2} = 1.5 ± 1.0 kcal mol^{-1} and log A_{1+2} = 10.3 ± 0.6. It should be noted that in this interpretation collisional deactivation of CℓOℓ* is restricted to $C\ell_2$.
1 + 2	[9.64]	-	-	352 K	[58] P.C.Beadle, J.H.Knox, F.Placido and K.C.Waugh, Trans.Faraday Soc., 65. 1571 (1969). A-factor calculated from transition state theory.
1 + 2	-	-	10.64	352 K 20-300 torr	[59] J.H.Knox and K.C.Waugh, Trans.Faraday Soc., 65, 1585 (1969). See Cℓ + C_2H_4, reference [64]. As for C_2H_4, no data on the rate of addition of $C\ell_2$ to CH_2=CHCℓ was obtained, owing to extensive surface reaction. The above value of k for Cℓ + CH_2=CHCℓ was calculated by Knox and Waugh from the data of Ayscough et al. [64], assuming the mechanism of Knox and Riddick (see Cℓ + cis-CHCℓ=CHCℓ system, reference [67]).

PREFERRED RATE PARAMETERS

log k_{1+2} = 10.64

COMMENTS

The preferred rate constant is that calculated by Knox and Waugh [59] from the data of reference [64] on the competitive chlorinations of trans-$CHC\ell=CHC\ell$ and $CH_2=CHC\ell$, by a re-interpretation involving the mechanism proposed by Knox and Riddick [Trans.Faraday Soc., 62, 1190 (1966)]. Thus Knox and Waugh calculated $k_{trans\text{-}CHC\ell=CHC\ell}/k_{CH_2=CHC\ell} = 1.1$, and taking log $k_{trans\text{-}CHC\ell=CHC\ell} = 10.60$, this gives log $k_{CH_2=CHC\ell} = 10.64$.

As for the $C\ell + C_2H_4$ system (see Comments), it should be pointed out that $k_{trans\text{-}CHC\ell=CHC\ell}$, upon which the preferred value depends, in its turn depends upon $k_{C_3H_8}$, and there is some recent evidence to suggest that the absolute values for k_{RH} may be too low.

The limited experimental evidence available indicates that $k_{CH_2=CHC\ell}$ is essentially temperature independent.

From the thermodynamic data and the rate constant of the reverse reaction listed by Benson and O'Neal [NSRDS-NBS 21, 1970], $\log(k_{-1}/s^{-1}) = 13 - 23.8/2.3RT$, it can be calculated that $\log k_1 = 9.6 - 0.4/2.3RT$ at 298 K. The agreement with the preferred rate parameters is not very good, but it is within the accumulated errors which are quite large in this case.

The enthalpy changes indicate that path 1 should be the favoured route for addition.

REACTION: CHLORINE ATOM + 1,1-DICHLOROETHYLENE

$$\dot{C\ell}(I) + CH_2{=}CC\ell_2(II) \xrightarrow{1} CC\ell_3\dot{C}H_2(III)$$
$$\xrightarrow{2} CH_2C\ell\dot{C}C\ell_2(IV)$$

	(I)	(II)	(III)	(IV)
$\Delta H^o_{f(300)}$	28.9 (121.0)	[4.3 (18.0)]	[15 (63)]	[7 (29)]
$S^o_{(300)}$	39.5 (165.4)	69.7 (291.7)	81.6 (341.7)	82.9 (346.9)
$C^o_{p(300)}$	5.2 (21.8)	16.5 (69.0)	21.5 (90.0)	20.0 (83.7)

	Path 1	Path 2
ΔH^o	[-18 (-75)]	[-26 (-109)]
ΔS^o	-27.6 (-115.6)	-26.3 (-110.1)
ΔC^o_p	-0.2 (-0.8)	-1.7 (-7.1)

Path	log A	E	log k	Conditions	Reference and Experimental Method
1 + 2	[9.3]	[0]	-	298-321 K 70-330 torr	[65] P.B.Ayscough, F.S.Dainton and B.E.Fleischfresser, Trans.Faraday Soc., 62, 1838 (1966). Static system. $C\ell$ from continuous or intermittent photolysis (λ = 365.5 nm) of $C\ell_2$. Stoichiometry, $C\ell_2 + CH_2{=}CC\ell_2 = CH_2C\ell CC\ell_3$, confirmed by analysis of product (gas chromatography). Reaction rates measured by decrease in pressure (Bourdon gauge). A mechanism involving vibrationally excited adduct radicals was proposed (see $C\ell + C_2H_4$ system,

ΔH^o_f, ΔH^o and E in kcal mol^{-1} (kJ mol^{-1}); S^o, C^o_p, ΔS^o and ΔC^o_p in cal mol^{-1} K^{-1} (J mol^{-1} K^{-1}); A and k in l mol^{-1} s^{-1}

Path	log A	E	log k	Conditions	Reference and Experimental Method
					reference [55]), but no rate constants for the addition, $C\ell$ to $CH_2{=}CC\ell_2$, were obtained. It is argued that addition occurs mainly to produce the $CH_2C\ell\dot{C}C\ell_2$ radical (path 2) on the basis of previous results. The above Arrhenius parameters are given in this paper as being calculated from previous results on the basis of $E_{CH_2=CHC\ell} = 0$.
1 + 2	[9.86]	-	-	352 K	[58] P.C.Beadle, J.H.Knox, F.Placido and K.C.Waugh, Trans.Faraday Soc., 65, 1571 (1969). A-factor calculated from transition state theory.

PREFERRED RATE PARAMETERS

$\log k_{1+2} = [10.5]$

COMMENTS

The above tentative value is an estimate based on the preferred rate constants for $C\ell$ additions to other chloroethylenes. The enthalpy changes indicate that path 2 should be the favoured route for addition.

REACTION: CHLORINE ATOM + *cis*-DICHLOROETHYLENE $\dot{C}\ell$(I) + *cis*-$CHC\ell{=}CHC\ell$(II) ⟶ $CHC\ell_2\dot{C}HC\ell$(III)

	(I)	(II)	(III)		
$\Delta H^\circ_{f(300)}$	28.9 (121.0)	1.0 (4.2)	[9 (38)]	ΔH°	[-21 (-88)]
$S^\circ_{(300)}$	39.5 (165.4)	70.8 (296.5)	82.9 (347.0)	ΔS°	-27.4 (-114.7)
$C^\circ_{p(300)}$	5.2 (21.8)	15.8 (66.2)	20.7 (86.6)	ΔC°_p	-0.3 (-1.3)

log A	E	log k	Conditions	Reference and Experimental Method
9.4	0.95 (3.97)	8.7(313 K)	298-328 K 50 torr	[63] P.B.Ayscough, A.J.Cocker, F.S.Dainton, S.Hirst and M.Weston, Proc.Chem.Soc., 244 (1961). Preliminary communication summarizing results, but with no experimental details.
10.3	1.2 (5.0)	9.47(323 K)	303-343 K 50-300 torr	[66] P.B.Ayscough, A.J.Cocker, F.S.Dainton and S.Hirst, Trans.Faraday Soc., 58, 295 (1962). Static system. $C\ell$ from continuous and intermittent photolysis (λ = 365.5 nm) of $C\ell_2$ in presence of *cis*-$CHC\ell{=}CHC\ell$. Stoichiometry was confirmed to be $C\ell_2 + CHC\ell{=}CHC\ell = CHC\ell_2CHC\ell_2$ by gas chromatographic analysis of the product. Reaction rate measured from rate of decrease of pressure in the system. A mechanism involving vibrationally excited adduct radicals was proposed (see $C\ell + C_2H_4$ system, reference [55]). The concentration of $C\ell$ atoms was determined from lifetime studies involving rotating sector measurements, and rate constants for the addition reaction, $C\ell + CHC\ell{=}CHC\ell \longrightarrow CHC\ell_2CHC\ell^*$, were calculated from the overall rates and the concentrations of $C\ell$ and $CHC\ell{=}CHC\ell$.
10.3	1.5±1.0 (6.3)	9.3(328 K)	303-353 K 65-350 torr	[64] P.B.Ayscough, A.J.Cocker, F.S.Dainton and S.Hirst, Trans.Faraday Soc., 58, 318 (1962). For details see $C\ell + CH_2{=}CHC\ell$.
10.90	0.19 (0.80)	10.78(352 K)	308-406 K 60-380 torr	[67] J.H.Knox and J.Riddick, Trans.Faraday Soc., 62, 1190 (1966). Static system. $C\ell$ from photolysis (W-filament lamp) of $C\ell_2$ in presence of olefin and propane with and without added CO_2. Competitive method; addition reaction measured relative to $C\ell$ attack on C_3H_8. Reaction rates measured by rates of formation of products (gas chromatography). The following reaction scheme was considered: $\dot{C}\ell + cis\text{-}O\ell \xrightarrow{1} C\ell\dot{O}\ell^*$ $C\ell\dot{O}\ell^* \xrightarrow{2} x\ cis\text{-}O\ell + (1-x)trans\text{-}O\ell + \dot{C}\ell$

ΔH^o_f, ΔH^o and E in kcal mol^{-1} (kJ mol^{-1}); S^o, C^o_p, ΔS^o and ΔC^o_p in cal mol^{-1} K^{-1} (J mol^{-1} K^{-1}); A and k in l mol^{-1} s^{-1}

log A	E	log k	Conditions	Reference and Experimental Method
				$Cl\dot{O}l^* + M_i \xrightarrow{3} Cl\dot{O}l + M_i$ $Cl\dot{O}l^* + Cl_2 \xrightarrow{4}$ y cis-Ol + (1-y)trans-Ol + $\dot{C}l_3$ $Cl\dot{O}l^* + Ol \xrightarrow{5}$ z cis-Ol + (1-z)trans-Ol + $Cl\dot{O}l$ $Cl\dot{O}l^* + Ol \xrightarrow{6}$ z'cis-Ol + (1-z')trans-Ol + $Cl\dot{O}l^*$ $Cl\dot{O}l^* \xrightarrow{7} Cl\dot{O}l$ $\dot{C}l + C_3H_8 \xrightarrow{8} HCl + n\text{-}\dot{C}_3H_7$ $n\text{-}\dot{C}_3H_7 + Cl_2 \xrightarrow{9} n\text{-}C_3H_7Cl + \dot{C}l$ The usual termination reactions were assumed, although the relative rates are independent of the exact termination reactions. The rate of formation of $n\text{-}C_3H_7Cl$ by reactions (8) and (9) was used as the internal standard. Reaction rate ratios were calculated from $R_{ClOlCl}/R_{C_3H_7Cl} = [ClOlCl][C_3H_8]/[nC_3H_7Cl][Ol]$, where [ClOlCl] and $[n\text{-}C_3H_7Cl]$ are the final concentrations of products and $[C_3H_8]$ and [Ol] initial concentrations of reactants. M_i in equation (3) refers to the various molecular species capable of undergoing energy transfer. Steady state expressions were derived and tested from which it was established that the overall rate of the exchange reactions (4), (5) and (6), is less than 5% of the deactivation rate. The following simplified expressions applied: $R_{ClOlCl}/R_{n\text{-}C_3H_7Cl} = \frac{k_1}{k_8} \times \frac{\Sigma k_3[M_i] + k_7}{k_2 + \Sigma k_3[M_i] + k_7}$ $R_{trans\text{-}Ol}/R_{n\text{-}C_3H_7Cl} = \frac{k_1}{k_8} \times \frac{(1-x)k_2}{k_2 + \Sigma k_3[M_i] + k_7}$ where $R_{trans\text{-}Ol}$ is the rate of formation of trans-CHCl=CHCl from reaction (2). Values of k_1/k_8 were obtained from the intercepts of plots of $R_{ClOlCl}/R_{n\text{-}C_3H_7Cl}$ against $R_{trans\text{-}Ol}/R_{n\text{-}C_3H_7Cl}$. It was shown that ClOl* can be deactivated with the same efficiency by Cl_2, ClOlCl, C_3H_8 and CO_2. Measurement of the temperature coefficient of the above rate ratios yielded the Arrhenius parameters given above, based on log k_8 = 11.01 - 0.98/2.3RT [G.C.Fettis and J.H.Knox, Prog.Reaction Kinetics, 2, 3 (1964)].
[9.78]	-	-	352 K	[58] P.C.Beadle, J.H.Knox, F.Placido and K.C.Waugh, Trans.Faraday Soc., 65, 1571 (1969). A-factor calculated from transition state theory.

PREFERRED RATE PARAMETERS

log k = 10.78

COMMENTS

The kinetic analysis of this system carried out by Knox and Riddick [67] is undoubtedly the most extensive and plausible, and the derived data is preferred here.

There still remains the controversy regarding the values of $k_{C_3H_8}$ on which $k_{cis\text{-}CHC\ell=CHC\ell}$ is based (see Comments on previous $C\ell$ + chloroethylene systems).

The small temperature coefficient found by Knox and Riddick [67] is not significant in relation to the errors involved, and the rate constant can be taken to be temperature independent.

From the thermodynamic data and the rate constant of the reverse reaction listed by Benson and O'Neal [NSRDS-NBS 21 (1970)], $\log(k_{-1}/s^{-1}) = 12 - 20.3/2.3RT$, it can be calculated that $\log k_1 = 7.8 + 0.1/2.3RT$ at 298 K. The calculated value of A_1 indicates that the experimental value of A_{-1} is much too low.

REACTION: CHLORINE ATOM + TRANS-DICHLOROETHYLENE

$\dot{C\ell}$(I) + trans-$CHC\ell=CHC\ell$(II) ⟶ $CHC\ell_2\dot{C}HC\ell$(III)

	(I)	(II)	(III)		
$\Delta H^\circ_{f(300)}$	28.9 (121.0)	1.2 (5.0)	[9 (38)]	ΔH°	[-21 (-88)]
$S^\circ_{(300)}$	39.5 (165.4)	70.8 (296.5)	82.9 (347.0)	ΔS°	-27.4 (-114.7)
$C^\circ_{p(300)}$	5.2 (21.8)	15.8 (66.2)	20.7 (86.6)	ΔC°_p	-0.3 (-1.3)

log A	E	log k	Conditions	Reference and Experimental Method
10.48	-0.17 (-0.7)	10.60(352 K)	308-406 K 60-380 torr	[67] J.H.Knox and J.Riddick, Trans.Faraday Soc., 62, 1190 (1966). The above Arrhenius parameters are based on $\log k_8 = 11.01 - 0.98/2.3RT$, as before. For details see $C\ell$ + cis-$CHC\ell=CHC\ell$. Experiments on trans-$CHC\ell=CHC\ell$ performed in conjunction with cis-$CHC\ell=CHC\ell$ experiments, and data analyzed in the same way.
[10.0]	-	-	352 K	[58] P.C.Beadle, J.H.Knox, F.C.Placido and K.C.Waugh, Trans.Faraday Soc., 65, 1571 (1969). A-factor calculated from transition state theory.
-	-	-	352 K 20-300 torr	[59] J.H.Knox and K.C.Waugh, Trans.Faraday Soc., 65, 1585 (1969). For details see $C\ell + C_2H_4$ system and $C\ell$ + cis-$CHC\ell=CHC\ell$ system [67].

ΔH^o_f, ΔH^o and E in kcal mol^{-1} (kJ mol^{-1}); S^o, C^o_p, ΔS^o and ΔC^o_p in cal mol^{-1} K^{-1} (J mol^{-1} K^{-1}); A and k in l mol^{-1} s^{-1}

log A	E	log k	Conditions	Reference and Experimental Method
				A few experiments on the $C\ell_2$/trans-$CHC\ell{=}CHC\ell$ system were carried out in the absence of C_3H_8, yielding data on the decomposition of the $CHC\ell CHC\ell_2$* radical in agreement with the data of Knox and Riddick [67], and confirming their mechanism and interpretations. No new data on $C\ell$ + trans-$CHC\ell{=}CHC\ell$.

PREFERRED RATE PARAMETERS

log k = 10.60

COMMENTS

See comments for $C\ell$ + cis-$CHC\ell{=}CHC\ell$ system. The rate constant is probably temperature independent. From the thermodynamic data and the rate constant for the reverse reaction listed by Benson and O'Neal [NSRDS-NBS, 21 (1970)], $\log(k_{-1}/s^{-1}) = 12 - 20.3/2.3RT$, it can be calculated that $\log k_1 = 7.8 + 0.1/2.3RT$ at 298 K. The calculated value of A_1 indicates that the experimental value of A_{-1} is much too low.

REACTION: CHLORINE ATOM + TRICHLOROETHYLENE

$$\dot{C\ell}(I) + CHC\ell{=}CC\ell_2(II) \xrightarrow{1} CHC\ell_2\dot{C}C\ell_2(III)$$
$$\xrightarrow{2} \dot{C}HC\ell CC\ell_3(IV)$$

	(I)	(II)	(III)	(IV)		Path 1	Path 2
$\Delta H^\circ_{f(300)}$	28.9 (121.0)	-2.0 (-8.4)	[7 (29)]	[10 (42)]	ΔH°	[-20 (-84)]	[-17 (-71)]
$S^\circ_{(300)}$	39.5 (165.4)	77.5 (324.3)	89.2 (373.0)	87.4 (365.9)	ΔS°	-27.8 (-116.3)	-29.6 (-123.9)
$C^\circ_{p(300)}$	5.2 (21.8)	19.3 (80.8)	23.9 (100.0)	24.9 (104.2)	ΔC°_p	-0.6 (-2.5)	0.4 (1.7)

Path	log A	E	log k	Conditions	Reference and Experimental Method
1 + 2	10.0	0	-	323-423 K 20-100 torr	[54] R.Eckling, P.Goldfinger, G.Huybrechts, G.Martens, L.Meyers and S.Smoes, Chem.Ber., 93, 3014 (1960). Calculated from data of various sources.
1 + 2	9.6	0.7 (2.9)	9.1(313 K)	298-328 K 50 torr	[63] P.B.Ayscough, A.J.Cocker, F.S.Dainton, S.Hirst and M.Weston, Proc.Chem. Soc., 244 (1961). Preliminary communication summarizing results but with no experimental details.

Path	log A	E	log k	Conditions	Reference and Experimental Method
1 + 2	10.3	1.5 (6.3)	9.3(328 K)	303-353 K 65-350 torr	[64] P.B.Ayscough, A.J.Cocker, F.S.Dainton and S.Hirst, Trans.Faraday Soc., 58, 318 (1962). For details see $C\ell + CH_2{=}CHC\ell$. From the experimental data in this paper, Knox and Waugh [59] have recalculated the ratio $k_{CHC\ell=CC\ell_2}/k_{trans\text{-}CHC\ell=CHC\ell} = 1.9$, based on the mechanism of Knox and Riddick [67]. Hence, taking $\log k_{trans\text{-}CHC\ell=CHC\ell} = 10.6$, it follows that $\log k_{CHC\ell=CC\ell_2} = 10.3$.
1 + 2	-	-	9.75(441 K)	430-452 K 100-200 torr	[68] G.Huybrechts, L.Meyers and G.Verbeke, Trans.Faraday Soc., 58, 1128 (1962). Static system. $C\ell$ from photolysis of $C\ell_2$. Reaction rates from rates of formation of products continuously monitored by mass spectrometry. $C\ell$ addition to $CHC\ell{=}CC\ell_2$ measured relative to reaction $\dot{C}\ell + C_2HC\ell_5 \longrightarrow HC\ell + \dot{C}_2C\ell_5$.
1 + 2	9.3	0	-	298-313 K 70-330 torr	[65] P.B.Ayscough, F.S.Dainton and B.E.Fleischfresser, Trans.Faraday Soc., 62, 1838 (1966). For details see $C\ell + CH_2{=}CC\ell_2$. No results on addition of $C\ell$ to $CHC\ell{=}CC\ell_2$ were obtained, and the above Arrhenius parameters are quoted in this paper as being calculated from previous results on the basis of $E_{CHC\ell=CC\ell_2} = 0$.
1 + 2	[9.25]	-	-	352 K	[58] P.C.Beadle, J.H.Knox, F.C.Placido and K.C.Waugh, Trans.Faraday Soc., 65, 1571 (1969). A-factor calculated by transition state theory.
1 + 2	10.54	0	10.54(350 K)	300-400 K 50 torr	[59] J.H.Knox and K.C.Waugh, Trans.Faraday Soc., 65, 1585 (1969). Above Arrhenius parameters based on $\log k_8 = 11.01 - 0.98/2.3RT$ [G.C.Fettis and J.H.Knox, Prog.Reaction Kinetics, 2, 3 (1964)]. For details see $C\ell + CH_2{=}CH_2$ and $C\ell$ + cis-$CHC\ell{=}CHC\ell$ system reference [67].

PREFERRED RATE PARAMETERS

$\log k_{1+2} = 10.54$

COMMENTS

See comments for $C\ell$ + cis-$CHC\ell{=}CHC\ell$ system.

ΔH_f^o, ΔH^o and E in kcal mol^{-1} (kJ mol^{-1}); S^o, C_p^o, ΔS^o and ΔC_p^o in cal mol^{-1} K^{-1} (J mol^{-1} K^{-1}); A and k in l mol^{-1} s^{-1}

From the thermodynamic data and the rate constant of the reverse reaction listed by Benson and O'Neal [NSRDS-NBS 21 (1970)], $\log(k_{-1}/s^{-1}) = 13.7 - 20.4/2.3RT$, it can be calculated that $\log k_1 = 9.4 - 1.0/2.3RT$ at 298 K. The agreement between the calculated value of k_1 and the preferred value of k_{1+2} is not very good, but is probably within the rather large accumulated errors.

The enthalpy changes indicate that addition probably occurs by both paths 1 and 2.

REACTION: CHLORINE ATOM + TETRACHLOROETHYLENE $\dot{C\ell}(I) + C\ell_2{=}CC\ell_2(II) \longrightarrow CC\ell_3\dot{C}C\ell_2(III)$

	(I)	(II)	(III)		
$\Delta H^\circ_{f(300)}$	28.9 (121.0)	[-2.7 (-11.3)]	[9 (38)]	ΔH°	[-17 (-71)]
$S^\circ_{(300)}$	39.5 (165.4)	81.3 (340.4)	95.5 (399.7)	ΔS°	-25.3 (-105.8)
$C^\circ_{p(300)}$	5.2 (21.8)	22.8 (95.5)	28.1 (117.6)	ΔC°_p	0.1 (0.4)

log A	E	log k	Conditions	Reference and Experimental Method
8.4	-0.5 (-2.1)	8.7(417 K)	358-475 K ? torr	[69] P.Goldfinger, M.Jeunehomme and G.Martens, J.Chem.Phys., 29, 456 (1958). Summary of early data on photochlorination reactions, but all data based on reaction schemes without participation of activated chloroalkyl radicals.
9.6	0	-	323-423 K 20-100 torr	[54] R.Eckling, P.Goldfinger, G.Huybrechts, G.Martens, L.Meyers and S.Smoes, Chem.Ber., 93, 3014 (1960). Calculated from data of various sources.
9.6	0	-	323-423 K 50 torr	[63] P.B.Ayscough, A.J.Cocker, F.S.Dainton, S.Hirst and M.Weston, Proc.Chem.Soc., 244 (1961). Preliminary communication summarizing results from various data sources with no experimental details.
9.36	0	9.36	360-475 K 140-310 torr	[70] P.Goldfinger, G.Huybrechts and G.Martens, Trans.Faraday Soc., 57, 2210 (1961). Static system. $C\ell$ from photolysis of $C\ell_2$. Competitive system $C\ell + CC\ell_2{=}CC\ell_2$ and one of CH_4, $CH_2C\ell_2$, $CHC\ell_3$ or $C_2HC\ell_5$. Overall reaction rates measured by decrease in pressure and [$C\ell$] determined photometrically. Extrapolation of rate ratio to infinite pressure gave values for k_1/k_2: $\dot{C\ell} + CC\ell_2{=}CC\ell_2 \xrightarrow{1} \dot{C}_2C\ell_5$ $\dot{C\ell} + RH \xrightarrow{2} \dot{R} + HC\ell$ Above result is average for all different RH molecules based on data for

log A	E	log k	Conditions	Reference and Experimental Method
				H-abstraction of Knox and Nelson [Trans.Faraday Soc., 55, 937 (1959)]. Knox and Waugh [59] suggest a value of log k_1 = 10.20, based on more complete data for competitive reaction (2).
(i)[10.28] (ii) [8.68]	-	-	352 K	[58] P.C.Beadle, J.H.Knox, F.C.Placido and K.C.Waugh, Trans.Faraday Soc., 65, 1571 (1969). A factors calculated from transition state theory, (i) based on same model as for other chloroethylene systems, (ii) based on different models both for the energized molecule and activated complex.
9.41	0	9.41(350 K)	300-400 K 7-230 torr	[59] J.Knox and K.C.Waugh, Trans.Faraday Soc., 65, 1585 (1969). Above result based on log k_8 = 11.01 - 0.98/2.3RT [G.C.Fettis and J.H.Knox, Prog. Reaction Kinetics, 2, 3 (1964)]. For details see $C\ell + CH_2{=}CH_2$ and $C\ell + CHC\ell{=}CC\ell_2$. The rate constant for $C\ell$ addition to $CC\ell_2{=}CC\ell_2$ is independent of pressure above 7 torr.
-	-	10.1	310-348 K 40-642 torr	[71] J.A.Franklin, G.Huybrechts and C.Cillien, Trans.Faraday Soc., 65, 2094 (1969). For details see $C\ell + CH_2{=}CH_2$ and reference [70] above. Addition reaction competed against H-abstraction from 1,2-$C_2H_4C\ell_2$ and C_2H_6. Using same notation as in [70], k_1/k_2 was obtained from a plot of the rate ratio against the reciprocal of the $C\ell_2$ pressure. The above result is an average value from separate studies with 1,2-$C_2H_4C\ell_2$ and C_2H_6 as the internal standards.
-	-	9.5	298 K 7-98 torr	[72] D.D.Davis, W.Braun and A.M.Bass, Int.J.Chem.Kinet., 2, 101 (1970). Static system. $C\ell$ generated from vacuum ultraviolet flash photolysis of $CC\ell_4$ moderated with He. $[C\ell]$ monitored by resonance fluorescence absorption. Reaction rates measured by $d[C\ell]/dt$ under conditions of low $[C\ell]$ (pseudo-first order). Absolute rate constants calculated from rate of change in $[C\ell]$. Bimolecular rate constants independent of total pressure (> 7 torr) and of pressure of $CC\ell_2{=}CC\ell_2$ (0.002-0.015 torr).

PREFERRED RATE PARAMETERS

log k = [9.5] at 298 K

ΔH_f^o, ΔH^o and E in kcal mol^{-1} (kJ mol^{-1}); S^o, C_p^o, ΔS^o and ΔC_p^o in cal mol^{-1} K^{-1} (J mol^{-1} K^{-1}); A and k in l mol^{-1} s^{-1}

COMMENTS

There is a very wide spread of measured rate constants for this reaction, even among the three most recent references [59,71,72]. The apparent agreement between references [59] and [72] is unreal, since the value log k_1 = 9.41 at 350 K, reported in reference [59], is based on log $k_{C_3H_8}$ = 10.5 at 350 K, which may be underestimated if the data on $k_{C_2H_6}$ in reference [72] is correct. At the same time, the new data [72] on $k_{C_2H_6}$ has the effect of increasing k_1 based on the results of reference [71].

It is difficult to accept log k_1 = 9.5, independent of temperature, in relation to the rate constants for the other chloroethylenes which are an order of magnitude higher.

A possible explanation is that $C\ell$ addition to $CC\ell_2$=$CC\ell_2$ may involve a small activation energy (1-2 kcal mol^{-1}). It would be useful if the absolute rate constants for $C\ell$ additions to other chloroethylenes were measured by the kinetic spectroscopic method [72].

From the thermodynamic data and the rate constant of the reverse reaction listed by Benson and O'Neal [NSRDS-NBS 21 (1970)] $\log(k_{-1}/s^{-1})$ = 12.8 - 16.8/2.3RT, it can be calculated that log k_1 = 8.6 - 0.4/2.3RT, which indicates that the Arrhenius parameters A_{-1} and E_{-1} have been considerably underestimated.

REACTION: CHLORINE ATOM + PERFLUOROCYCLOBUTENE

$\dot{C}l$ (I) + (II) ⟶ (III)

	(I)	(II)	(III)	
$\Delta H^\circ_{f(300)}$	28.9 (121.0)			ΔH°
$S^\circ_{(300)}$	39.5 (165.4)			ΔS°
$C^\circ_{p(300)}$	5.2 (21.9)			ΔC°_p

log A	E	log k	Conditions	Reference and Experimental Method
-	-	-	283-343 K 50-700 torr	[73] Z.R.Alberto, J.J.Cosa, C.A.Vallana and E.H.Staricco, Can.J.Chem., 49, 1252 (1971). Same method, rate law and mechanism as for $C\ell$ + CH_2=CF_2 reaction. Overall activation energy (4.6 kcal mol^{-1}) was equated to E_3 for the rate-controlling step: $C\ell\dot{O}\ell + C\ell_2 \xrightarrow{3} C\ell O\ell C\ell + C\ell$.

PREFERRED RATE PARAMETERS

No quantitative data available

COMMENTS

The reaction is clearly fast with E close to 0 kcal mol^{-1}.

REACTION: CHLORINE ATOM + PERFLUOROCYCLOHEXENE

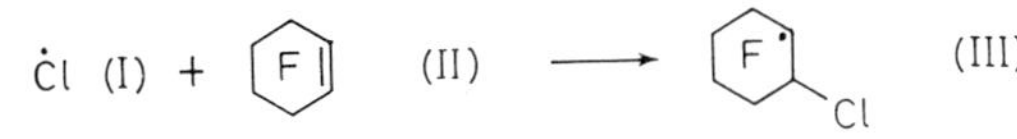

	(I)	(II)	(III)	
$\Delta H^\circ_{f(300)}$	28.9 (121.0)	-462.7 (-1938.0)		ΔH°
$S^\circ_{(300)}$	39.5 (165.4)			ΔS°
$C^\circ_{p(300)}$	5.2 (21.9)			ΔC°_p

log A	E	log k	Conditions	Reference and Experimental Method
-	-	-	283-323 K 50-300 torr	[74] J.J.Cosa, C.A.Vallana and E.H.Staricco, Can.J.Chem., 47, 1067 (1969). Same method, rate law and mechanism as for $Cl + CH_2{=}CF_2$ reaction. Overall activation energy (5.1 kcal mol^{-1}) was equated to E_3 for the rate-controlling step: $Cl\dot{O}l + Cl_2 \xrightarrow{3} ClOlCl + \dot{C}l$

PREFERRED RATE PARAMETERS

No quantitative data available

COMMENTS

The reaction is clearly fast, with E close to 0 kcal mol^{-1}.

ΔH^o_f, ΔH^o and E in kcal mol^{-1} (kJ mol^{-1}); S^o, C^o_p, ΔS^o and ΔC^o_p in cal mol^{-1} K^{-1} (J mol^{-1} K^{-1}); A and k in l mol^{-1} s^{-1}

REACTION: SELENIUM ATOM + ETHYLENE $Se(I) + CH_2{=}CH_2(II) \longrightarrow [\overline{CH_2CH_2Se}](III)$

	(I)	(II)	(III)	
$\Delta H^\circ_{f(300)}$	49.2 (206.0)	12.5 (52.3)		ΔH°
$S^\circ_{(300)}$	42.2 (176.7)	52.4 (219.4)		ΔS°
$C^\circ_{p(300)}$	5.0 (20.9)	10.3 (43.1)		ΔC°_p

log A	E	log k	Conditions	Reference and Experimental Method
-	-	-	298 K	[75] A.B.Callear and W.J.R.Tyerman, Proc.Chem.Soc., 296 (1964). Static system. $Se(^3P)$ from flash photolysis of CSe_2 moderated with N_2. Photolysis in presence of alkenes causes an increase in the rate of decay of Se, and strong new bands appear in far ultraviolet spectrum, ascribed to ethylene selenide. Relative rate constants for Se additions to alkenes were derived by measuring the intensities of the ethylene selenide spectra produced by photolyses of CSe_2 in mixtures with C_2H_4 and a second alkene.
10.03	2.81 (11.8)	8.41(357 K)	302-412 K 55-105 torr	[76] A.B.Callear and W.J.R.Tyerman, Trans.Faraday Soc., 62, 371 (1966). Static system. $Se(^3P)$ from flash photolysis of CSe_2 moderated with N_2. [Se] determined from absorption $5^3S \leftarrow 4^3P$ at 196 nm, and rate constants were obtained from -d[Se]/dt or from the rate of formation of the ethylene selenide bands.
-	-	-	298 K 50-95 torr	[77] A.B.Callear and W.J.R.Tyerman, Trans.Faraday Soc., 62, 2760 (1966). Same system and results as reference [75].

PREFERRED RATE PARAMETERS

Reliable as given above

COMMENTS

The absorption spectra of flashed CSe_2 + olefin mixtures show band systems in the 2000-2300 Å region which have been ascribed to olefin selenides [76]. These adduct structures also seem likely by analogy with O and S atom reactions with olefins.

REACTION: SELENIUM ATOM + PROPENE $Se(I) + CH_3CH{=}CH_2(II) \longrightarrow [\overline{CH_3CHCH_2Se}](III)$

	(I)	(II)	(III)	
$\Delta H^\circ_{f(300)}$	49.2 (206.0)	4.9 (20.5)		ΔH°
$S^\circ_{(300)}$	42.2 (176.7)	63.8 (267.0)		ΔS°
$C^\circ_{p(300)}$	5.0 (20.9)	15.3 (64.0)		ΔC°_p

log A	E	log k	Conditions	Reference and Experimental Method
10.11	2.35 (9.84)	8.68(360 K)	302-418 K 50 torr	[75] A.B.Callear and W.J.R.Tyerman, Trans.Faraday Soc., 62, 2760 (1966); Proc.Chem. Soc., 296 (1964). For details see $Se + CH_2{=}CH_2$. Relative rate data were checked by measuring the intensities of the propylene selenide absorptions as well as those for ethylene selenide. Above Arrhenius parameters were calculated from $E_{C_2H_4} - E_{C_3H_6}$ = 0.46 kcal mol^{-1} and log $(A_{C_3H_6}/A_{C_2H_4})$ = 0.077, and the equation log $k_{C_2H_4}$ = 10.03 - 2.81/2.3RT.
-	-	8.56	301 K	[76] A.B.Callear and W.J.R.Tyerman, Trans.Faraday Soc., 62, 371 (1966). For details see $Se + CH_2{=}CH_2$. Absolute rate constant determined from the intensities of the propylene selenide absorptions only.

PREFERRED RATE PARAMETERS

log k = 10.11 - 2.35/2.3RT

COMMENTS

There is a slight discrepancy between the absolute value k(301 K) from reference [76] and the value calculated from the above Arrhenius parameters (log k = 8.40 at 301 K) based on the relative data. See also comments for $Se + CH_2{=}CH_2$.

ΔH^o_f, ΔH^o and E in kcal mol^{-1} (kJ mol^{-1}); S^o, C^o_p, ΔS^o and ΔC^o_p in cal mol^{-1} K^{-1} (J mol^{-1} K^{-1}); A and k in l mol^{-1} s^{-1}

REACTION: SELENIUM ATOM + BUTENE-1   Se(I) + $CH_3CH_2CH{=}CH_2$(II) ⟶ [$CH_3CH_2\overline{CHCH_2Se}$](III)

	(I)	(II)	(III)	
$\Delta H^\circ_{f(300)}$	49.2 (206.0)	0.0 (0.0)		ΔH°
$S^\circ_{(300)}$	42.2 (176.7)	73.6 (308.2)		ΔS°
$C^\circ_{p(300)}$	5.0 (20.9)	20.6 (86.3)		ΔC°_p

log A	E	log k	Conditions	Reference and Experimental Method
10.49	2.25 (9.42)	9.12(360 K)	302-418 K 100 torr	[75] A.B.Callear and W.J.R.Tyerman, Proc.Chem.Soc., 296 (1964); Trans.Faraday Soc., 62, 2760 (1966). For details see Se + $CH_2{=}CH_2$. Relative rate data were checked by measuring the intensities of the butene-1 selenide absorptions as well as those for ethylene selenide. The above Arrhenius parameters were calculated from $E_{C_2H_4} - E_{C_4H_8\text{-}1}$ = 0.56 kcal mol^{-1} and $\log (A_{C_4H_8\text{-}1}/A_{C_2H_4})$ = 0.45 and the equation $\log k_{C_2H_4}$ = 10.03 - 2.81/2.3RT.

PREFERRED RATE PARAMETERS

Reliable as given above

COMMENTS

See comments for Se + $CH_2{=}CH_2$.

REACTION: SELENIUM ATOM + cis-BUTENE-2

Se(I) + cis-$CH_3CH{=}CHCH_3$(II) ⟶ [$CH_3\overline{CHCH(CH_3)Se}$](III)

	(I)	(II)	(III)	
$\Delta H^\circ_{f(300)}$	49.2 (206.0)	-1.9 (-7.9)		ΔH°
$S^\circ_{(300)}$	42.2 (176.7)	73.6 (308.2)		ΔS°
$C^\circ_{p(300)}$	5.0 (20.9)	19.4 (81.2)		ΔC°_p

log A	E	log k	Conditions	Reference and Experimental Method
10.28	1.21 (5.07)	9.55(360 K)	302-418 K 50 torr	[75] A.B.Callear and W.J.R.Tyerman, Proc.Chem.Soc., 296 (1964); Trans.Faraday Soc., 62, 2760 (1966). For details see Se + $CH_2{=}CH_2$. The above Arrhenius parameters were calculated from $E_{C_2H_4} - E_{cis\text{-}C_4H_8\text{-}2} = 1.60$ kcal mol^{-1} and log $(A_{cis\text{-}C_4H_8\text{-}2}/A_{C_2H_4}) = 0.257$, and the equation log $k_{C_2H_4} = 10.03 - 2.81/2.3RT$.

PREFERRED RATE PARAMETERS

Reliable as given above

COMMENTS

See comments for Se + $CH_2{=}CH_2$.

ΔH^o_f, ΔH^o and E in kcal mol^{-1} (kJ mol^{-1}); S^o, C^o_p, ΔS^o and ΔC^o_p in cal mol^{-1} K^{-1} (J mol^{-1} K^{-1}); A and k in l mol^{-1} s^{-1}

REACTION: SELENIUM ATOM + TRANS-BUTENE-2 $Se(I) + trans\text{-}CH_3CH{=}CHCH_3(II) \longrightarrow [CH_3CHCH(CH_3)Se](III)$

	(I)	(II)	(III)	
$\Delta H^\circ_{f(300)}$	49.2 (206.0)	-2.9 (-12.1)		ΔH°
$S^\circ_{(300)}$	42.2 (176.7)	72.4 (303.1)		ΔS°
$C^\circ_{p(300)}$	5.0 (20.9)	20.7 (86.7)		ΔC°_p

log A	E	log k	Conditions	Reference and Experimental Method
10.24	0.59 (2.47)	9.88(360 K)	302-418 K 50 torr	[75] A.B.Callear and W.J.R.Tyerman, Proc.Chem.Soc., 296 (1964); Trans.Faraday Soc., 62, 2760 (1966). For details see Se + $CH_2{=}CH_2$. The above Arrhenius parameters were calculated from $E_{C_2H_4} - E_{trans\text{-}C_4H_8\text{-}2} = 2.22$ kcal mol^{-1} and log $(A_{trans\text{-}C_4H_8\text{-}2}/A_{C_2H_4}) = 0.17$, and the equation log $k_{C_2H_4} = 10.03 - 2.81/2.3RT$.

PREFERRED RATE PARAMETERS

Reliable as given above

COMMENTS

See comments for Se + $CH_2{=}CH_2$.

REACTION: SELENIUM ATOM + 2-METHYLPROPENE (ISOBUTENE) $Se(I) + (CH_3)_2C{=}CH_2(II) \longrightarrow [(CH_3)_2\overline{CCH_2Se}](III)$

	(I)	(II)	(III)	
$\Delta H^\circ_{f(300)}$	49.2 (206.0)	-3.6 (-15.1)		ΔH°
$S^\circ_{(300)}$	42.2 (176.7)	71.3 (298.5)		ΔS°
$C^\circ_{p(300)}$	5.0 (20.9)	21.6 (90.4)		ΔC°_p

log A	E	log k	Conditions	Reference and Experimental Method
10.38	1.01 (4.23)	9.77(360 K)	302-418 K 50 torr	[75] A.B.Callear and W.J.R.Tyerman, Proc.Chem.Soc., 296 (1964); Trans.Faraday Soc., 62, 2760 (1966). For details see $Se + CH_2{=}CH_2$. Relative rate data were checked by measuring the intensities of the isobutene selenide absorptions as well as those for ethylene selenide. The above Arrhenius parameters were calculated from $E_{C_2H_4} - E_{iso\text{-}C_4H_8} = 1.80$ kcal mol^{-1} and $\log (A_{iso\text{-}C_4H_8}/A_{C_2H_4}) = 0.34$, and the equation $\log k_{C_2H_4} = 10.03 - 2.81/2.3RT$.

PREFERRED RATE PARAMETERS

Reliable as given above

COMMENTS

See comments for $Se + CH_2{=}CH_2$.

ΔH^o_f, ΔH^o and E in kcal mol^{-1} (kJ mol^{-1}); S^o, C^o_p, ΔS^o and ΔC^o_p in cal mol^{-1} K^{-1} (J mol^{-1} K^{-1}); A and k in l mol^{-1} s^{-1}

REACTION: SELENIUM ATOM + PENTENE-1 $Se(I) + CH_3CH_2CH_2CH{=}CH_2(II) \longrightarrow [CH_3CH_2CH_2\overline{CHCH_2Se}](III)$

	(I)	(II)	(III)	
$\Delta H^\circ_{f(300)}$	49.2 (206.0)	-4.9 (-20.5)		ΔH°
$S^\circ_{(300)}$	42.2 (176.7)	83.0 (347.5)		ΔS°
$C^\circ_{p(300)}$	5.0 (20.9)	26.1 (109.3)		ΔC°_p

log A	E	log k	Conditions	Reference and Experimental Method
∿10.38	2.21 (9.25)	∿9.04(360 K)	302-418 K 50 torr	[77] A.B.Callear and W.J.R.Tyerman, <u>Trans.Faraday Soc.</u>, <u>62</u>, 2760 (1966). For details see $Se + CH_2{=}CH_2$. The above Arrhenius parameters were calculated from $E_{C_2H_4} - E_{C_5H_{10}-1} = 0.60$ kcal mol^{-1} and $\log (A_{C_5H_{10}-1}/A_{C_2H_4}) \simeq 0.36$, and the equation $\log k_{C_2H_4} = 10.03 - 2.81/2.3RT$.

PREFERRED RATE PARAMETERS

Reliable as given above

COMMENTS

See comments for $Se + CH_2{=}CH_2$.

REACTION: SELENIUM ATOM + VINYL CHLORIDE $Se(I) + CH_2{=}CHCl(II) \longrightarrow [\overline{CH_2CHClSe}](III)$

	(I)	(II)	(III)	
$\Delta H^\circ_{f(300)}$	49.2 (206.0)	8.6 (36.0)		ΔH°
$S^\circ_{(300)}$	42.2 (176.7)	63.0 (263.6)		ΔS°
$C^\circ_{p(300)}$	5.0 (20.9)	13.0 (54.4)		ΔC°_p

log A	E	log k	Conditions	Reference and Experimental Method
9.89	2.44 (10.22)	8.41(360 K)	302-418 K 50 torr	[75] A.B.Callear and W.J.R.Tyerman, Proc.Chem.Soc., 296 (1964); Trans.Faraday Soc., 62, 2760 (1966). For details see Se + $CH_2{=}CH_2$. The above Arrhenius parameters were calculated from $E_{C_2H_4} - E_{C_2H_3Cl} = 0.37$ kcal mol^{-1} and $\log (A_{C_2H_3Cl}/A_{C_2H_4}) = -0.15$, and the equation $\log k_{C_2H_4} = 10.03 - 2.81/2.3RT$.

PREFERRED RATE PARAMETERS

Reliable as given above

COMMENTS

See comments for Se + $CH_2{=}CH_2$.

ΔH^o_f, ΔH^o and E in kcal mol^{-1} (kJ mol^{-1}); S^o, C^o_p, ΔS^o and ΔC^o_p in cal mol^{-1} K^{-1} (J mol^{-1} K^{-1}); A and k in l mol^{-1} s^{-1}

REACTION: SELENIUM ATOM + BUTADIENE-1,3 $Se(I) + CH_2{=}CHCH{=}CH_2(II) \longrightarrow [CH_2{=}CH\overline{CHCH_2Se}](III)$

	(I)	(II)	(III)	
$\Delta H^{\circ}_{f(300)}$	49.2 (206.0)	26.1 (109.2)		ΔH°
$S^{\circ}_{(300)}$	42.2 (176.7)	68.0 (284.5)		ΔS°
$C^{\circ}_{p(300)}$	5.0 (20.9)	19.1 (80.0)		ΔC°_{p}

log A	E	log k	Conditions	Reference and Experimental Method
-	-	9.96	298 K 50 torr	[75] A.B.Callear and W.J.R.Tyerman, Proc.Chem.Soc., 296 (1964). For details see $Se + CH_2{=}CH_2$. Recalculated from relative rate constants $k_{C_4H_6}/k_{C_2H_4} = 98.3$ and preferred value for $Se + CH_2{=}CH_2$ ($\log k_{C_2H_4} = 7.97$ at 298 K).
10.72	0.88 (3.68)	10.19(360 K)	302-418 K 50 torr	[77] A.B.Callear and W.J.R.Tyerman, Trans.Faraday Soc., 62, 2760 (1966). For details see $Se + CH_2{=}CH_2$. Arrhenius parameters calculated from $E_{C_2H_4} - E_{C_4H_6} = 1.93$ kcal mol^{-1} and $\log (A_{C_4H_6}/A_{C_2H_4}) = 0.69$, and the equation $\log k_{C_2H_4} = 10.03 - 2.81/2.3RT$.

PREFERRED RATE PARAMETERS

$\log k = 10.72 - 0.88/2.3RT$

COMMENTS

See comments for $Se + CH_2{=}CH_2$.

REACTION: SELENIUM ATOM + ACRYLONITRILE $Se(I) + CH_2{=}CHCN(II) \longrightarrow [\overline{SeCH_2CHCN}](III)$

	(I)	(II)	(III)	
$\Delta H^\circ_{f(300)}$	49.2 (206.0)	43.7 (183.0)		ΔH°
$S^\circ_{(300)}$	42.2 (176.7)	64.2 (268.8)		ΔS°
$C^\circ_{p(300)}$	5.0 (20.9)	14.9 (62.4)		ΔC°_p

log A	E	log k	Conditions	Reference and Experimental Method
-	-	8.51	298 K 50 torr	[75] A.B.Callear and W.J.R.Tyerman, Proc.Chem.Soc., 296 (1964). For details see Se + $CH_2{=}CH_2$. Recalculated from relative rate constants $k_{CH_2=CHCN}/k_{C_2H_4}$ = 3.5 and preferred value for Se + $CH_2{=}CH_2$ (log $k_{C_2H_4}$ = 7.97 at 298 K).
~9.08	0.72 (3.01)	8.64(360 K)	302-418 K 50 torr	[77] A.B.Callear and W.J.R.Tyerman, Trans.Faraday Soc., 62, 2760 (1966). For details see Se + $CH_2{=}CH_2$. Arrhenius parameters calculated from $E_{C_2H_4} - E_{CH_2=CHCN}$ = 2.19 kcal mol^{-1} and log $(A_{CH_2=CHCN}/A_{C_2H_4}) \simeq -0.97$, and the equation log $k_{C_2H_4}$ = 10.03 - 2.81/2.3RT.

PREFERRED RATE PARAMETERS

log k = 9.08 - 0.7/2.3RT

COMMENTS

See comments for Se + $CH_2{=}CH_2$.

ΔH^o_f, ΔH^o and E in kcal mol^{-1} (kJ mol^{-1}); S^o, C^o_p, ΔS^o and ΔC^o_p in cal mol^{-1} K^{-1} (J mol^{-1} K^{-1}); A and k in l mol^{-1} s^{-1}

REACTION: SELENIUM ATOM + ALLYL CHLORIDE $Se(I) + CH_2{=}CHCH_2Cl(II) \longrightarrow [\overline{SeCH_2C}HCH_2Cl](III)$

	(I)	(II)	(III)	
$\Delta H^\circ_{f(300)}$	49.2 (206.0)	-1.3 (-5.4)		ΔH°
$S^\circ_{(300)}$	42.2 (176.7)	73.4 (307.3)		ΔS°
$C^\circ_{p(300)}$	5.0 (20.9)	18.2 (76.2)		ΔC°_p

log A	E	log k	Conditions	Reference and Experimental Method
-	-	8.33	302 K 50 torr	[77] A.B.Callear and W.J.R.Tyerman, Trans.Faraday Soc., 62, 2760 (1966). For details see $Se + CH_2{=}CH_2$. Recalculated from relative rate constants $k_{C_3H_5Cl}/k_{C_2H_4} = 2.3$, and preferred value for $Se + CH_2{=}CH_2$ ($\log k_{C_2H_4} = 7.97$ at 298 K).

PREFERRED RATE PARAMETERS

Reliable as given above

COMMENTS

See comments for $Se + CH_2{=}CH_2$.

REACTION: SELENIUM ATOM + ACETYLENE $\qquad Se(I) + CH{\equiv}CH(II) \longrightarrow [\overline{CH{=}CHSe}](III)$

	(I)	(II)	(III)	
$\Delta H^\circ_{f(300)}$	49.2 (206.0)	54.2 (226.6)		ΔH°
$S^\circ_{(300)}$	42.2 (176.7)	48.0 (200.7)		ΔS°
$C^\circ_{p(300)}$	5.0 (20.9)	10.5 (43.9)		ΔC°_p

log A	E	log k	Conditions	Reference and Experimental Method
-	-	<6.97	298 K 50 torr	[75] A.B.Callear and W.J.R.Tyerman, Proc.Chem.Soc., 296 (1964). For details see $Se + CH_2{=}CH_2$. Recalculated from relative rate constants $k_{C_2H_2}/k_{C_2H_4} < 0.1$ and preferred value for $Se + CH_2{=}CH_2$ ($\log k_{C_2H_4} = 7.97$ at 298 K).

PREFERRED RATE PARAMETERS

Reliable as given above

COMMENTS

See comments for $Se + CH_2{=}CH_2$.

ΔH^o_f, ΔH^o and E in kcal mol^{-1} (kJ mol^{-1}); S^o, C^o_p, ΔS^o and ΔC^o_p in cal mol^{-1} K^{-1} (J mol^{-1} K^{-1}); A and k in l mol^{-1} s^{-1}

REACTION: BROMINE ATOM + ETHYLENE $\dot{Br}(I) + CH_2{=}CH_2(II) \longrightarrow BrCH_2\dot{C}H_2(III)$

	(I)	(II)	(III)		
$\Delta H^\circ_{f(300)}$	26.7 (111.7)	12.5 (52.3)	30.4 (127.2)	ΔH°	-8.8 (-36.8)
$S^\circ_{(300)}$	41.8 (174.9)	52.4 (219.4)	70.2 (294.0)	ΔS°	-24.0 (-100.4)
$C^\circ_{p(300)}$	5.0 (20.9)	10.3 (43.1)	15.0 (62.7)	ΔC°_p	-0.3 (-1.3)

log A	E	log k	Conditions	Reference and Experimental Method
-	-	7.60	333 K 600 torr	[78] D.A.Armstrong and J.W.T.Spinks, Can.J.Chem., 37, 1210 (1959). Static system. Br from Co^{60} γ-radiation of HBr. Reaction rates derived from measurements of pressure decrease recorded by spoon gauge. Rate constants derived from rate data on basis of assumed diffusion rates of Br atoms in system.
-	-	-	313-393 K 5-240 torr	[79] P.I.Abell, Trans.Faraday Soc., 60, 2214 (1964). Static system. Photolysis of HBr (full Hg arc) in presence of pairs of olefins. Relative rates from analysis of alkyl bromides (gas chromatography) and rate constants, relative to ethylene as standard, for a series of olefins. It is not possible to derive absolute rate constants for Br + C_2H_4 reaction from this system.
-	-	-	228-318 K 360 torr	[80] K.T.Wong and D.A.Armstrong, Chem.Comm., 353 (1966). As for reference [83].
-	-	-	288-293 K 120-270 torr	[81] T.Berces and Z.G.Szabo, Acta.Chim.Sci.Hung.Tomus., 52, 397 (1967); 52, 405 (1967). Static system. Thermal reaction between Br_2 and ethylene. Overall reaction rates measured by pressure decrease (quartz spiral manometer). The effect of inert gases, N_2 and C_2H_6, was studied, and the overall rate expression derived, but no mechanism was suggested and no rate constants obtained. Earlier references to this system are given in this paper.
-	-	-	323-423 K 25-125 torr	[82] R.J.Field and P.I.Abell, Trans.Faraday Soc., 65, 743 (1969). Static system. Photolysis of HBr (λ > 220 nm) in presence of C_2H_4 plus added SF_6 Reaction rates from decrease in total pressure, measured by quartz spiral Bourdon gauge. Proposed mechanism: $HBr + h\nu \xrightarrow{I} \dot{H} + \dot{Br}$ $\dot{H} + HBr \xrightarrow{2} H_2 + \dot{Br}$ $\dot{Br} + CH_2{=}CH_2 \xrightarrow{1} BrCH_2\dot{C}H_2^*$ $BrCH_2\dot{C}H_2^* + M \xrightarrow{3} BrCH_2\dot{C}H_2 + M$

log A	E	log k	Conditions	Reference and Experimental Method
				$Br\dot{C}H_2\dot{C}H_2{}^* + HBr \xrightarrow{4} CH_3CH_2Br + \dot{B}r$ $\dot{B}r + \dot{B}r + M \xrightarrow{5} Br_2 + M$ $\dot{B}r + wall \xrightarrow{6} \frac{1}{2}Br_2 + wall$ Rate expression of the form: $\text{Initial Rate} = \frac{k_1k_4(kI_0)}{k_6k_{-1}}[C_2H_4][HBr]^2[M]^x$ and measured overall activation energy $E_{expt.} = -5.7$ kcal mol^{-1}. With C_2H_5Br as added gas, reaction (3) is rapid, and mechanism has to include: $BrCH_2CH_2 \xrightarrow{7} CH_2{=}CH_2 + Br$ $BrCH_2CH_2 + HBr \xrightarrow{8} CH_3CH_2Br + Br$ and the rate expression becomes: $\text{Rate} = \frac{k_1k_3k_8(kI_0)^{\frac{1}{2}}}{k_{-1}k_7k_5}[C_2H_4][HBr]^{3/2}[C_2H_5Br]^{\frac{1}{2}}$ and $E_{expt.} = -2.8$ kcal mol^{-1}. The decrease in $E_{expt.}$ with C_2H_5Br as third body was rationalized in terms of the recombination of Br atoms on the walls requiring an activation energy of 5-6 kcal mol^{-1}. It was not possible to obtain E_1, although the data are consistent with a low activation energy for the addition reaction.
6.96	0	6.96	288-318 K 600 torr	[83] K.T.Wong and D.A.Armstrong, Can.J.Chem., 47, 4183 (1969); 48, 2426 (1970). Static system. Br from rotating sector photolysis (305-325 nm) of acetone in presence of HBr with C_3H_8 as third body. Reaction rate from formation of C_2H_5Br. Rate constants obtained relative to reaction $Br + Br + C_3H_8 \longrightarrow Br_2 + C_3H_8$, for which rate constants were measured.
[9.5]	∿2.9	7.60	(ref.[78])	[84] S.W.Benson and H.E.O'Neal, NSRDS-NBS, 21 (1970). Approximate activation energy calculated from rate constant from ref.[78] and an assumed value of the A-factor, log A = 9.5.

PREFERRED RATE PARAMETERS

log k = [9.5 - 2.9/2.3RT]

ΔH_f^o, ΔH^o and E in kcal mol^{-1} (kJ mol^{-1}); S^o, C_p^o, ΔS^o and ΔC_p^o in cal mol^{-1} K^{-1} (J mol^{-1} K^{-1}); A and k in l mol^{-1} s^{-1}

COMMENTS

The above tentative Arrhenius parameters should be regarded as upper limits. The most recent measurement of the rate constant, reference [83], for Br addition to C_2H_4 is suspect, since the results are based on measurements of the rate constant for the reaction $Br + Br + C_3H_8 \xrightarrow{k_r} Br_2 + C_3H_8$, which are at least a power of ten too high. If, however, the results are recalculated on the basis of the more reasonable value $\log k_r(\ell^2/mol^2 s) = 10$, then, from the data for Br addition, $\log k_{C_2H_4}$ at 300 K is calculated to be ~ 6, which is even worse. The evidence from reference [80] for a low E for Br addition to C_2H_4 is reasonably convincing, and consequently the estimate of reference [84] is considered to be an upper limit. On the other hand, if the relative rates of addition of Br to ethylene and isobutene are correct, $k_{iso\text{-}C_4H_8}/k_{C_2H_4} = 384$ (see relative rates of Br addition, p.363), and if it is assumed that $E_{iso\text{-}C_4H_8} = 0$, then $E_{C_2H_4} - E_{iso\text{-}C_4H_8} \simeq 3$ on the basis of $A_{iso\text{-}C_4H_8} \simeq 2.5\ A_{C_2H_4}$.

There is no reliable data on the reverse reaction, $CH_2CH_2Br \longrightarrow Br + C_2H_4$, to permit calculation of the addition reaction via the equilibrium constant, but the above tentative Arrhenius parameters in conjunction with the thermodynamic data lead to $\log (k_{-1}/s^{-1}) = 12.9 - 11.1/2.3RT$, which seems reasonable [S.W.Benson and H.E.O'Neal, NSRDS-NBS 21, 1970].

Obviously, further measurements on the $Br + C_2H_4$ reaction are required to obtain a more accurate rate constant. For a review of Br additions to olefins, see R.J.Cvetanović, Adv.Photochem., 1, 173 (1963).

REACTION: BROMINE ATOM + PROPENE

$$\dot{Br}(I) + CH_3CH{=}CH_2(II) \xrightarrow{1} CH_3\dot{C}HCH_2Br(III)$$
$$\xrightarrow{2} CH_3CHBr\dot{C}H_2(IV)$$

	(I)	(II)	(III)	(IV)		Path 1	Path 2
$\Delta H^\circ_{f(300)}$	26.7 (111.8)	4.9 (20.5)	22.0 (92.1)	22.3 (93.4)	ΔH°	-9.6 (-40.2)	-9.3 (-38.9)
$S^\circ_{(300)}$	41.8 (175.0)	63.8 (267.1)	79.8 (334.1)		ΔS°	-25.8 (-108.0)	
$C^\circ_{p(300)}$	5.0 (20.9)	15.3 (64.1)	20.5 (85.8)		$\Delta C^\circ_{p(300)}$	0.2 (0.8)	

Path	log A	E	log k	Conditions	Reference and Experimental Method
1 + 2	-	-	[8.85](298 K)	313-393 K 5-240 torr	[79] P.I.Abell, Trans.Faraday Soc., 60, 2214 (1964). For details see $Br + CH_2{=}CH_2$. Recalculated from relative rate constants $k_a/k_{C_2H_4} = 17.7$ and preferred value for $Br + CH_2{=}CH_2$ ($\log k = [7.6]$ at 298 K).

PREFERRED RATE PARAMETERS

Tentatively as given above

COMMENTS

More accurate value must await further work on $Br + C_2H_4 \longrightarrow BrCH_2CH_2$ reaction.

REACTION: BROMINE ATOM + BUTENE-1

$$\dot{B}r(I) + CH_2{=}CHCH_2CH_3(II) \xrightarrow{1} CH_2Br\dot{C}HCH_2CH_3(III)$$
$$\xrightarrow{2} \dot{C}H_2CHBrCH_2CH_3(IV)$$

	(I)	(II)	(III)	(IV)		Path 1	Path 2
$\Delta H^o_{f(300)}$	26.7 (111.8)	0.0 (0.0)	17.0 (71.2)	17.4 (72.9)	$\Delta H°$	-9.7 (-40.6)	-9.3 (38.9)
$S°_{(300)}$	41.8 (175.0)	73.6 (308.0)	89.2 (373.5)		$\Delta S°$	-26.2 (-109.7)	
$C°_{p(300)}$	5.0 (20.9)	20.6 (86.2)	26.0 (108.9)		$\Delta C°_{p(300)}$	0.4 (1.7)	
$C°_{p(800)}$	5.0 (20.9)	42.0 (175.8)	47.9 (200.6)		$\Delta C°_{p(800)}$	0.9 (3.8)	
					$<\Delta C°_p>$	0.7 (2.9)	

Path	log A	E	log k	Conditions	Reference and Experimental Method
1 + 2	-	-	[8.95](298 K)	313-393 K 5-240 torr	[79] P.I.Abell, Trans.Faraday Soc., 60, 2214 (1964). For details see $Br + CH_2{=}CH_2$. Recalculated from relative rate constants $k_{C_4H_8\text{-}1}/k_{C_2H_4}$ = 22.6 and preferred value for $Br + CH_2{=}CH_2$ (log k = [7.6] at 298 K).

PREFERRED RATE PARAMETERS

Tentatively as given above

COMMENTS

More accurate value must await further work on $Br + C_2H_4 \longrightarrow BrCH_2CH_2$ reaction.

ΔH^o_f, ΔH^o and E in kcal mol^{-1} (kJ mol^{-1}); S^o, C^o_p, ΔS^o and ΔC^o_p in cal mol^{-1} K^{-1} (J mol^{-1} K^{-1}); A and k in l mol^{-1} s^{-1}

REACTION: BROMINE ATOM + cis-BUTENE-2

$\dot{B}r(I) + cis\text{-}CH_3CH{=}CHCH_3(II) \longrightarrow CH_3CHBr\dot{C}HCH_3(III)$

	(I)	(II)	(III)		
$\Delta H^\circ_{f(300)}$	26.7 (111.8)	-1.9 (-7.9)	13.9 (58.2)	ΔH°	-10.9 (-45.6)
$S^\circ_{(300)}$	41.8 (175.0)	73.6 (308.2)		ΔS°	
$C^\circ_{p(300)}$	5.0 (20.9)	19.4 (81.2)		ΔC°_p	

log A	E	log K	Conditions	Reference and Experimental Method
-	-	[9.58](298 K)	313-393 K 5-240 torr	[79] P.I.Abell, Trans.Faraday Soc., 60, 2214 (1964). For details see Br + $CH_2{=}CH_2$. Recalculated from relative rate constants $k_{cis\text{-}C_4H_8\text{-}2}/k_{C_2H_4}$ = 95.0 and preferred value for Br + $CH_2{=}CH_2$ (log k = [7.6] at 298 K).

PREFERRED RATE PARAMETERS

Tentatively as given above

COMMENTS

More accurate value must await further work on Br + $C_2H_4 \longrightarrow BrCH_2CH_2$ reaction.

REACTION: BROMINE ATOM + TRANS-BUTENE-2

$\dot{B}r(I) + trans\text{-}CH_3CH{=}CHCH_3(II) \longrightarrow CH_3CHBr\dot{C}HCH_3(III)$

	(I)	(II)	(III)		
$\Delta H^\circ_{f(300)}$	26.7 (111.8)	-2.9 (-12.1)	13.9 (58.2)	ΔH°	-9.9 (-41.5)
$S^\circ_{(300)}$	41.8 (175.0)	72.4 (302.8)		ΔS°	
$C^\circ_{p(300)}$	5.0 (20.9)	20.7 (86.6)		$\Delta C^\circ_{p(300)}$	
$C^\circ_{p(800)}$	5.0 (20.9)	41.3 (172.9)		$\Delta C^\circ_{p(800)}$	

log A	E	log k	Conditions	Reference and Experimental Method
-	-	[9.59](298 K)	313-393 K 5-240 torr	[79] P.I.Abell, Trans.Faraday Soc., 60, 2214 (1964). For details see $Br + CH_2{=}CH_2$. Recalculated from relative rate constants $k_{trans\text{-}C_4H_8\text{-}2}/k_{C_2H_4}$ = 98.5 and preferred value for $Br + CH_2{=}CH_2$ (log k = [7.6] at 298 K).

PREFERRED RATE PARAMETERS

Tentatively as given above

COMMENTS

More accurate value must await further work on $Br + C_2H_4 \longrightarrow BrCH_2CH_2$ reaction.

ΔH^o_f, ΔH^o and E in kcal mol^{-1} (kJ mol^{-1}); S^o, C^o_p, ΔS^o and ΔC^o_p in cal mol^{-1} K^{-1} (J mol^{-1} K^{-1}); A and k in l mol^{-1} s^{-1}

REACTION: **BROMINE ATOM + VINYL FLUORIDE**

$$\dot{B}r(I) + CH_2{=}CHF(II) \xrightarrow{1} CH_2Br\dot{C}HF(III)$$
$$\xrightarrow{2} \dot{C}H_2CHFBr(IV)$$

	(I)	(II)	(III)	(IV)		Path 1	Path 2
$\Delta H^\circ_{f(300)}$	26.7 (111.7)	-31.6 (-132.3)	[-9 (-38)]		ΔH°	[-4 (-17)]	
$S^\circ_{(300)}$	41.8 (175.0)	60.4 (252.9)	77.6 (324.9)		ΔS°	-24.6 (-103.0)	
$C^\circ_{p(300)}$	5.0 (20.9)	11.9 (49.8)	17.0 (71.2)		ΔC°_p	0.1 (0.4)	

Path	log A	E	log k	Conditions	Reference and Experimental Method
1 + 2	-	[$\geqslant$0]	-	323 K 36-117 torr	[85] G.J.Kavarnos and P.I.Abell, Trans.Faraday Soc., 66, 2022 (1970).

For details see $Br + C_2F_4$ system. Proposed mechanism:

$$HBr + h\nu \xrightarrow{I} \dot{H} + \dot{B}r$$
$$\dot{B}r + O\ell \xrightarrow{1+2} Br\dot{O}\ell$$
$$Br\dot{O}\ell \xrightarrow{-1,-2} \dot{B}r + O\ell$$
$$Br\dot{O}\ell + HBr \xrightarrow{3} BrO\ell H + \dot{B}r$$
$$2Br\dot{O}\ell \xrightarrow{4} \text{dimer}$$
$$\dot{B}r + Br\dot{O}\ell \xrightarrow{5} BrO\ell Br$$
$$2\dot{B}r + M \xrightarrow{6} Br_2 + M$$

Experimental results indicated rate law of the form:

$$\text{Initial Rate} = k_3[HBr]^{\frac{1}{2}}(KI_0[O\ell]/2k_5)^{\frac{1}{2}}$$

where $K = k_{1+2}/k_{-1-2}$ and termination is by reaction (5)

Overall activation energy = -2.0 kcal mol^{-1}, thus

$E_3 + (E_{1+2} - E_{-1-2} - E_5)/2 = -2.0$, and assuming $E_3 = 2.3$, $E_5 = 0$ and $E_{1+2} \geqslant 0$ kcal mol^{-1}, it follows that $E_{-1-2} \geqslant 9$ kcal mol^{-1}.

PREFERRED RATE PARAMETERS

Insufficient data to make a meaningful selection

COMMENTS

The overall rate of addition of HBr to $CH_2{=}CHF$ is slower than the rate of HBr addition to $CH_2{=}CH_2$, indicating a slower rate of addition of Br to $CH_2{=}CHF$ than to $CH_2{=}CH_2$, consistent with E > 0 for Br addition to $CH_2{=}CHF$. No thermochemical data are available for the system.

REACTION: BROMINE ATOM + cis-DIFLUOROETHYLENE $\dot{Br}(I) + cis\text{-}CHF{=}CHF \longrightarrow CHBrF\dot{C}HF(III)$

	(I)	(II)	(III)	
$\Delta H^\circ_{f(300)}$	26.7 (111.7)	-75.2 (-314.9)		ΔH°
$S^\circ_{(300)}$	41.8 (175.0)	65.6 (274.7)		ΔS°
$C^\circ_{p(300)}$	5.0 (20.9)	13.6 (56.9)		ΔC°_p

log A	E	log k	Conditions	Reference and Experimental Method
-	[⩾0]	-	323 K 150 torr	[85] G.J.Kavarnos and P.I.Abell, Trans.Faraday Soc., 66, 2022 (1970).

For details see $Br + C_2F_4$ system. Proposed mechanism:

$$HBr + h\nu \xrightarrow{I} \dot{H} + \dot{Br}$$
$$\dot{Br} + Ol \xrightarrow{1} Br\dot{O}l$$
$$Br\dot{O}l \xrightarrow{-1} \dot{Br} + Ol$$
$$Br\dot{O}l + HBr \xrightarrow{3} BrOlH + \dot{Br}$$
$$2Br\dot{O}l \xrightarrow{4} dimer$$
$$\dot{Br} + Br\dot{O}l \xrightarrow{5} BrOlBr$$
$$2\dot{Br} + M \xrightarrow{6} Br_2 + M$$

Experimental results indicate rate laws of the form:

$$\text{Initial Rate} = k_5[HBr]^{3/2}(KI_0[Ol]/2k_7)^{\frac{1}{2}}$$

or

$$\text{Initial Rate} = k_3K[HBr]^{3/2}[Ol](I_0/k_6[M])^{\frac{1}{2}}$$

depending upon conditions and where $K = k_1/k_{-1}$. Under conditions of high olefin concentration, the first rate law applies, and the overall activation energy was found to be -1.7 kcal mol^{-1}; hence, $E_3 + (E_1 - E_{-1} - E_5)/2 = -1.7$, and assuming $E_3 = 2.3$, $E_5 = 0$ and $E_1 \geqslant 0$, it follows that $E_{-1} \geqslant 8$ kcal mol^{-1}.

PREFERRED RATE PARAMETERS

Insufficient data to make a meaningful selection

COMMENTS

The overall rate of addition of HBr to cis-CHF=CHF is slower than the rate of HBr addition to $CH_2{=}CH_2$, indicating a slower rate of addition of Br to cis-CHF=CHF than to $CH_2{=}CH_2$, consistent with $E > 0$ for Br addition to cis-CHF=CHF. No thermochemical data are available for the system.

ΔH^o_f, ΔH^o and E in kcal mol^{-1} (kJ mol^{-1}); S^o, C^o_p, ΔS^o and ΔC^o_p in cal mol^{-1} K^{-1} (J mol^{-1} K^{-1}); A and k in l mol^{-1} s^{-1}

REACTION: BROMINE ATOM + TRANS-DIFLUOROETHYLENE $\dot{B}r(I)$ + trans-CHF=CHF(II) ⟶ CHBrF$\dot{C}$HF(III)

	(I)	(II)	(III)	
$\Delta H^\circ_{f(300)}$	26.7 (111.7)	-75.2 (-314.9)		ΔH°
$S^\circ_{(300)}$	41.8 (175.0)	65.6 (274.7)		ΔS°
$C^\circ_{p(300)}$	5.0 (20.9)	13.6 (56.9)		ΔC°_p

log A	E	log k	Conditions	Reference and Experimental Method
-	[≥0]	-	323 K 150 torr	[85] G.J.Kavarnos and P.I.Abell, Trans.Faraday Soc., 66, 2022 (1970). As for Br + cis-CHF=CHF system.

PREFERRED RATE PARAMETERS

Insufficient data to make a meaningful selection

COMMENTS

The overall rate of HBr addition to trans-CHF=CHF is slower than the rate of HBr addition to $CH_2=CH_2$, indicating a slower rate of addition of Br to trans-CHF=CHF than to $CH_2=CH_2$, consistent with $E > 0$ for Br addition to trans-CHF=CHF. No thermochemical data are available for the system.

REACTION: BROMINE ATOM + TRIFLUOROETHYLENE

$$\dot{B}r(I) + CF_2{=}CHF(II) \xrightarrow{1} CF_2Br\dot{C}HF(III)$$
$$\xrightarrow{2} \dot{C}F_2CHFBr(IV)$$

	(I)	(II)	(III)	(IV)		Path 1	Path 2
$\Delta H^\circ_{f(300)}$	26.7 (111.7)	-117.3 (-491.1)			ΔH°		
$S^\circ_{(300)}$	41.8 (175.0)	70.1 (293.5)			ΔS°		
$C^\circ_{p(300)}$	5.0 (20.9)	16.5 (69.1)			ΔC°_p		

Path	log A	E	log k	Conditions	Reference and Experimental Method
1 + 2	-	0	-	323 K 27-92 torr	[85] G.J.Kavarnos and P.I.Abell, Trans.Faraday Soc., 66, 2022 (1970). See $Br + C_2F_4$ system. Proposed mechanism:

$$HBr + h\nu \xrightarrow{I} \dot{H} + \dot{B}r$$
$$\dot{B}r + O\ell \xrightarrow{1+2} Br\dot{O}\ell$$
$$Br\dot{O}\ell \xrightarrow{-1-2} \dot{B}r + O\ell$$
$$Br\dot{O}\ell + HBr \xrightarrow{3} BrO\ell H + \dot{B}r$$
$$2\ Br\dot{O}\ell \xrightarrow{4} \text{dimer}$$
$$\dot{B}r + Br\dot{O}\ell \xrightarrow{5} BrO\ell Br$$
$$2\ \dot{B}r + M \xrightarrow{6} Br_2 + M$$

Experimental results conform to a rate law of the form:

$$\text{Initial Rate} = k_3[HBr]^{\frac{1}{2}}(KI_0[O\ell]/2k_5)^{\frac{1}{2}}$$

where $K = k_{1+2}/k_{-1-2}$ and termination is by reaction (5). The overall experimental activation energy was -3.7 kcal mol^{-1}; thus $E_3 + (E_{1+2} - E_{-1-2} - E_5)/2 = -3.7$, and assuming $E_3 = 2.3$, $E_5 = 0$ and $E_{1+2} = 0$, it follows that $E_{-1-2} = 12$ kcal mol^{-1}.

PREFERRED RATE PARAMETERS

Insufficient data to make a meaningful selection

COMMENTS

The overall rate of addition of HBr to $CHF{=}CF_2$ is comparable to the rate of addition of HBr to C_2H_4, and both indicate a rapid rate of addition of Br to $CHF{=}CF_2$ and to $CH_2{=}CH_2$ consistent with a low or zero activation energy.

ΔH°_f, ΔH° and E in kcal mol^{-1} (kJ mol^{-1}); S°, C°_p, ΔS° and ΔC°_p in cal mol^{-1} K^{-1} (J mol^{-1} K^{-1}); A and k in l mol^{-1} s^{-1}

REACTION: BROMINE ATOM + TETRAFLUOROETHYLENE $\dot{B}r(I) + CF_2{=}CF_2(II) \longrightarrow CF_2Br\dot{C}F_2(III)$

	(I)	(II)	(III)	
$\Delta H^\circ_{f(300)}$	26.7 (111.7)	-157.9 (-661.1)		ΔH°
$S^\circ_{(300)}$	41.8 (175.0)	74.6 (312.4)		ΔS°
$C^\circ_{p(300)}$	5.0 (20.9)	19.4 (81.2)		ΔC°_p

log A	E	log k	Conditions	Reference and Experimental Method
-	[0]	-	323-403 K 18-107 torr	[85] G.J.Kavarnos and P.I.Abell, Trans.Faraday Soc., 65, 2089 (1969); 66, 2022 (1970). Static system. Photolysis of HBr (full Hg arc) in presence of olefin plus added SF_6 or C_2F_6. Reaction rates from decrease in total pressure measured by quartz spiral Bourdon gauge. Major product of the reaction was CF_2HCF_2Br (gas chromatography and proton magnetic resonance), and the stoichiometry corresponded to the reaction $HBr + C_2F_4 \rightarrow CF_2HCF_2Br$. Proposed mechanism:

$$HBr + h\nu \xrightarrow{I} \dot{H} + \dot{B}r$$
$$\dot{B}r + CF_2{=}CF_2 \xrightarrow{1} BrCF_2\dot{C}F_2$$
$$BrCF_2\dot{C}F_2 \xrightarrow{-1} \dot{B}r + CF_2{=}CF_2$$
$$BrCF_2\dot{C}F_2 + HBr \xrightarrow{3} CF_2HCF_2Br + \dot{B}r$$
$$2\ BrCF_2\dot{C}F_2 \xrightarrow{4} \text{dimer}$$
$$\dot{B}r + BrCF_2\dot{C}F_2 \xrightarrow{5} BrCF_2CF_2Br$$
$$2\ \dot{B}r + M \xrightarrow{6} Br_2 + M$$

This is consistent with rate law:

$$\text{Initial Rate} = (k_3/k_4^{\frac{1}{2}})I_0^{\frac{1}{2}}[HBr]^{3/2}$$

assuming termination by reaction (4), which was confirmed experimentally. The lack of a pressure effect by added gases indicates that the $BrCF_2CF_2$ radicals were not vibrationally excited. The measured overall activation energy was $E = 2.3$ kcal mol^{-1}. Thus, $E_3 - \frac{1}{2}E_4 = 2.3$, and assuming $E_4 = 0$, it follows that $E_3 = 2.3$ kcal mol^{-1}. This is consistent with an assumed value of $E_1 = 0$ kcal mol^{-1}. The overall reaction between HBr and C_2F_4 is much more rapid than between HBr and C_2H_4 since the adduct radical decomposes more slowly in the former reaction.

PREFERRED RATE PARAMETERS

Insufficient data to make a meaningful selection

COMMENTS

The kinetic data on the overall addition of HBr to C_2F_4 are consistent with a rapid addition of Br to C_2F_4, indicating a low activation energy. The activation energy for the reverse reaction, $BrCF_2CF_2 \xrightarrow{4} Br + C_2F_4$ has been estimated [85] to be $E_4 \simeq 23$ kcal mol^{-1}, but thermochemical data for the system are lacking.

REACTION: BROMINE ATOM + ALLENE

$\dot{B}r(I) + CH_2{=}C{=}CH_2(II) \xrightarrow{1} CH_2Br\dot{C}{=}CH_2(III)$

$\xrightarrow{2} \dot{C}H_2CBr{=}CH_2(IV)$

	(I)	(II)	(III)	(IV)		Path 1	Path 2
$\Delta H^\circ_{f(300)}$	26.7 (111.7)	45.9 (192.2)	-66.8 (-279.7)		ΔH°	-5.8 (-24.3)	
$S^\circ_{(300)}$	41.8 (174.9)	58.3 (244.1)	[78 (327)]		ΔS°	[-22 (-92)]	
$C^\circ_{p(300)}$	5.0 (20.9)	14.2 (59.5)	[18 (75)]		ΔC°_p	[-1.2 (-5.0)]	

Path	log A	E	log k	Conditions	Reference and Experimental Method
1 + 2	-	-	-	323-423 K 10-480 torr	[85a] P.I.Abell and R.S.Anderson, Tetrahedron Lett., 49, 3727 (1964). For details see Br + $CF_2{=}CF_2$, reference [85]. The overall activation energy for HBr addition to allene was determined to be 4.5 kcal mol^{-1}.

PREFERRED RATE PARAMETERS

No meaningful estimate possible

COMMENTS

It is argued in reference [85a] that addition is likely to occur via path 1, and although there is no analytical data in the paper to support this conclusion, it seems likely that path 1 will be the favoured, though not exclusive, route for addition (see E-A.I.Heiba and W.O.Haag, J.Org.Chem., 31, 3814 (1966)].

ΔH^o_f, ΔH^o and E in kcal mol^{-1} (kJ mol^{-1}); S^o, C^o_p, ΔS^o and ΔC^o_p in cal mol^{-1} K^{-1} (J mol^{-1} K^{-1}); A and k in l mol^{-1} s^{-1}

REACTION: TELLURIUM ATOM + ETHYLENE $\quad Te(I) + CH_2{=}CH_2(II) \longrightarrow [\overline{CH_2CH_2Te}](III)$

	(I)	(II)	(III)	
$\Delta H^\circ_{f(300)}$		12.5 (52.3)		ΔH°
$S^\circ_{(300)}$		52.4 (219.4)		ΔS°
$C^\circ_{p(300)}$		10.3 (43.1)		ΔC°_p

log A	E	log k	Conditions	Reference and Experimental Method
-	-	7.30	298 K (?) 30-300 torr	[86] J.Connor, G.Greig and O.P.Strausz, J.Amer.Chem.Soc., 91, 5695 (1969). Static system. Te(^{3}P) from flash photolysis of Me_2Te moderated with CO_2. [Te] from absorption at 214.3 nm and 225.9 nm. Rate constants calculated from -d[Te]/dt in presence of the olefin, with allowance made for the fast reaction between Te and Me_2Te. Absorptions which appear when olefins are present have been ascribed to epitelluride molecules, which were confirmed by kinetic mass spectrometry.
8.94	2.5 (10.5)	7.11(298 K)	298-353 K ? torr	[52] J.Connor, A.Van Roodselaar, R.W.Fair and O.P.Strausz, J.Amer.Chem.Soc., 93, 560 (1971). Same method as for reference [86].

PREFERRED RATE PARAMETERS

$\log k = 8.94 - 2.5/2.3RT$

COMMENTS

There is reasonable agreement between the room temperature rate constants of the above two references. The data of reference [52] are preferred on the grounds that they are more recent measurements by the same group.

REACTION: TELLURIUM ATOM + PROPENE

$Te(I) + CH_3CH{=}CH_2(II) \longrightarrow [CH_3\dot{C}HCH_2\dot{T}e](III)$

	(I)	(II)	(III)	
$\Delta H^\circ_{f(300)}$		4.9 (20.5)		ΔH°
$S^\circ_{(300)}$		63.8 (267.0)		ΔS°
$C^\circ_{p(300)}$		15.3 (64.0)		ΔC°_p

log A	E	log k	Conditions	Reference and Experimental Method
-	-	8.30	298 K (?) 30-300 torr	[86] J.Connor, G.Greig and O.P.Strausz, J.Amer.Chem.Soc., 91, 5695 (1969). For details see Te + $CH_2{=}CH_2$.
8.52	0.6 (2.5)	8.08 (298 K)	298-353 K ? torr	[52] J.Connor, A.Van Roodselaar, R.W.Fair and O.P.Strausz, J.Amer.Chem.Soc., 93, 560 (1971) Same method as for reference [86].

PREFERRED RATE PARAMETERS

log k = 8.52 - 0.6/2.3RT

COMMENTS

There is reasonable agreement between the room temperature rate constants of the above two references. The data of reference [52] are preferred on the grounds that they are more recent measurements by the same group.

ΔH^o_f, ΔH^o and E in kcal mol^{-1} (kJ mol^{-1}); S^o, C^o_p, ΔS^o and ΔC^o_p in cal mol^{-1} K^{-1} (J mol^{-1} K^{-1}); A and k in l mol^{-1} s^{-1}

REACTION: TELLURIUM ATOM + 2,3-DIMETHYLBUTENE-2 (TETRAMETHYLETHYLENE) $Te(I) + (CH_3)_2C{=}C(CH_3)_2(II) \longrightarrow [(CH_3)_2\overline{CC(CH_3)_2Te}](III)$

	(I)	(II)	(III)	
$\Delta H^\circ_{f(300)}$		-19.6 (-81.9)		ΔH°
$S^\circ_{(300)}$		87.4 (365.3)		ΔS°
$C^\circ_{p(300)}$		33.0 (137.9)		ΔC°_p

log A	E	log k	Conditions	Reference and Experimental Method
8.42	-1.6 (-6.7)	9.59(298 K)	298-353 K ? torr	[52] J.Connor, A.Van Roodselaar, R.W.Fair and O.P.Strausz, J.Amer.Chem.Soc., 93, 560 (1971). For details see $Te + CH_2{=}CH_2$.

PREFERRED RATE PARAMETERS

log k = 8.42 + 1.6/2.3RT

COMMENTS

The negative activation energy reported in reference [52] for this reaction appears to be real, but is difficult to explain. The obvious explanation that the reaction was in a pressure sensitive region does not apply as the reactions were apparently carried out with substantial pressures of CO_2 diluent.

REACTION: METHYL RADICAL + ETHYLENE $\dot{C}H_3(I) + CH_2{=}CH_2(II) \longrightarrow CH_3CH_2\dot{C}H_2(III)$

	(I)	(II)	(III)		
$\Delta H^o_{f(300)}$	34.0 (142.3)	12.5 (52.3)	21.0 (87.8)	ΔH^o	-25.5 (-106.7)
$S^o_{(300)}$	46.1 (193.0)	52.4 (219.4)	68.5 (286.5)	ΔS^o	-30.0 (-125.5)
$C^o_{p(300)}$	8.8 (36.8)	10.3 (43.1)	17.1 (71.5)	$\Delta C^o_{p(300)}$	-2.0 (-8.4)
$C^o_{p(600)}$	11.5 (48.2)	16.6 (69.5)	28.5 (119.3)	$\Delta C^o_{p(600)}$	0.4 (1.7)
				$<\Delta C^o_p>$	-0.8 (-3.3)

log A	E	log k	Conditions	Reference and Experimental Method
-	6.8 (28.5)	-	573 K 50-450 torr	[87] F.A.Raal and C.J.Danby, J.Chem.Soc., 2219 (1949). Static system. CH_3 from photolysis (λ = 360.0 nm) of CH_3CHO. Reaction rates measured by rate of decrease in pressure (mercury manometer). Concentrations of unreacted CH_3CHO were analysed chemically. The temperature coefficient of the overall rate was found to be E = -2.9 kcal mol^{-1} and from the proposed mechanism this gave $E_1 - E_2 = -2.9$ for the reactions $CH_3 + C_2H_4 \xrightarrow{1} C_3H_7$ and $CH_3 + CH_3CHO \xrightarrow{2} CH_4 + CH_3CO$. $E_1 = 6.8$ was calculated on the basis of $E_2 = 9.7$, which is now known to be wrong.
8.25	7.0 (29.3)	4.95(463 K)	413-513 K 50 torr	[88] L.Mandelcorn and E.W.R.Steacie, Can.J.Chem., 32, 474 (1954). Arrhenius parameters relative to log k_c = 10.34 for CH_3 combination; $\log (k_1/k_c^{\frac{1}{2}})(\ell^{\frac{1}{2}}mol^{-\frac{1}{2}}s^{-\frac{1}{2}}) = 3.08 - 7.0/2.3RT$. Static system. CH_3 from photolysis (λ > 230 nm) of CH_3COCH_3, and the reaction rates determined by mass balance method involving analyses of CH_4, C_2H_6 and CO. This method cannot distinguish between the formation of different adduct radicals in addition to asymmetrical alkenes.
-	8.2 (34.3)	-	403-438 K 10-380 torr	[89] L.C.Landers and D.H.Volman, J.Amer.Chem.Soc., 79, 2996 (1957). Static system. CH_3 from pyrolysis of di-t-butyl peroxide. Reaction rates

ΔH^o_f, ΔH^o and E in kcal mol^{-1} (kJ mol^{-1}); S^o, C^o_p, ΔS^o and ΔC^o_p in cal mol^{-1} K^{-1} (J mol^{-1} K^{-1}); A and k in l mol^{-1} s^{-1}

log A	E	log k	Conditions	Reference and Experimental Method
				measured by rate of decrease in pressure (mercury manometer). Overall activation energy was determined to be E = 27.4 kcal mol^{-1}, and from the proposed mechanism this was assigned to $\frac{1}{2}E_3 + E_1 - \frac{1}{2}E_C$ where the reactions are $((CH_3)_3CO)_2 \xrightarrow{3} 2CH_3 + 2CH_3COCH_3$, $CH_3 + C_2H_4 \xrightarrow{1} C_3H_7$, and $CH_3 + CH_3 \xrightarrow{c} C_2H_6$. Taking $E_3 = 38.4$ and $E_C = 0$, the above value of E_1 was deduced.
9.08	8.86 (36.3)	4.47(415 K)	397-432 K 15-34 torr	[90] R.K.Brinton, J.Chem.Phys., 29, 781 (1958). Arrhenius parameters relative to log k_C = 10.34 for CH_3 combination; log $(k_1/k_C^{\frac{1}{2}})(\ell^{\frac{1}{2}}mol^{-\frac{1}{2}}s^{-\frac{1}{2}})$ = 3.91 - 8.86/2.3RT. Static system. CH_3 from pyrolysis of di-t-butyl peroxide. Reaction rates determined by analysis of products (low temperature distillation followed by mass spectrometry) of the reactions of the adduct radicals under conditions of low $[C_2H_4]$ to limit polymerisation reactions.
8.52	7.82 (32.7)	4.33(418 K)	393-443 K 13-400 torr	[91] A.M.Hogg and P.Kebarle, J.Amer.Chem.Soc., 86, 4558 (1964). Arrhenius parameters relative to log k_C = 10.34 for CH_3 combination; log $(k_1/k_C^{\frac{1}{2}})(\ell^{\frac{1}{2}}mol^{-\frac{1}{2}}s^{-\frac{1}{2}})$ = 3.35 - 7.82/2.3RT. As for reference [90] above, with gas chromatographic analysis of products.
8.15	6.84 (28.6)	4.85(453 K)	403-503 K 100 torr	[92] L.Endrenyi and D.J.LeRoy, J.Phys.Chem., 71, 1334 (1967). Arrhenius parameters relative to log k_C = 10.34 for CH_3 combination; log $(k_1/k_C^{\frac{1}{2}})(\ell^{\frac{1}{2}}mol^{-\frac{1}{2}}s^{-\frac{1}{2}})$ = 3.25 - 9.43/2.3RT. Static system. CH_3 from photolysis (λ = 313 nm) of CH_3COCH_3. Reaction rates determined by analysis of products. Higher molecular weight products determined than in reference [90].
8.63	7.9 (33.6)	4.33(403 K)	353-453 K 3000 torr	[93] R.J.Cvetanović and R.S.Irwin, J.Chem.Phys., 46, 1694 (1967). Arrhenius parameters derived from $E_1 - E_3 = 0.35$ and $A_1/A_3 = 12.6$, where log k_3 = 7.53 - 7.6/2.3RT for $CH_3 + iso\text{-}C_4H_{10} \xrightarrow{3} CH_4 + C_4H_9$, based on log k_C = 10.34 for CH_3 combination. Static system. Improved mass balance method based on photolysis of biacetyl whereby every CH_3 in system is accounted for by a molecule of CO, even if all CH_3CO radicals do not decompose. Rates determined from mass balance of products, CO, CH_4, from photolyses in presence of isobutene and olefin. The preferred value of log A_3 is 0.3 lower than the value used by Cvetanović and Irwin and was selected to give agreement between the absolute value of k_1 from reference [93] with the absolute values of k_1 from references [90,91,92]. There is ample justification for the preferred value of A_1 [V.N.Kondratiev, Handbook on Kinetic

log A	E	log k	Conditions	Reference and Experimental Method
				Constants of Gaseous Reactions, "Nauka", Moscow, 1970]. Arrhenius parameters for addition reaction calculated from $E_1 - E_3 = 0.35$ and $A_1/A_3 = 12.6$. It should be noted that since this is a mass balance method, it does not distinguish between different adduct radicals in additions to unsymmetrical alkenes, but gives composite rate constants.
[8.82]	-	-	414 K	[94] B.Yu.Ainbinder and M.V.Bazilevskii, Zh.Strukt.Khim., 9, 513 (1968). Calculated A-factor based on transition-state theory.
-	[7 to 10]	-	-	[95] M.V.Basilevskii and I.E.Chlenov, Theoret.Chim.Acta (Berl.), 15, 174 (1969). Theoretical measurement obtained by construction of a potential energy surface model for $CH_3 + CH_2{=}CH_2$.

PREFERRED RATE PARAMETERS

log k = 8.52 - 7.7/2.3RT

COMMENTS

The data of references [90,91,92,93] all seem reliable, and there is good agreement in the absolute rate constants of references [90,91,92]. The absolute rate constants of reference [93] were adjusted to agree with the other data by choosing a reasonable value of the A-factor for the reference reaction.
The above Arrhenius parameters were then calculated from a least-mean-squares treatment of the absolute rate constants of references [90,91,92,93] and an Arrhenius plot of the data is shown overleaf. (Figure 1).
The evaluated parameters for the reverse radical decomposition reaction, log (k_{-1}/s^{-1}) = 13.8 - 33.2/2.3RT at 600 K [Benson and O'Neal, NSRDS-NBS 21 (1967)], together with the thermodynamic data, yield a calculated rate constant given by log k_1 = 8.9 - 8.1/2.3RT at 600 K in good agreement with the above preferred parameters.
It must be pointed out, however, that the preferred value of k_{-1} is partly based on the data for k_1 as well as transition state theory, and consequently is not a true test for consistency. The most recent experimental value for the rate constant of the reverse reaction is given by log (k_{-1}/s^{-1}) = 14.40 - 32.6/2.3RT at 600 K, [M.M.Papic and K.J.Laidler, Can.J.Chem., 49, 549 (1971)], which, in conjunction with the thermodynamic data, yields log k_1 = 9.54 - 7.5/2.3RT at 600 K.
Although the calculated value of E_1 agrees well with the preferred experimental value, there is a serious discrepancy between the calculated and experimental values of A_1. This discrepancy has already been pointed out by Benson and O'Neal. We feel the most likely source of error is in the experimental value of A_{-1} which is obtained from an extrapolation.
There is no evidence in any of the above studies of the addition reaction to indicate that k_1 is pressure dependent above about 15 torr and at temperatures below about 500 K, and this is substantiated by the above study of the n-propyl decomposition reaction.
See also summary of alkyl radical additions, p.219, and comments.

ΔH_f^o, ΔH^o and E in kcal mol^{-1} (kJ mol^{-1}); S^o, C_p^o, ΔS^o and ΔC_p^o in cal mol^{-1} K^{-1} (J mol^{-1} K^{-1}); A and k in l mol^{-1} s^{-1}

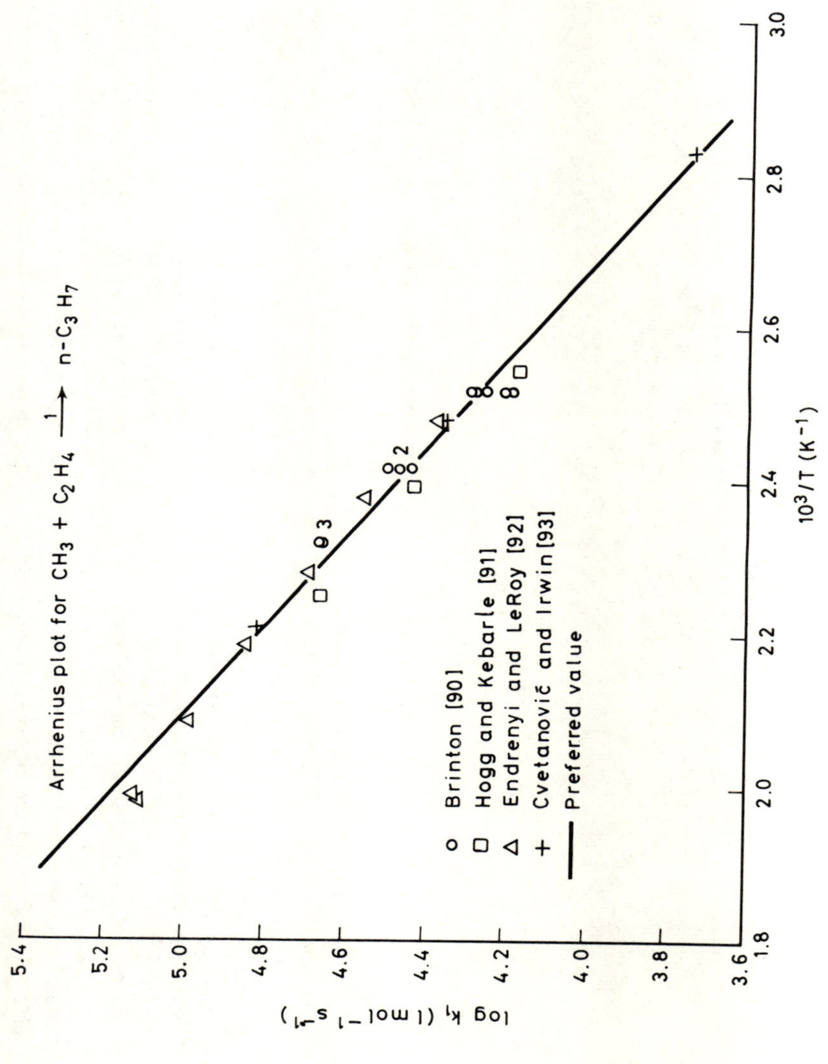

FIGURE 1. Arrhenius plot for $CH_3 + C_2H_4 \xrightarrow{1} n\text{-}C_3H_7$

REACTION: METHYL RADICAL + PROPENE

$\dot{C}H_3(I) + CH_3CH{=}CH_2(II) \xrightarrow{1} CH_3\dot{C}HCH_2CH_3(III)$

$\xrightarrow{2} (CH_3)_2CH\dot{C}H_2(IV)$

	(I)	(II)	(III)	(IV)
$\Delta H^\circ_{f(300)}$	34.0 (142.2)	4.9 (20.5)	12.3 (51.5)	13.8 (57.7)
$S^\circ_{(300)}$	46.1 (192.8)	63.8 (267.0)	76.7 (320.8)	75.6 (316.2)
$C^\circ_{p(300)}$	8.8 (36.8)	15.3 (64.0)	23.0 (96.2)	22.8 (95.4)
$C^\circ_{p(600)}$	11.5 (48.2)	25.8 (108.0)	38.3 (160.4)	38.8 (162.5)

	Path 1	Path 2
ΔH°	-26.6 (-111.3)	-25.1 (-105.0)
ΔS°	-33.2 (-138.8)	-34.3 (-143.4)
$\Delta C^\circ_{p(300)}$	-1.1 (-4.6)	-1.3 (-5.4)
$\Delta C^\circ_{p(600)}$	1.0 (4.2)	1.5 (6.3)
$<\Delta C^\circ_p>$	0.0 (0.0)	0.1 (0.4)

Path	log A	E	log k	Conditions	Reference and Experimental Method
1 + 2	7.78	6.0 (25.2)	5.01(473 K)	413-533 K 50 torr	[88] L.Mandelcorn and E.W.R.Steacie, Can.J.Chem., 32, 474 (1954). Arrhenius parameters relative to log k_c = 10.34 for CH_3 combination; $\log (k_{1+2}/k_c^{\frac{1}{2}})(\ell^{\frac{1}{2}}mol^{-\frac{1}{2}}s^{-\frac{1}{2}}) = 2.61 - 6.0/2.3RT$. For details see $CH_3 + CH_2{=}CH_2$.
1 + 2	-	5.6 (23.4)	-	403-438 K 10-350 torr	[89] L.C.Landers and D.H.Volman, J.Amer.Chem.Soc., 79, 2996 (1957). For details see $CH_3 + CH_2{=}CH_2$. Overall activation energy E = 24.8 = $E_{1+2} + \frac{1}{2}E_3 - \frac{1}{2}E_c$, hence from E_3 = 38.4 and E_c = 0, it follows that E_{1+2} = 5.6 kcal mol^{-1}.
1 + 2	8.88	8.8 (36.8)	4.19(410 K)	380-441 K 15 torr	[96] M.Miyoshi and R.K.Brinton, J.Chem.Phys., 36, 3019 (1962). Arrhenius parameters relative to log k_c = 10.34 for CH_3 combination; $\log (k_{1+2}/k_c^{\frac{1}{2}})(\ell^{\frac{1}{2}}mol^{-\frac{1}{2}}s^{-\frac{1}{2}}) = 3.67 - 8.8/2.3RT$. Static system. CH_3 from pyrolysis of di-t-butyl peroxide. Reaction rates determined by analysis of products (low-temperature distillation followed by mass spectrometry and gas chromatography) of the reactions of the adduct radicals under conditions of low $[C_3H_6]$ to limit polymerization reactions, but at low temperatures, as many as 20% of the butyl adduct radicals added to propene to form heptyl radicals and the further products of the C_7 radicals were analyzed or were estimated. Analysis of the addition products from the adduct radicals, sec-C_4H_9 and iso-C_4H_9, indicated that approximately 90% of the CH_3 radicals add to the terminal C atom in propene at 380 K.

ΔH^o_f, ΔH^o and E in kcal mol^{-1} (kJ mol^{-1}); S^o, C^o_p, ΔS^o and ΔC^o_p in cal mol^{-1} K^{-1} (J mol^{-1} K^{-1}); A and k in l mol^{-1} s^{-1}

Path	log A	E	log k	Conditions	Reference and Experimental Method
1 + 2	8.22	7.4	4.51(403 K)	353-453 K 3000 torr	[93] R.J.Cvetanović and R.S.Irwin, J.Chem.Phys., 46, 1694 (1967). Arrhenius parameters derived from $E_{1+2} - E_3 = -0.17$ and $A_{1+2}/A_3 = 4.8$ where $\log k_3 = 7.53 - 7.6/2.3RT$ for $CH_3 + iso\text{-}C_4H_{10} \xrightarrow{3} CH_4 + C_4H_9$ [A.F.Trotman-Dickenson and G.S.Milne, NSRDS-NBS 9, 1967]. For details see $CH_3 + CH_2{=}CH_2$.

PREFERRED RATE PARAMETERS

$\log k_{1+2} = 8.22 - 7.4/2.3RT$

COMMENTS

The preferred rate constant is that obtained in reference [93]. The rate constants of reference [96] are in reasonable agreement with the preferred data at the highest temperatures, but are significantly smaller at the lower temperatures, as shown in Fig.2. In reference [93] the rate constants were determined by analysis of the products of the reactions of the adduct radicals, but at the lower temperatures a significant proportion of the initial adduct radicals added further to propene to form C_7 products which may not all have been accounted for. It should be noted that the preferred data refer to total addition, i.e. at the central as well as the terminal C atom. The data of reference [96] indicate that about 90% of the addition occurs at the terminal C atom, but the data are insufficiently precise to warrant a calculation of k_1 and k_2 from the composite k_{1+2}. It would be well worth repeating the experiments of reference [93] in the presence of a radical scavenger, as for reference [96] of the CH_3 + cis-C_4H_8-2 system.

The most recent study of the reverse decomposition reaction s-$C_4H_9 \xrightarrow{-1} CH_3 + C_3H_6$ reports $\log (k_{-1}/s^{-1}) = 14.8 - 32.6/2.3RT$ at 575 K [M.C.Lin and K.J.Laidler, Can.J.Chem., 45, 1315 (1967)], and in conjunction with the thermodynamic data, for path 1, leads to $\log k_1 = 9.4 - 6.6/2.3RT$ at 600 K. The calculated value of E_1 is in good agreement with the above preferred E_{1+2}, but the calculated log A_1 is much too high. The most likely cause of the discrepancy appears to be in the experimental value of A_{-1}, which seems to be an over-estimate. See also summary of alkyl radical additions (page 219) and comments.

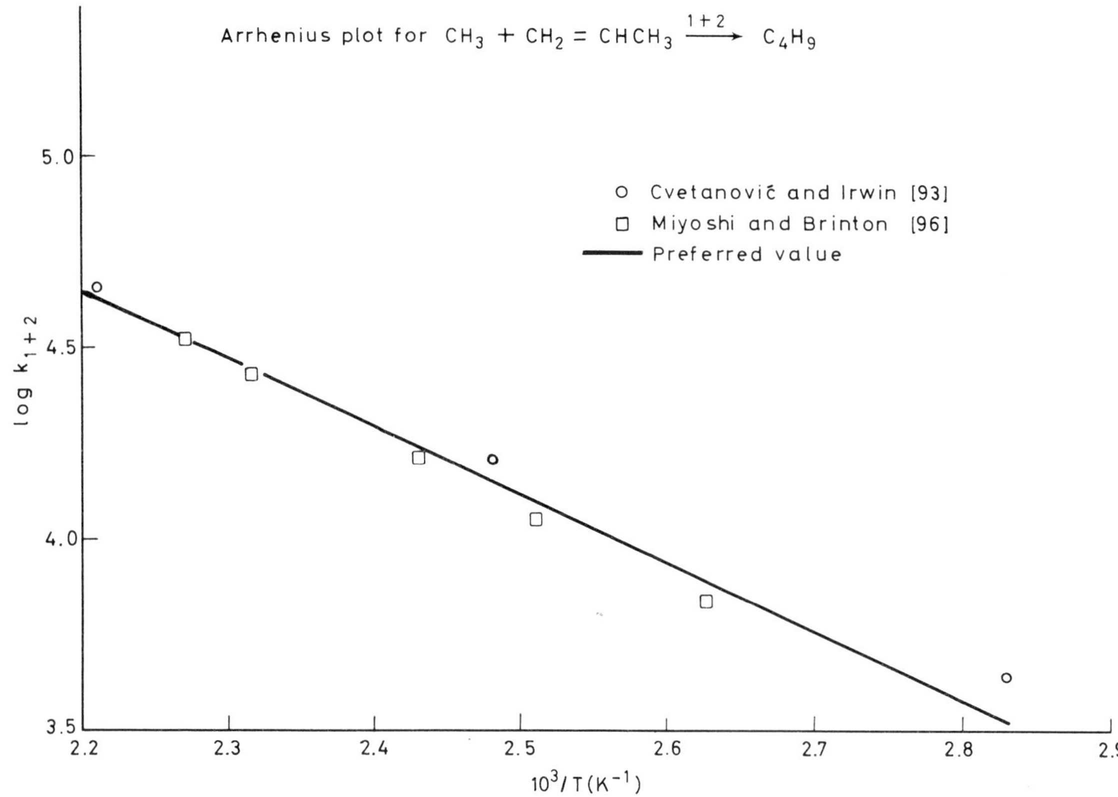

FIGURE 2. Arrhenius plot for $CH_3 + CH_2{=}CHCH_3 \xrightarrow{1+2} C_4H_9$

REACTION: METHYL RADICAL + BUTENE-1

$\dot{C}H_3(I) + CH_2{=}CHCH_2CH_3(II) \xrightarrow{1} CH_3CH_2\dot{C}HCH_2CH_3(III)$

$\xrightarrow{2} CH_2\dot{C}H(CH_3)CH_2CH_3(IV)$

	(I)	(II)	(III)	(IV)		Path 1	Path 2
$\Delta H^\circ_{f(300)}$	34.0 (142.3)	0.0 (0.0)	7.4 (31.0)	8.9 (37.3)	ΔH°	-26.6 (-111.4)	-25.1 (-105.1)
$S^\circ_{(300)}$	46.1 (193.0)	73.6 (308.2)	86.0 (360.0)	85.0 (355.9)	ΔS°	-33.7 (-141.1)	-34.7 (-145.3)
$C^\circ_{p(300)}$	8.8 (36.8)	20.6 (86.3)	28.5 (119.3)	28.4 (118.9)	ΔC°_p	-0.9 (-3.8)	-1.0 (-4.2)

Path	log A	E	log k	Conditions	Reference and Experimental Method
1	-	-	3.87(673 K)	623-773 K 30-100 torr	[97] J.R.McNesby and A.S.Gordon, J.Amer.Chem.Soc., 79, 5902 (1957). Rate constant relative to log k_4 = 2.73 at 400 K for $CD_3 + (CD_3)_2CO \xrightarrow{4} CD_4 + CD_3COCD_2$ [J.A.Kerr and D.Timlin, J.Chem.Soc.(A), 1241 (1969)], and parameters refer to addition of CD_3 to butene-1.
2	-	-	2.91(673 K)	623-773 K 30-100 torr	Static system. CD_3 from photolysis of $(CD_3)_2CO$. Reaction rates determined by analysis of products of reactions of adduct radicals using mass spectrometry and gas chromatography.
1 + 2	8.01	7.2 (30.1)	4.11(403 K)	353-453 K 3000 torr	[93] R.J.Cvetanović and R.S.Irwin, J.Chem.Phys., 46, 1694 (1967). Arrhenius parameters derived from $E_{1+2} - E_3$ = -0.44 and A_{1+2}/A_3 = 3.0 where log k_3 = 7.53 - 7.6/2.3RT for $CH_3 + i\text{-}C_4H_{10} \xrightarrow{3} CH_4 + C_4H_9$. For details see $CH_3 + CH_2{=}CH_2$.

PREFERRED RATE PARAMETERS

log k_{1+2} = 8.01 - 7.2/2.3RT

COMMENTS

The composite rate constant (path 1 + 2) of reference [93] is preferred to the absolute rate constants derived from the data of reference [97], on the grounds that the former involve a technique which has yielded data for other CH_3 additions which are reliable and consistent with independent results. The relative rate constants for terminal and non-terminal addition (paths 1 and 2) from reference [97] are probably reliable, i.e. 12.0% non-terminal addition at 673 K. From the preferred parameters and the thermodynamic data, we calculate for the reverse reaction (path 1) log (k_{-1}/s^{-1}) = 13.5 - 33.2/2.3RT at 298 K, which is reasonable.
See also summary of alkyl additions (page 219) and comments.

REACTION: METHYL RADICAL + cis-BUTENE-2 $\dot{C}H_3(I) + cis\text{-}CH_3CH{=}CHCH_3(II) \longrightarrow CH_3\dot{C}HCH(CH_3)_2(III)$

	(I)	(II)	(III)		
$\Delta H^\circ_{f(300)}$	34.0 (142.3)	-2.0 (-8.4)	5.3 (22.2)	ΔH°	-26.7 (-111.8)
$S^\circ_{(300)}$	46.1 (193.0)	73.6 (308.2)	83.3 (348.8)	ΔS°	-36.4 (-152.4)
$C^\circ_{p(300)}$	8.8 (36.8)	19.4 (81.2)	28.3 (118.5)	$\Delta C^\circ_{p(300)}$	0.1 (0.4)
$C^\circ_{p(500)}$	10.6 (44.3)	29.6 (123.6)	41.2 (172.1)	$\Delta C^\circ_{p(500)}$	1.0 (4.2)

log A	E	log k	Conditions	Reference and Experimental Method
-	-	3.45(673 K)	623-773 K 30-100 torr	[97] J.R.McNesby and A.S.Gordon, J.Amer.Chem.Soc., 79, 5902 (1957). Arrhenius parameters relative to $\log k_4 = 9.07 - 11.6/2.3RT$ for $CD_3 + (CD_3)_2CO \xrightarrow{4} CD_4 + CD_2COCD_3$ [J.A.Kerr and D.Timlin, J.Chem.Soc.(A), 1241 (1969)]. For details see $CH_3 + CH_2{=}CHCH_2CH_3$. Parameters are for CD_3 addition.
7.64	7.5 (31.4)	3.58(403 K)	353-453 K 300 torr	[93] R.J.Cvetanović and R.S.Irwin, J.Chem.Phys., 46, 1694 (1967). Arrhenius parameters derived from $E_1 - E_3 = -0.12$ and $A_1/A_3 = 1.3$, where $\log k_3 = 7.53 - 7.6/2.3RT$ for $CH_3 + iso\text{-}C_4H_{10} \xrightarrow{3} CH_4 + C_4H_9$. For details see $CH_3 + CH_2{=}CH_2$.
-	7.85[a] (32.9)	-	298 K	[98] C.W.Larson, B.S.Rabinovitch and D.C.Tardy, J.Chem.Phys., 47, 4570 (1967). Static system. The study involved measuring the relative rates of decomposition of chemically activated alkyl radicals produced from the addition of H atoms to olefins: e.g. $H + CH_2{=}CHCH(CH_3)_2 \longrightarrow CH_3\dot{C}HCH(CH_3)_2^*$ $CH_3\dot{C}HCH(CH_3)_2^* \xrightarrow{-1} \dot{C}H_3 + trans\text{-}C_4H_8\text{-}2$ $\xrightarrow{-1'} \dot{C}H_3 + cis\text{-}C_4H_8\text{-}2$ The relative rates of decomposition of the radical were determined by gas-chromatographic analyses of the olefins. Relative critical energies for the reverse alkyl radical additions to the olefins (reactions (-1) and (-1')) were calculated from an RRKM treatment of the activated radical data for a series of olefins. The above critical energy is based on an assumed value of 8.00 kcal mol^{-1} for the critical energy of CH_3 addition to trans-butene-2. For this type of reaction, the critical energy (E_c) of the reaction is < 1 kcal mol^{-1} smaller than the experimental activation energy (E), but in most cases the values of

[a] critical energy of the reaction.

ΔH^o_f, ΔH^o and E in kcal mol^{-1} (kJ mol^{-1}); S^o, C^o_p, ΔS^o and ΔC^o_p in cal mol^{-1} K^{-1} (J mol^{-1} K^{-1}); A and k in l mol^{-1} s^{-1}

log A	E	log k	Conditions		Reference and Experimental Method
					E_c reported by the above method should be close to E since the standard E_c values upon which they are based are approximately equal to the measured E for the reference reaction.
7.65	7.00 (29.2)	3.98(417 K)	399-436 K 60 torr	[99]	N.Yokoyama and R.K.Brinton, Can.J.Chem., 47, 2987 (1969). Arrhenius parameters relative to log k_c = 10.34 for CH_3 combination; $\log(k_1/k_c^{\frac{1}{2}})(\ell^{\frac{1}{2}}mol^{-\frac{1}{2}}s^{-\frac{1}{2}}) = 2.48 - 7.0/2.3RT$. Static system. CH_3 from pyrolysis of di-t-butyl peroxide. Adduct radicals scavenged by large concentration of CH_3CHO. Reaction rates from analyses of products of adduct radicals by low-temperature distillation and gas chromatography. Good kinetic analyses of comprehensive data.

PREFERRED RATE PARAMETERS

$\log k = 7.65 - 7.3/2.3RT$

COMMENTS

The preferred parameters are the average of references [93] and [99]. The preferred data and the thermodynamic data yield $\log (k_{-1}/s^{-1}) = 13.7 - 33.4/2.3RT$ at 298 K, which is reasonable for a radical decomposition. See summary of alkyl radical additions (page 219) and comments.

REACTION: METHYL RADICAL + TRANS-BUTENE-2

$\dot{C}H_3(I) + trans\text{-}CH_3CH{=}CHCH_3(II) \longrightarrow CH_3\dot{C}HCH(CH_3)_2(III)$

	(I)	(II)	(III)		
$\Delta H^\circ_{f(300)}$	34.0 (142.3)	-3.0 (-12.6)	5.3 (22.2)	ΔH°	-25.7 (-107.6)
$S^\circ_{(300)}$	46.1 (193.0)	72.4 (303.1)	83.3 (348.8)	ΔS°	-35.2 (-147.4)
$C^\circ_{p(300)}$	8.8 (36.8)	20.7 (86.7)	28.3 (118.5)	$\Delta C^\circ_{p(300)}$	-1.2 (-5.0)
$C^\circ_{p(500)}$	10.6 (44.3)	30.4 (127.1)	41.2 (172.1)	$\Delta C^\circ_{p(500)}$	0.2 (0.8)

log A	E	log k	Conditions	Reference and Experimental Method
-	-	3.56(673 K)	623-773 K 30-100 torr	[97] J.R.McNesby and A.S.Gordon, J.Amer.Chem.Soc., 79, 5902 (1957). Rate constant relative to log k_4 = 9.07 - 11.6/2.3RT for $CD_3 + (CD_3)_2CO \xrightarrow{4} CD_4 + CD_3COCD_2$ [J.A.Kerr and D.Timlin, J.Chem.Soc.(A), 1241 (1969)]. For details see $CH_3 + CH_2{=}CHCH_2CH_3$.
8.15	8.1 (33.9)	3.76(403 K)	353-453 K 3000 torr	[93] R.J.Cvetanović and R.S.Irwin, J.Chem.Phys., 46, 1694 (1967). Arrhenius parameters derived from $E_1 - E_3$ = 0.49 and A_1/A_3 = 4.2, where log k_3 = 7.53 - 7.6/2.3RT for $CH_3 + iso\text{-}C_4H_{10} \xrightarrow{3} CH_4 + C_4H_9$. For details see $CH_3 + CH_2{=}CH_2$.

PREFERRED RATE PARAMETERS

log k = 8.15 - 8.1/2.3RT

COMMENTS

The large discrepancy between the rate constant of reference [97] and the Arrhenius parameters of reference [93] cannot be explained, but the latter are preferred on the basis that they involve a technique which has yielded data for other CH_3 additions which are reliable and consistent with independent results. From the preferred parameters and the thermodynamic data, it can be calculated that log (k_{-1}/s^{-1}) = 14.0 - 33.2/2.3RT at 298 K, which is reasonable for a radical decomposition. See summary of alkyl radical additions (page 219) and comments.

ΔH^o_f, ΔH^o and E in kcal mol^{-1} (kJ mol^{-1}); S^o, C^o_p, ΔS^o and ΔC^o_p in cal mol^{-1} K^{-1} (J mol^{-1} K^{-1}); A and k in l mol^{-1} s^{-1}

REACTION: METHYL RADICAL + 2-METHYLPROPENE (ISOBUTENE)

$\dot{C}H_3(I) + (CH_3)_2C{=}CH_2(II) \xrightarrow{1} (CH_3)_3C\dot{C}H_2(III)$

$\xrightarrow{2} (CH_3)_2\dot{C}CH_2CH_3(IV)$

	(I)	(II)	(III)	(IV)		Path 1	Path 2
$\Delta H^\circ_{f(300)}$	34.0 (142.4)	-3.6 (-15.1)	7.1 (29.7)	1.8 (7.5)	ΔH°	-23.3 (-97.6)	-28.6 (-119.7)
$S^\circ_{(300)}$	46.1 (193.0)	71.3 (298.5)	80.7 (337.9)	83.3 (348.8)	ΔS°	-36.7 (-153.7)	-34.1 (-142.8)
$C^\circ_{p(300)}$	8.8 (36.8)	21.6 (90.4)	28.9 (121.0)	28.1 (117.7)	ΔC°_p	-1.5 (-6.3)	-2.3 (-9.6)

Path	log A	E	log k	Conditions	Reference and Experimental Method
1	-	-	2.88(673 K)	623-773 K 30-100 torr	[97] J.R.McNesby and A.S.Gordon, J.Amer.Chem.Soc., 79, 5902 (1957). Rate constants relative to $\log k_4 = 9.07 - 11.6/2.3RT$ for $CD_3 + (CD_3)_2CO \xrightarrow{4} CD_4 + CD_3COCD_2$ [J.A.Kerr and D.Timlin, J.Chem.Soc.(A), 1241 (1969)]. For details see $CH_3 + CH_2{=}CHCH_2CH_3$.
2	-	-	-	As above	
1 + 2	8.15	6.9 (28.9)	4.41(403 K)	353-453 K 3000 torr	[93] R.J.Cvetanović and R.S.Irwin, J.Chem.Phys., 46, 1694 (1967). Arrhenius parameters derived from $E_{1+2} - E_3 = -0.66$ and $A_{1+2}/A_3 = 4.2$, where $\log k_3 = 7.53 - 7.6/2.3RT$ for $CH_3 + iso\text{-}C_4H_{10} \xrightarrow{3} CH_4 + C_4H_9$.

PREFERRED RATE PARAMETERS

$\log k_{1+2} = 8.15 - 6.9/2.3RT$

COMMENTS

See comments for CH_3 + Butene-1. The Preferred parameters for k_{1+2} and the thermodynamic data (path 1) yield $\log (k_{-1}/s^{-1}) = 14.4 - 29.6/2.3RT$ at 298 K, which is reasonable for a radical decomposition. See summary of alkyl radical additions (page 219) and comments. It seems likely that at least 95% of the addition occurs at the terminal C atom in isobutene (path 2); see H + isobutene system.

REACTION: METHYL RADICAL + *cis*-PENTENE-2

$\dot{C}H_3(I) + C\text{-}CH_3CH{=}CHCH_2CH_3(II) \xrightarrow{1} (CH_3)_2CH\dot{C}HCH_2CH_3(III)$

$\xrightarrow{2} CH_3\dot{C}HCH(CH_3)CH_2CH_3(IV)$

	(I)	(II)	(III)	(IV)		Path 1	Path 2
$\Delta H^\circ_{f(300)}$	34.0 (142.3)	-6.6 (-27.6)	0.4 (1.7)	0.4 (1.7)	ΔH°	-27.0 (-113.0)	-27.0 (-113.0)
$S^\circ_{(300)}$	46.1 (193.0)	82.3 (344.6)	92.7 (388.1)	92.7 (388.1)	ΔS°	-35.7 (-149.5)	-35.7 (-149.5)
$C^\circ_{p(300)}$	8.8 (36.8)	25.8 (108.0)	33.8 (141.5)	33.8 (141.5)	ΔC°_p	-0.8 (-3.3)	-0.8 (-3.3)

Path	log A	E	log k	Conditions	Reference and Experimental Method
1	-	8.07[a] (33.79)	-	298 K	[98] C.W.Larson, B.S.Rabinovitch and D.C.Tardy, J.Chem.Phys., 47, 4570 (1967). For details see CH_3 + *cis*-butene-2. Radicals (III) and (IV) were generated from H atom additions to *cis*- and *trans*-4-methylpentene-2 and 3-methylpentene-1. The above critical energies are relative to an assumed value of 8.00 kcal mol^{-1} for the critical energy of the CH_3 addition to *trans*-pentene-2.
2	-	8.20[a] (34.33)	-	298 K	

[a] critical energy of the reaction.

PREFERRED RATE PARAMETERS

Reasonable as given above

ΔH^o_f, ΔH^o and E in kcal mol^{-1} (kJ mol^{-1}); S^o, C^o_p, ΔS^o and ΔC^o_p in cal mol^{-1} K^{-1} (J mol^{-1} K^{-1}); A and k in l mol^{-1} s^{-1}

REACTION: METHYL RADICAL + 3-METHYLBUTENE-1

$\dot{C}H_3(I) + (CH_3)_2CHCH{=}CH_2(II) \xrightarrow{1} (CH_3)_2CH\dot{C}HCH_2CH_3(III)$

$\xrightarrow{2} (CH_3)_2CHCH(CH_3)\dot{C}H_2(IV)$

	(I)	(II)	(III)	(IV)		Path 1	Path 2
$\Delta H^\circ_{f(300)}$	34.0 (142.3)	-6.8 (-28.5)	0.4 (1.7)	1.8 (7.5)	ΔH°	-26.8 (-112.2)	-25.4 (-106.3)
$S^\circ_{(300)}$	46.1 (193.0)	[80 (335)]	92.7 (388.1)	91.7 (383.9)	ΔS°	[-33 (-138)]	[-34 (-142)]
$C^\circ_{p(300)}$	8.8 (36.8)	[26.8 (112.2)]	33.8 (141.5)	33.6 (140.7)	ΔC°_p	[-1.8 (-7.5)]	[-2.0 (-8.3)]

Path	log A	E	log k	Conditions	Reference and Experimental Method
1	-	7.00[a] (29.31)	-	298 K	[98] C.W.Larson, B.S.Rabinovitch and D.C.Tardy, J.Chem.Phys., 47, 4570 (1967). For details see CH_3 + cis-butene-2. Radical (III) generated from H atom addition to cis- and trans-4-methylpentene-2. The above critical energy is relative to an assumed value of 8.00 kcal mol^{-1} for the critical energy of the CH_3 addition to trans-pentene-2.
2	-	-	-	-	

[a] critical energy of the reaction

PREFERRED RATE PARAMETERS

Reasonable as given above

REACTION: METHYL RADICAL + 2-METHYLBUTENE-2 (TRIMETHYLETHYLENE) $\dot{C}H_3(I) + (CH_3)_2C{=}CHCH_3(II) \xrightarrow{1} (CH_3)_3C\dot{C}HCH_3(III)$

$\xrightarrow{2} (CH_3)_2\dot{C}CH(CH_3)_2(IV)$

	(I)	(II)	(III)	(IV)		Path 1	Path 2
$\Delta H^\circ_{f(300)}$	34.0 (142.3)	-10.3 (-43.1)	-1.4 (-5.9)	-5.2 (-21.8)	ΔH°	-25.1 (-105.1)	-28.9 (-121.0)
$S^\circ_{(300)}$	46.1 (193.0)	79.9 (334.5)	88.5 (370.5)	90.0 (376.8)	ΔS°	-37.5 (-157.0)	-36.0 (-150.7)
$C^\circ_{p(300)}$	8.8 (36.8)	25.5 (106.8)	34.3 (143.6)	33.4 (139.8)	$\Delta C^\circ_{p(300)}$	0 (0)	-0.9 (-3.8)
$C^\circ_{p(500)}$	10.6 (44.3)	38.2 (159.8)	51.8 (216.8)	49.8 (208.6)	$\Delta C^\circ_{p(500)}$	3.0 (12.6)	1.0 (4.2)

Path	log A	E	log k	Conditions	Reference and Experimental Method
1 + 2	[7.13]	[6.1] [(25.5)]	3.83(403 K)	403-455 K 3000 torr	[93] R.J.Cvetanović and R.S.Irwin, J.Chem.Phys., 46, 1694 (1967). Arrhenius parameters derived from $E_{1+2} - E_3$ = [-1.55] and A_{1+2}/A_3 = [0.4], where log k_3 = 7.53 - 7.6/2.3RT for CH_3 + iso-$C_4H_{10} \xrightarrow{3} CH_4 + C_4H_9$. For details see $CH_3 + CH_2{=}CH_2$. The results were considered tentative as the rate constants were measured at two temperatures only.

PREFERRED RATE PARAMETERS

Tentatively as given

COMMENTS

From the above results for the addition reaction and the thermodynamic data, it can be calculated that for the reverse radical decomposition reaction (path -2), log $k_{-1}(s^{-1})$ = 13.2 - 34.5/2.3RT at 298 K, which is reasonable. See also summary of alkyl radical additions (page 219) and comments.

ΔH°_f, ΔH° and E in kcal mol^{-1} (kJ mol^{-1}); S°, C°_p, ΔS° and ΔC°_p in cal mol^{-1} K^{-1} (J mol^{-1} K^{-1}); A and k in l mol^{-1} s^{-1}

REACTION: METHYL RADICAL + cis-HEXENE-2

$\dot{C}H_3(I) + C\text{-}CH_3CH{=}CHCH_2CH_2CH_3 \xrightarrow{1} CH_3CH(CH_3)\dot{C}HCH_2CH_2CH_3(III)$

$\xrightarrow{2} CH_3\dot{C}HCH(CH_3)CH_2CH_2CH_3(IV)$

	(I)	(II)	(III)	(IV)		Path 1	Path 2
$\Delta H^\circ_{f(300)}$	34.0 (142.3)	-11.7 (-49.0)	-4.6 (-19.3)	-4.6 (-19.3)	ΔH°	-27.3 (-114.3)	-27.3 (-114.3)
$S^\circ_{(300)}$	46.1 (193.0)	91.6 (383.6)	101.6 (425.4)	101.6 (425.4)	ΔS°	-36.1 (-151.2)	-36.1 (-151.2)
$C^\circ_{p(300)}$	8.8 (36.8)	30.0 (125.6)	39.3 (164.5)	39.3 (164.5)	ΔC°_p	0.5 (2.1)	0.5 (2.1)

Path	log A	E	log k	Conditions	Reference and Experimental Method
1	-	8.07[a] (33.79)	-	298 K	[98] C.W.Larson, B.S.Rabinovitch and D.C.Tardy, J.Chem.Phys., 47, 4570 (1967); C.W.Larson and B.S.Rabinovitch, ibid., 50, 871 (1969).
2	-	-	-	-	For details see CH_3 + cis-butene-2. Radical (III) was generated from H atom addition to trans-2-methylhexene-3. The above critical energy is relative to an assumed value of 8.00 kcal mol^{-1} for the critical energy of the CH_3 addition to trans-hexene-2.
1	-	-	-	-	[100] C.W.Larson, D.C.Tardy and B.S.Rabinovitch, J.Chem.Phys., 49, 299 (1968).
2		8.25[a] (34.4)	-	298 K	For details see CH_3 + cis-butene-2. Radical (IV) generated from H atom addition to 3-methylhexene-1. The above critical energy is relative to an assumed value of 8.00 kcal mol^{-1} for the critical energy of the CH_3 addition to trans-hexene-2.

[a] critical energy of the reaction

PREFERRED RATE PARAMETERS

Reasonable as given above

REACTION: METHYL RADICAL + 2-METHYLPENTENE-1

$\dot{C}H_3(I) + CH_2{=}C(CH_3)CH_2CH_2CH_3(II) \xrightarrow{1} CH_3CH_2\dot{C}(CH_3)CH_2CH_2CH_3(III)$

$\xrightarrow{2} \dot{C}H_2C(CH_3)_2CH_2CH_2CH_3(IV)$

	(I)	(II)	(III)	(IV)
$\Delta H^\circ_{f(300)}$	34.0 (142.3)	-13.3 (-55.7)	-8.1 (-33.9)	-3.8 (-15.9)
$S^\circ_{(300)}$	46.1 (193.0)	90.6 (379.3)	102.2 (427.9)	99.6 (417.0)
$C^\circ_{p(300)}$	8.8 (36.8)	32.2 (134.8)	39.1 (163.7)	39.9 (167.1)

	Path 1	Path 2
ΔH°	-28.8 (-120.6)	-24.5 (-102.6)
ΔS°	-34.5 (-144.5)	-37.1 (-155.3)
ΔC°_p	-1.9 (-8.0)	-1.1 (-4.6)

Path	log A	E	log k	Conditions	Reference and Experimental Method
1	-	6.86[a] (28.72)	-	298 K	[98] C.W.Larson, B.S.Rabinovitch and D.C.Tardy, J.Chem.Phys., 47, 4570 (1967). For details see CH_3 + cis-butene-2. Radical (III) generated from H atom addition to cis-3-methylhexene-2. The above critical energy is relative to an assumed value of 7.00 kcal mol^{-1} for the critical energy of the C_2H_5 addition to 2-methylbutene-1.
2	-	-	-	-	

[a] critical energy of the reaction

PREFERRED RATE PARAMETERS

Reasonable as given above

ΔH^o_f, ΔH^o and E in kcal mol^{-1} (kJ mol^{-1}); S^o, C^o_p, ΔS^o and ΔC^o_p in cal mol^{-1} K^{-1} (J mol^{-1} K^{-1}); A and k in l mol^{-1} s^{-1}

REACTION: METHYL RADICAL + cis-4-METHYLPENTENE-2

$\dot{C}H_3(I)$ + cis-$CH_3CH{=}CHCH(CH_3)_2(II) \xrightarrow{1} (CH_3)_2CH\dot{C}HCH(CH_3)_2(III)$

$\xrightarrow{2} CH_3\dot{C}HCH(CH_3)CH(CH_3)_2(IV)$

	(I)	(II)	(III)	(IV)		Path 1	Path 2
$\Delta H^\circ_{f(300)}$	34.0 (142.3)	-13.5 (-56.5)	-6.7 (-28.1)	-6.7 (-28.1)	ΔH°	-27.2 (-113.9)	-27.2 (-113.9)
$S^\circ_{(300)}$	46.1 (193.0)	[89 (373)]	99.4 (416.2)	99.4 (416.2)	ΔS°	[-36 (-151)]	[-36 (-151)]
$C^\circ_{p(300)}$	8.8 (36.8)	[29.8 (124.8)]	39.0 (163.3)	39.0 (163.3)	ΔC°_p	[0.4 (1.7)]	[0.4 (1.7)]

Path	log A	E	log k	Conditions	Reference and Experimental Method
1	-	-	-	298 K	[101] C.W.Larson and B.S.Rabinovitch, J.Chem.Phys., 50, 871 (1969).
2	-	8.72[a] (36.51)	-	-	For details see CH_3 + cis-butene-2. Radical (IV) generated from H atom addition to 3,4-dimethylpentene-1. The above critical energy is relative to an assumed value of 8.00 kcal mol^{-1} for the critical energy of the CH_3 addition to trans-4-methylpentene-2.

[a] critical energy of the reaction

PREFERRED RATE PARAMETERS

Reasonable as given above

REACTION: METHYL RADICAL + 2,3-DIMETHYLBUTENE-2 (TETRAMETHYLETHYLENE) $\dot{C}H_3(I) + (CH_3)_2C{=}C(CH_3)_2(II) \longrightarrow (CH_3)_3C\dot{C}(CH_3)_2(III)$

	(I)	(II)	(III)		
$\Delta H^\circ_{f(300)}$	34.0 (142.3)	-16.6 (-69.4)	-11.9 (-49.8)	ΔH°	-29.3 (-122.6)
$S^\circ_{(300)}$	46.1 (193.0)	87.4 (365.9)	95.2 (398.6)	ΔS°	-38.3 (-160.4)
$C^\circ_{p(300)}$	8.8 (36.8)	30.3 (126.8)	39.4 (165.0)	$\Delta C^\circ_{p(300)}$	0.3 (1.3)
$C^\circ_{p(500)}$	10.6 (44.3)	46.0 (192.4)	59.8 (250.3)	$\Delta C^\circ_{p(500)}$	3.2 (13.4)

log A	E	log k	Conditions	Reference and Experimental Method
[7.01]	[6.8] [(28.5)]	3.33(403 K)	403-453 K 3000 torr	[93] R.J.Cvetanović and R.S.Irwin, J.Chem.Phys., 46, 1694 (1967). Arrhenius parameters derived from $E_1 - E_3$ = [-0.84] and A_1/A_3 = [0.3], where $\log k_3 = 7.53 - 7.6/2.3RT$ for CH_3 + iso-$C_4H_{10} \xrightarrow{3} CH_4 + C_4H_9$. For details see $CH_3 + CH_2{=}CH_2$. The results were considered tentative as the rate constants were measured at two temperatures only.

PREFERRED RATE PARAMETERS

Tentatively as given

COMMENTS

From the above results for the addition reaction and the thermodynamic data, it can be shown that for the reverse radical decomposition, $\log k_{-1}(s^{-1}) = 13.5 - 35.5$ at 298 K, which is reasonable. See also summary of alkyl additions (page 219) and comments.

ΔH°_f, ΔH° and E in kcal mol^{-1} (kJ mol^{-1}); S°, C°_p, ΔS° and ΔC°_p in cal mol^{-1} K^{-1} (J mol^{-1} K^{-1}); A and k in l mol^{-1} s^{-1}

REACTION: METHYL RADICAL + TETRAFLUOROETHYLENE $\dot{C}H_3(I) + CF_2{=}CF_2(II) \longrightarrow CH_3CF_2\dot{C}F_2(III)$

	(I)	(II)	(III)		
$\Delta H^\circ_{f(300)}$	34.0 (142.3)	-157.9 (-661.1)	-159.5 (-667.9)	ΔH°	-35.6 (-149.0)
$S^\circ_{(300)}$	46.1 (193.0)	74.6 (312.4)	86.3 (361.3)	ΔS°	-34.4 (-144.0)
$C^\circ_{p(300)}$	8.8 (36.8)	19.4 (81.2)	25.8 (108.0)	$\Delta C^\circ_{p(300)}$	-2.4 (-10.0)
$C^\circ_{p(500)}$	10.6 (44.3)	24.0 (100.4)	36.2 (151.4)	$\Delta C^\circ_{p(500)}$	1.6 (6.7)

log A	E	log k	Conditions	Reference and Experimental Method
8.05	5.2 (21.8)	5.23(403 K)	353-453 K 750 torr	[102] J.M.Sangster and J.C.J.Thynne, Int.J.Chem.Kinet., 1, 571 (1969). Arrhenius parameters derived from $E_1 - E_3 = -2.42$ and $A_1/A_3 = 3.3$, where $\log k_3 = 7.53 - 7.6/2.3RT$ for $CH_3 + iso\text{-}C_4H_{10} \xrightarrow{3} CH_4 + C_4H_9$. Same experimental method as that for $CH_3 + CH_2{=}CH_2$ system, reference [93].

PREFERRED RATE PARAMETERS

Reliable as given

COMMENTS

There are no data on the reverse decomposition reaction $CH_3CF_2\dot{C}F_2 \xrightarrow{-1} \dot{C}H_3 + CF_2{=}CF_2$, but the above preferred value of k_1 and the thermodynamic data yield $\log k_{-1}(s^{-1}) = 13.8 - 33.7/2.3RT$ at 500 K, which seems reasonable in relation to the decomposition of n-propyl (see $CH_3 + CH_2{=}CH_2$ system).

REACTION: METHYL RADICAL + HEXAFLUOROPROPENE

$\dot{C}H_3(I) + CF_3CF{=}CF_2(II) \xrightarrow{1} CF_3\dot{C}FCF_2CH_3(III)$

$\xrightarrow{2} CF_3(CH_3)CF\dot{C}F_2(IV)$

	(I)	(II)	(III)	(IV)		Path 1	Path 2
$\Delta H^\circ_{f(300)}$	34.0 (142.3)		[-269 (-1126)]	[-265 (-1110)]	ΔH°		
$S^\circ_{(300)}$	46.1 (193.0)		[102 (427)]		ΔS°		
$C^\circ_{p(300)}$	8.8 (36.8)				ΔC°_p		

Path	log A	E	log k	Conditions	Reference and Experimental Method
1 + 2	8.14	5.7 (23.9)	5.14(415 K)	354-476 K 750 torr	[103] J.C.J.Thynne, Int.J.Chem.Kinet., 3, 155 (1970). Arrhenius parameters derived from $E_{1+2} - E_3 = -1.9$ kcal mol^{-1} and $A_{1+2}/A_3 = 4.1$, where log k = 7.53 - 7.6/2.3RT for $CH_3 + iso\text{-}C_4H_{10} \xrightarrow{3} CH_4 + C_4H_9$. Same experimental method as that for $CH_3 + CH_2{=}CH_2$ system, reference [93].

PREFERRED RATE PARAMETERS

Reasonable as given above

COMMENTS

It is not possible to distinguish the favoured path for addition from the experiments of reference [103], although it would be surprising if path 1 was not faster than path 2.

ΔH^o_f, ΔH^o and E in kcal mol^{-1} (kJ mol^{-1}); S^o, C^o_p, ΔS^o and ΔC^o_p in cal mol^{-1} K^{-1} (J mol^{-1} K^{-1}); A and k in l mol^{-1} s^{-1}

REACTION: METHYL RADICAL + VINYL CHLORIDE

$$\dot{C}H_3(I) + CH_2{=}CHC\ell(II) \xrightarrow{1} CH_3CH_2\dot{C}HC\ell(III)$$
$$\xrightarrow{2} \dot{C}H_2CHC\ell CH_3(IV)$$

	(I)	(II)	(III)	(IV)		Path 1	Path 2
$\Delta H^\circ_{f(300)}$	34.0 (142.3)	8.6 (36.0)	13.5 (56.5)	12.9 (54.0)	ΔH°	-29.1 (-121.8)	-29.7 (-123.3)
$S^\circ_{(300)}$	46.1 (193.0)	63.0 (263.6)	76.9 (321.7)	78.4 (328.1)	ΔS°	-32.2 (-134.7)	-30.7 (-128.5)
$C^\circ_{p(300)}$	8.8 (36.8)	13.0 (54.4)	20.0 (83.7)	21.0 (87.8)	$\Delta C^\circ_{p(300)}$	-1.8 (-7.5)	-0.8 (-3.3)
$C^\circ_{p(500)}$	10.6 (44.3)	17.8 (74.5)	29.4 (123.0)	28.2 (117.9)	$\Delta C^\circ_{p(500)}$	1.0 (4.2)	-0.2 (-0.8)

Path	log A	E	log k	Conditions	Reference and Experimental Method
1 + 2	9.74	9.05 (37.89)	4.95(417 K)	393-433 K 6-24 torr	[91] A.M.Hogg and P.Kebarle, J.Amer.Chem.Soc., 86, 4558 (1964). Arrhenius parameters relative to log k_c = 10.34 for CH_3 combination; log $(k_{1+2}/k_c^{\frac{1}{2}})(\ell^{\frac{1}{2}}mol^{-\frac{1}{2}}s^{-\frac{1}{2}})$ = 4.57 - 9.05/2.3RT. For details see CH_3 + $CH_2{=}CH_2$, references [91,92]. Rate constants for addition to $CH_2{=}CHC\ell$ determined by mass balance method, based on analysis of the products CH_3COCH_3, CH_4 and C_2H_6. The authors of this paper assumed that the addition followed path 1 exclusively, but in fact the mass balance method gives composite rate constants k_{1+2}, although it seems likely that path 1 is the favoured route, (see comments).

PREFERRED RATE PARAMETERS

[8.3 - 6.2/2.3RT]

COMMENTS

The Arrhenius parameters appear to be over-estimates. Assuming the data of reference [91] refer to path 1, it can be calculated from the thermodynamic data that for the reverse radical decomposition, log (k_{-1}/s^{-1}) = 14.9 - 36.0/2.3RT. Both the calculated A_{-1} and E_{-1} seem excessively large in relation to the n-C_3H_7 decomposition, and again probably indicate that k_1 has been over-estimated. The preferred Arrhenius parameters are based on an assumed value of A_1, and the measured rate constant at the middle of the experimental temperature range.

It should be noted that the data of reference [91] refer to a very short temperature range. It is possible that some addition occurs via path 2 leading to a vibrationally excited adduct radical $\dot{C}H_2CHC\ell CH_3$*, which would readily decompose, $\dot{C}H_2CHC\ell CH_3^* \longrightarrow CH_2{=}CHCH_3 + C\ell$, to yield C_3H_6, which, if present in sufficient quantity, could upset the mass balance upon which the data for k_1 are based.

REACTION: METHYL RADICAL + ALLENE

$$\dot{C}H_3(I) + CH_2{=}C{=}CH_2(II) \xrightarrow{1} CH_3CH_2\dot{C}{=}CH_2(III)$$
$$\xrightarrow{2} \dot{C}H_2C(CH_3){=}CH_2(IV)$$

	(I)	(II)	(III)	(IV)
$\Delta H^o_{f(300)}$	34.0 (142.3)	45.9 (192.2)	52.1 (218.1)	30.7 (128.5)
$S^o_{(300)}$	46.1 (193.0)	58.3 (244.1)	75.0 (313.8)	71.2 (298.0)
$C^o_{p(300)}$	8.8 (36.8)	14.2 (59.5)	20.3 (85.0)	20.8 (87.0)
$C^o_{p(500)}$	10.6 (44.3)	19.8 (82.9)	30.7 (128.4)	31.1 (130.2)

	Path 1	Path 2
ΔH^o	-27.8 (-116.4)	-49.2 (-205.8)
ΔS^o	-29.4 (-123.0)	-33.2 (-138.9)
$\Delta C^o_{p(300)}$	-2.7 (-11.3)	-2.2 (-9.2)
$\Delta C^o_{p(500)}$	0.3 (1.2)	0.7 (2.9)

Path	log A	E	log k	Conditions	Reference and Experimental Method
1	8.3	8.1 (33.9)	4.16(428 K)	373-483 K 40-100 torr	[104] R.R.Getty, J.A.Kerr and A.F.Trotman-Dickenson, J.Chem.Soc.(A), 979 (1967).

Arrhenius parameters relative to log k_c = 10.34 for CH_3 combination; log $(k_{1+2}/k_c^{\frac{1}{2}})(\ell^{\frac{1}{2}}mol^{-\frac{1}{2}}s^{-\frac{1}{2}})$ = 3.13 - 8.1/2.3RT.

Static system. Alkyl radicals from photolysis (full Hg arc) of corresponding aldehyde. Aldehyde used as scavenger for adduct radicals ($\dot{A}$):

$$\dot{R} + O\ell \xrightarrow{1,2} \dot{A}$$
$$\dot{A} + RCHO \xrightarrow{3} AH + R\dot{C}O$$
$$R\dot{C}O \xrightarrow{4} \dot{R} + CO$$
$$\dot{R} + RCHO \xrightarrow{5} RH + R\dot{C}O$$
$$2\dot{R} \xrightarrow{C} R_2$$

This is the 'aldehyde' method which has been used to study the additions of several alkyl radicals to alkenes and alkynes [J.A.Kerr and A.F.Trotman-Dickenson, Prog.Reaction Kinetics, 1, 108 (1961)]. Rate constants were calculated from the equation $k_1/k_c^{\frac{1}{2}} = R_{AH}/R_{R_2}^{\frac{1}{2}}[O\ell]$, in which R_{AH} and R_{R_2} are the rates of formation of the abstraction product (reaction (3)) and the radical dimer, which were analyzed by gas chromatography. It was thus assumed that the adduct radicals only form the product AH, and in some cases analysis revealed very little product from further additions of $\dot{A}$ to the unsaturate ($O\ell$). This method appears to give reasonable results, but, without a more complete kinetic analysis involving larger variations in the concentrations of starting materials,

ΔH^o_f, ΔH^o and E in kcal mol^{-1} (kJ mol^{-1}); S^o, C^o_p, ΔS^o and ΔC^o_p in cal mol^{-1} K^{-1} (J mol^{-1} K^{-1}); A and k in l mol^{-1} s^{-1}

Path	log A	E	log k	Conditions	Reference and Experimental Method
					the derived Arrhenius parameters should be assigned considerable errors as the assumption of complete removal of adduct radicals by reaction (3) may not be entirely justified. It was shown by product analysis that the addition occurred exclusively by path 1, i.e. by addition to the terminal C atom of allene. On thermochemical grounds, the addition to the central C atom should be favoured, but the geometry of the orbitals in allene is such that bonding by the adding radical at the central C atom precludes any overlap of the resulting p-orbital of the terminal C atom with the p-orbitals of the remaining double bond; although, once the σ bond is formed, and rotation about the C-C σ bond produces the resonance-stabilized allylic radical, the effect of stabilization does not influence the transition state of the addition reaction.

PREFERRED RATE PARAMETERS

Reasonable as given

COMMENTS

See comments on 'aldehyde method' in experimental section.
See also summary of alkyl radical additions (page 219)
and comments.

REACTION: METHYL RADICAL + BUTADIENE-1,3

$\dot{C}H_3(I) + CH_2{=}CHCH{=}CH_2(II) \xrightarrow{1} CH_3CH_2\dot{C}HCH{=}CH_2(III)$

$\xrightarrow{2} \dot{C}H_2CH(CH_3)CH{=}CH_2(IV)$

	(I)	(II)	(III)	(IV)		Path 1	Path 2
$\Delta H^\circ_{f(300)}$	34.0 (142.3)	26.1 (109.2)	25.3 (105.9)	38.7 (162.0)	ΔH°	-34.8 (-145.6)	-21.4 (-89.6)
$S^\circ_{(300)}$	46.1 (192.9)	68.0 (284.5)	80.2 (335.7)	82.9 (346.9)	ΔS°	-33.9 (-141.8)	-31.2 (-130.6)
$C^\circ_{p(300)}$	8.8 (36.8)	19.1 (79.9)	25.5 (106.7)	26.0 (108.8)	ΔC°_p	-2.4 (-10.0)	-1.9 (-8.0)

Path	log A	E	log k	Conditions	Reference and Experimental Method
1 + 2	∿6.8	∿2.5 (∿10.5)	∿5.6(473 K)	413-533 K 50 torr	[88] L.Mandelcorn and E.W.R.Steacie, Can.J.Chem., 32, 474 (1954). Arrhenius parameters relative to log k_c = 10.34 for CH_3 combination; log $(k_{1+2}/k_c^{\frac{1}{2}})(\ell^{\frac{1}{2}}mol^{-\frac{1}{2}}s^{-\frac{1}{2}})$ = ∿1.6 - (∿2.5/2.3RT). For details see $CH_3 + CH_2{=}CH_2$.
1 + 2	-	4.9 (20.5)	-	403-438 K 10-350 torr	[89] L.C.Landers and D.H.Volman, J.Amer.Chem.Soc., 79, 2996 (1957). For details see $CH_3 + CH_2{=}CH_2$. $\frac{1}{2}E_3 + E_{1+2} - \frac{1}{2}E_c = 24.1$ and taking $E_3 = 38.4$ and $E_c = 0$ gives $E_{1+2} = 4.9$ kcal mol^{-1}.
1 + 2	7.91	4.1 (19.3)	5.69(403 K)	353-453 K 3000 torr	[93] R.J.Cvetanović and R.S.Irwin, J.Chem.Phys., 46, 1694 (1967). Arrhenius parameters derived from $E_{1+2} - E_3 = -3.46$ and $A_{1+2}/A_3 = 2.4$, where log k_3 = 7.53 - 7.6/2.3RT for $CH_3 + iso\text{-}C_4H_{10} \xrightarrow{3} CH_4 + C_4H_9$.

PREFERRED RATE PARAMETERS

log k_1 = 7.91 - 4.1/2.3RT

COMMENTS

The data of reference [93] seem likely to refer to path 1 exclusively since there is a large difference between ΔH_1° and ΔH_2° ; the large exothermicity of path 1, due to the formation of the resonance stabilized adduct radical, also explains why the E for addition to butadiene is significantly lower than E for mono-alkenes. See also summary of alkyl additions (page 219) and comments.

ΔH_f°, ΔH° and E in kcal mol^{-1} (kJ mol^{-1}); S°, C_p°, ΔS° and ΔC_p° in cal mol^{-1} K^{-1} (J mol^{-1} K^{-1}); A and k in l mol^{-1} s^{-1}

REACTION: METHYL RADICAL + ACETYLENE $\dot{C}H_3(I) + CH{\equiv}CH(II) \longrightarrow CH_3CH{=}\dot{C}H(III)$

	(I)	(II)	(III)		
$\Delta H^\circ_{f(300)}$	34.0 (142.3)	54.2 (226.9)	57.7 (241.6)	ΔH°	-30.5 (-127.7)
$S^\circ_{(300)}$	46.1 (193.0)	48.0 (201.0)	68.3 (286.0)	ΔS°	-25.8 (-108.0)
$C^\circ_{p(300)}$	8.8 (36.8)	10.5 (44.0)	15.2 (63.6)	$\Delta C^\circ_{p(300)}$	-4.1 (-17.2)
$C^\circ_{p(500)}$	10.6 (44.3)	13.1 (54.8)	22.5 (94.2)	$\Delta C^\circ_{p(500)}$	-1.8 (-7.5)

log A	E	log k	Conditions	Reference and Experimental Method
7.88	5.5 (23.0)	5.34(473 K)	417-514 K 50 torr	[88] L.Mandelcorn and E.W.R.Steacie, Can.J.Chem., 32, 474 (1954). Arrhenius parameters relative to log k_C = 10.34 for CH_3 combination; $\log (k_1/k_C^{\frac{1}{2}})(\ell^{\frac{1}{2}}mol^{-\frac{1}{2}}s^{-\frac{1}{2}}) = 2.71 - 5.5/2.3RT$. For details see $CH_3 + CH_2{=}CH_2$.
-	5.1 (21.4)	-	403-438 K 10-350 torr	[89] L.C.Landers and D.H.Volman, J.Amer.Chem.Soc., 79, 2996 (1957). For details see $CH_3 + CH_2{=}CH_2$. $\frac{1}{2}E_3 + E_1 - \frac{1}{2}E_C = 24.3$, and taking $E_3 = 38.4$ and $E_C = 0$ gives $E_1 = 5.1$ kcal mol^{-1}.
8.4	7.7 (32.2)	4.44(425 K)	371-479 K 60 torr	[105] J.A.Garcia-Dominguez and A.F.Trotman-Dickenson, J.Chem.Soc., 940 (1962). Arrhenius parameters relative to log k_C = 10.34 for CH_3 combination; $\log (k_1/k_C^{\frac{1}{2}})(\ell^{\frac{1}{2}}mol^{-\frac{1}{2}}s^{-\frac{1}{2}}) = 3.25 - 7.7/2.3RT$. Details as for $CH_3 + CH_2{=}C{=}CH_2$, reference [104].

PREFERRED RATE PARAMETERS

log k = 8.4 - 7.7/2.3RT

COMMENTS

See comments on 'aldehyde method' in experimental section of $CH_3 + CH_2{=}C{=}CH_2$. There is a very large discrepancy between the rate constants of reference [88] and reference [105] which is difficult to explain, particularly since the mass balance method of reference [88] gave rate constants for $k_{C_2H_4}$ in reasonable accord with the preferred $k_{C_2H_4}$. We favour the data of reference [105], since the experimental method is more direct.

See also summary of alkyl additions (page 219) and comments.

REACTION: METHYL RADICAL + PROPYNE

$$\dot{C}H_3(I) + CH_3C{\equiv}CH(II) \xrightarrow{1} (CH_3)_2C{=}\dot{C}H(III)$$
$$\xrightarrow{2} CH_3\dot{C}{=}CHCH_3(IV)$$

	(I)	(II)	(III)	(IV)		Path 1	Path 2
$\Delta H^\circ_{f(300)}$	34.0 (142.3)	44.0 (184.2)	49.3 (206.4)	49.9 (208.9)	ΔH°	-28.7 (-120.2)	-28.1 (-117.7).
$S^\circ_{(300)}$	46.1 (193.0)	59.2 (247.9)	72.6 (304.0)	75.2 (314.8)	ΔS°	-32.7 (-136.9)	-30.1 (-126.0)
$C^\circ_{p(300)}$	8.8 (36.8)	14.6 (61.1)	21.6 (90.4)	20.7 (86.7)	$\Delta C^\circ_{p(300)}$	-1.8 (-7.5)	-2.7 (-11.3)
$C^\circ_{p(500)}$	10.6 (44.3)	19.7 (82.5)	31.3 (131.1)	30.4 (127.2)	$\Delta C^\circ_{p(500)}$	1.0 (4.2)	0.1 (0.4)

Path	log A	E	log k	Conditions	Reference and Experimental Method
1 + 2	8.7	8.8 (36.8)	4.14(322 K)	379-465 K 40-100 torr	[106] R.R.Getty, J.A.Kerr and A.F.Trotman-Dickenson, J.Chem.Soc.(A), 1360 (1967). Arrhenius parameters relative to log k_c = 10.34 for CH_3 combination; log $(k_{1+2}/k_c^{\frac{1}{2}})(\ell^{\frac{1}{2}}mol^{-\frac{1}{2}}s^{-\frac{1}{2}})$ = 3.53 - 8.8/2.3RT. For details see $CH_3 + CH_2{=}CH_2$. Product analysis showed that addition occurred predominantly at the terminal C atom, path 1 : path 2 ≃ 0.12.

PREFERRED RATE PARAMETERS

Tentatively as given above

COMMENTS

See comments on 'aldehyde method' in experimental section of $CH_3 + CH_2{=}C{=}CH_2$ system.
By comparison with results on CH_3 addition to $CH{\equiv}CH$, the above Arrhenius parameters seem slightly high. See summary of alkyl radical additions (page 219).

ΔH°_f, ΔH° and E in kcal mol^{-1} (kJ mol^{-1}); S°, C°_p, ΔS° and ΔC°_p in cal mol^{-1} K^{-1} (J mol^{-1} K^{-1}); A and k in l mol^{-1} s^{-1}

REACTION: ETHYL RADICAL + ETHYLENE $CH_3\dot{C}H_2$(I) + CH_2=CH_2(II) ⟶ $CH_3CH_2CH_2\dot{C}H_2$(III)

	(I)	(II)	(III)		
$\Delta H^\circ_{f(300)}$	26.0 (108.8)	12.5 (52.3)	15.8 (66.2)	ΔH°	-22.7 (-95.0)
$S^\circ_{(300)}$	59.8 (250.4)	52.4 (219.3)	78.3 (327.7)	ΔS°	-33.9 (-141.8)
$C^\circ_{p(300)}$	11.1 (46.5)	10.3 (43.1)	23.2 (97.1)	$\Delta C^\circ_{p(300)}$	1.8 (7.6)
$C^\circ_{p(500)}$	16.3 (68.2)	14.9 (62.4)	34.2 (143.2)	$\Delta C^\circ_{p(500)}$	3.0 (12.6)

log A	E	log k	Conditions	Reference and Experimental Method
6.75	5.5 (23.3)	3.5(363 K)	331-396 K 660 torr	[107] J.A.Pinder and D.J.LeRoy, Can.J.Chem., 35, 588 (1957). Arrhenius parameters relative to log k_c = 10.4 for C_2H_5 combination; $\log (k_1/k_c^{\frac{1}{2}})(\ell^{\frac{1}{2}}mol^{-\frac{1}{2}}s^{-\frac{1}{2}}) = 1.55 - 5.5/2.3RT$. Static system. Radicals from the addition of H atoms (Hg photosensitized decomposition H_2) to C_2H_4. Reaction rates based on analysis of products of reaction of adduct radicals. Analysis by low-temperature distillation and gas chromatography.
7.01	5.5 (23.3)	4.2(423 K)	398-448 K 45-65 torr	[108] F.W.Lampe and F.H.Field, Can.J.Chem., 37, 995 (1959). Arrhenius parameters relative to log k_c = 10.4 for C_2H_5 combination; $\log (k_1/k_c^{\frac{1}{2}})(\ell^{\frac{1}{2}}mol^{-\frac{1}{2}}s^{-\frac{1}{2}}) = 1.81 - 5.5/2.3RT$. Static system. Radicals from photolysis (λ = 366 nm) of azoethane. Reaction rates based on analysis of products of reactions of adduct radicals. Analysis by low-temperature distillation and gas chromatography.
8.8	8.6 (36.0)	3.9(428 K)	364-482 K 55-130 torr	[109] J.A.Kerr and A.R.Trotman-Dickenson, J.Chem.Soc., 1611 (1960). Arrhenius parameters relative to log k_c = 10.4 for C_2H_5 combination; $\log (k_1/k_c^{\frac{1}{2}})(\ell^{\frac{1}{2}}mol^{-\frac{1}{2}}s^{-\frac{1}{2}}) = 3.5 - 8.6/2.3RT$. 'Aldehyde method'. See experimental section of CH_3 + CH_2=C=CH_2.
8.17	7.56 (31.6)	3.75(374 K)	348-405 K 6-110 torr	[110] K.W.Watkins and L.A.O'Deen, J.Phys.Chem., 73, 4094 (1969). Arrhenius parameters relative to log k_c = 10.4 for C_2H_5 combination; $\log (k_1/k_c^{\frac{1}{2}})(\ell^{\frac{1}{2}}mol^{-\frac{1}{2}}s^{-\frac{1}{2}}) = 2.97 - 7.56/2.3RT$. Static system. Radicals from high intensity photolysis of corresponding azo compound (λ = full Hg arc, Pyrex reaction vessel). Reaction rates based on analysis of products of reactions of adduct radicals. Analysis by gas chromatography.

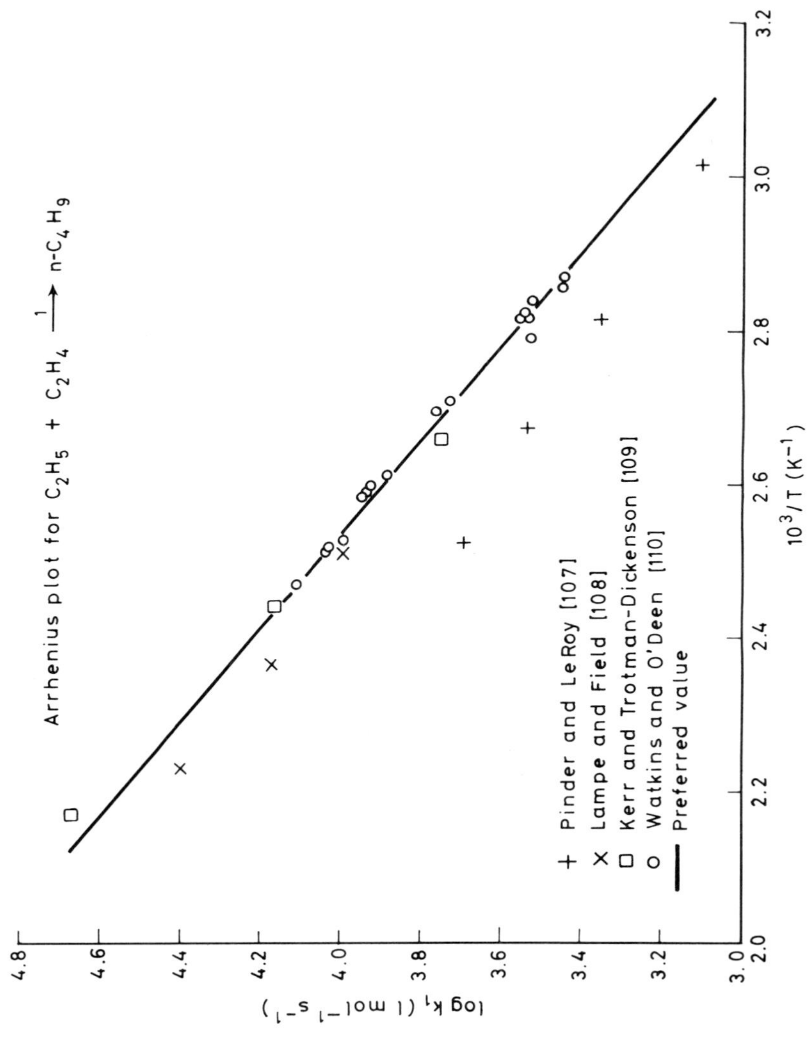

FIGURE 3. Arrhenius plot for $C_2H_5 + C_2H_4 \xrightarrow{1} n\text{-}C_4H_9$

log A	E	log k	Conditions	Reference and Experimental Method
				This study is the most comprehensive so far attempted and the rate constants are probably the most reliable. Unfortunately the temperature range is rather narrow.

PREFERRED RATE PARAMETERS

log k = 8.20 - 7.3/2.3RT

COMMENTS.

The preferred Arrhenius parameters were obtained from a least-mean-squares treatment of the rate constants of references [109,110]. All the data are shown in the plot (Fig.3). The rate constants of references [107,108] are distinctly lower than the preferred line, the most likely cause of the discrepancy being failure to account for all the products of the reactions of the adduct radicals. The most recent study of the reverse decomposition reaction, $n\text{-}C_4H_9 \xrightarrow{-1} C_2H_5 + C_2H_4$, reports $\log (k_{-1}/s^{-1}) = 13.6 - 28.7/2.3RT$ at a mean temperature of ∿ 500 K [W.E.Morganroth and J.G.Calvert, J.Amer.Chem.Soc., 88, 5387 (1966)], which, with the above thermodynamic data, leads to a calculated rate constant, $\log k_1 = 8.3 - 7.1/2.3RT$ at 500 K, in excellent agreement with the above preferred result. See also summary of alkyl radical additions (page 219) and comments.

REACTION: ETHYL RADICAL + BUTENE-1

$$CH_3\dot{C}H_2(I) + CH_2{=}CHCH_2CH_3(II) \xrightarrow{1} C_2H_5CH_2\dot{C}HCH_2CH_3(III)$$
$$\xrightarrow{2} \dot{C}H_2CH(C_2H_5)CH_2CH_3(IV)$$

	(I)	(II)	(III)	(IV)		Path 1	Path 2
$\Delta H^\circ_{f(300)}$	26.0 (108.8)	0.0 (0.0)	2.4 (10.0)	3.7 (15.5)	ΔH°	-23.6 (-98.8)	-22.3 (-93.4)
$S^\circ_{(300)}$	59.8 (250.4)	73.6 (308.0)	95.4 (399.4)	94.2 (394.4)	ΔS°	-38.0 (-159.1)	-39.2 (-164.2)
$C^\circ_{p(300)}$	11.1 (46.5)	20.6 (86.2)	34.0 (142.4)	33.9 (141.9)	ΔC°_p	2.3 (9.6)	2.2 (9.2)

Path	log A	E	log k	Conditions	Reference and Experimental Method
1	-	7.30[a] (30.57)	-	298 K	[98] C.W.Larson, B.S.Rabinovitch and D.C.Tardy, J.Chem.Phys., 47, 4570 (1967).
2	-	-	-	-	For details see CH_3 + cis-butene-2. Radical (III) generated from

[a] critical energy of the reaction

Path	log A	E	log k	Conditions	Reference and Experimental Method
					H atom addition to trans-hexene-3. The above critical energy is relative to an assumed value of 7.00 kcal mol^{-1} for the critical energy of the CH_3 addition to pentene-1.

PREFERRED RATE PARAMETERS

Reasonable as given above

REACTION: ETHYL RADICAL + CIS-BUTENE-2

$CH_3\dot{C}H_2(I) + cis\text{-}CH_3CH{=}CHCH_3(II) \longrightarrow CH_3CH(C_2H_5)\dot{C}HCH_3(III)$

	(I)	(II)	(III)		
$\Delta H^\circ_{f(300)}$	26.0 (108.8)	-1.9 (-7.9)	0.4 (1.7)	ΔH°	-23.7 (-99.2)
$S^\circ_{(300)}$	59.8 (250.4)	73.6 (308.2)	92.7 (388.1)	ΔS°	-40.7 (-170.4)
$C^\circ_{p(300)}$	11.1 (46.5)	19.4 (81.2)	33.8 (141.5)	ΔC°_p	3.3 (13.8)

log A	E	log k	Conditions	Reference and Experimental Method
-	8.47[a] (35.46)	-	298 K	[98] C.W.Larson, B.S.Rabinovitch and D.C.Tardy, J.Chem.Phys., 47, 4570 (1967). For details see CH_3 + cis-butene-2 system. Radical (III) generated from H atom addition to 3-methylpentene-1. The above critical energy is relative to an assumed value of 8.00 kcal mol^{-1} for the critical energy of the CH_3 addition to trans-pentene-2.

[a] critical energy of the reaction

PREFERRED RATE PARAMETERS

Reasonable as given above

ΔH^o_f, ΔH^o and E in kcal mol^{-1} (kJ mol^{-1}); S^o, C^o_p, ΔS^o and ΔC^o_p in cal mol^{-1} K^{-1} (J mol^{-1} K^{-1}); A and k in l mol^{-1} s^{-1}

REACTION: ETHYL RADICAL + TRANS-BUTENE-2

$CH_3\dot{C}H_2(I)$ + trans-$CH_3CH{=}CHCH_3(II) \longrightarrow CH_3CH(C_2H_5)\dot{C}HCH_3(III)$

	(I)	(II)	(III)		
$\Delta H^\circ_{f(300)}$	26.0 (108.8)	-2.9 (-12.1)	0.4 (1.7)	ΔH°	-22.7 (-95.0)
$S^\circ_{(300)}$	59.8 (250.4)	72.4 (302.8)	92.7 (388.1)	ΔS°	-39.5 (-165.4)
$C^\circ_{p(300)}$	11.1 (46.5)	20.7 (86.6)	33.8 (141.5)	ΔC°_p	2.0 (8.3)

log A	E	log k	Conditions	Reference and Experimental Method
-	8.64[a] (36.18)	-	298 K	[98] C.W.Larson, B.S.Rabinovitch and D.C.Tardy, J.Chem.Phys., 47, 4570 (1967). For details see CH_3 + cis-butene-2 system. Radical (III) generated from H atom addition to 3-methylpentene-1. The above critical energy is relative to an assumed value of 8.00 kcal mol^{-1} for the critical energy of the CH_3 addition to trans-pentene-2.

[a] critical energy of the reaction

PREFERRED RATE PARAMETERS

Reasonable as given above

REACTION: ETHYL RADICAL + CIS-PENTENE-2

$CH_3\dot{C}H_2(I)$ + cis-$CH_3CH{=}CHCH_2CH_3(II) \xrightarrow{1} CH_3\dot{C}HCH(C_2H_5)CH_2CH_3(III)$

$\xrightarrow{2} CH_3CH(C_2H_5)\dot{C}HCH_2CH_3(IV)$

	(I)	(II)	(III)	(IV)		Path 1	Path 2
$\Delta H^\circ_{f(300)}$	26.0 (108.8)	-6.6 (-27.6)	-4.6 (-19.3)	-4.6 (-19.3)	ΔH°	-24.0 (-100.5)	-24.0 (-100.5)
$S^\circ_{(300)}$	59.8 (250.4)	82.3 (344.6)	102.1 (427.5)	102.1 (427.5)	ΔS°	-40.0 (-167.5)	-40.0 (-167.5)
$C^\circ_{p(300)}$	11.1 (46.5)	25.8 (108.0)	39.3 (164.5)	39.3 (164.5)	ΔC°_p	2.4 (10.1)	2.4 (10.1)

Path	log A	E	log k	Conditions	Reference and Experimental Method
1	-	8.50[a] (35.59)	-	298 K	[98] C.W.Larson, B.S.Rabinovitch and D.C.Tardy, J.Chem.Phys., 47, 4570 (1967).

[a] critical energy of the reaction

Path	log A	E	log k	Conditions	Reference and Experimental Method
2	-	-	-	-	For details see CH_3 + cis-butene-2 system. Radical (III) generated from H atom addition to 3-ethylpentene-1. The above critical energy is relative to an assumed value of 8.00 kcal mol^{-1} for the critical energy of the C_2H_5 addition to trans-pentene-2.

PREFERRED RATE PARAMETERS

Reasonable as given above

REACTION: ETHYL RADICAL + 3-METHYLBUTENE-1

$$CH_3\dot{C}H_2(I) + CH_2{=}CHCH(CH_3)_2(II) \xrightarrow{1} C_2H_5CH_2\dot{C}HCH(CH_3)_2(III)$$
$$\xrightarrow{2} \dot{C}H_2CH(C_2H_5)CH(CH_3)_2(IV)$$

	(I)	(II)	(III)	(IV)		Path 1	Path 2
$\Delta H^\circ_{f(300)}$	26.0 (108.8)	-6.8 (-28.5)	-4.6 (-19.3)	-3.2 (-13.4)	ΔH°	-23.8 (-99.7)	-22.4 (-93.8)
$S^\circ_{(300)}$	59.8 (250.4)	[80 (335)]	102.1 (427.5)	101.1 (423.3)	ΔS°	[-38 (-159)]	[-39 (-163)]
$C^\circ_{p(300)}$	11.1 (46.5)	[26.8 (112.2)]	39.3 (164.5)	39.1 (163.7)	ΔC°_p	[1.4 (5.9)]	[1.2 (5.0)]

Path	log A	E	log k	Conditions	Reference and Experimental Method
1	-	7.19[a] (30.10)	-	298 K	[98] C.W.Larson, B.S.Rabinovitch and D.C.Tardy, J.Chem.Phys., 47, 4570 (1967).
2	-	-	-	-	For details see CH_3 + cis-butene-2 system. Radical (III) generated from H atom addition to trans-2-methylhexene-3. The above critical energy is relative to an assumed value of 8.00 kcal mol^{-1} for the critical energy of the CH_3 addition to trans-hexene-2.
1	-	7.19[a] (30.10)	-	298 K	[101] C.W.Larson and B.S.Rabinovitch, J.Chem.Phys., 50, 871 (1969).
2	-	-	-	-	For details see CH_3 + cis-butene-2 system, reference [98]. Radical (III) generated from H atom addition to trans-2-methylhexene-2. The above critical energy is relative to an assumed value of 7.00 kcal mol^{-1} for the critical energy of the CH_3 addition to 4-methylpentene-1.

[a] critical energy of the reaction

ΔH^o_f, ΔH^o and E in kcal mol^{-1} (kJ mol^{-1}); S^o, C^o_p, ΔS^o and ΔC^o_p in cal mol^{-1} K^{-1} (J mol^{-1} K^{-1}); A and k in l mol^{-1} s^{-1}

PREFERRED RATE PARAMETERS

Reasonable as given above

REACTION: ETHYL RADICAL + HEXENE-1

$$CH_3\dot{C}H_2(I) + CH_2{=}CH(CH_2)_3CH_3(II) \xrightarrow{1} C_2H_5CH_2\dot{C}H(CH_2)_3CH_3(III)$$
$$\xrightarrow{2} \dot{C}H_2CH(C_2H_5)(CH_2)_3CH_3(IV)$$

	(I)	(II)	(III)	(IV)		Path 1	Path 2
$\Delta H^\circ_{f(300)}$	26.0 (108.9)	-9.8 (-41.0)	-7.4 (-31.0)	-6.2 (-26.0)	ΔH°	-23.6 (-98.8)	-22.4 (-93.8)
$S^\circ_{(300)}$	59.8 (250.4)	94.7 (396.5)	114.2 (478.2)	114.1 (477.7)	ΔS°	-40.3 (-168.7)	-40.4 (-169.2)
$C^\circ_{p(300)}$	11.1 (46.5)	31.6 (132.3)	45.0 (188.4)	44.9 (188.0)	ΔC°_p	2.3 (9.6)	2.2 (9.2)

Path	log A	E	log k	Conditions	Reference and Experimental Method
1 + 2	7.59	6.8 (28.5)	3.74(386 K)	338-435 K 20 torr	[111] D.G.L.James and E.W.R.Steacie, Proc.Roy.Soc.(London), Ser.A., 244, 297 (1958). Arrhenius parameters relative to log k_c = 10.4 for C_2H_5 combination; log $(k_{1+2}/k_c^{\frac{1}{2}})(\ell^{\frac{1}{2}}mol^{-\frac{1}{2}}s^{-\frac{1}{2}})$ = 2.39 - 6.8/2.3RT. Static system. C_2H_5 from photolysis of diethylketone. Reaction rates calculated by a mass balance for C_2H_5 radicals, based on the analysis of the products CO, C_2H_6 and C_4H_{10} by low-temperature distillation and mass spectrometry.
1	-	6.78[a] (28.4)	-	298 K	[100] C.W.Larson, D.C.Tardy and B.S.Rabinovitch, J.Chem.Phys., 49, 299 (1968) For details see CH_3 + cis-butene-2, reference [98]. Radical (III) generated from H atom addition to trans-heptene-3. The above critical energy is relative to an assumed value of 7.00 kcal mol^{-1} for the critical energy of the CH_3 addition to hexene-1.
2	-	-	-	-	

[a] critical energy of the reaction

PREFERRED RATE PARAMETERS

log k_{1+2} = 7.59 - 6.8/2.3RT

COMMENTS

The experiments of reference [111] refer to the sum of the rate constants of paths 1 + 2, although path 1 will be the favoured route. Assuming that the preferred parameters approximate to k_1 it can be calculated from the thermodynamic data for path 1 that $\log (k_{-1}/s^{-1}) = 14.6 - 29.8/2.3RT$ at 298 K, which is reasonable for a radical decomposition. See also summary of alkyl additions (page 219) and comments.

REACTION: ETHYL RADICAL + HEPTENE-1

$CH_3\dot{C}H_2(I) + CH_2{=}CH(CH_2)_4CH_3(II) \xrightarrow{1} C_2H_5CH_2\dot{C}H(CH_2)_4CH_3(III)$

$\xrightarrow{2} \dot{C}H_2CH(C_2H_5)(CH_2)_4CH_3(IV)$

	(I)	(II)	(III)	(IV)		Path 1	Path 2
$\Delta H^\circ_{f(300)}$	26.0 (108.9)	-14.8 (-62.0)	-12.4 (-51.9)	-11.0 (-46.1)	ΔH°	-23.6 (-98.8)	-22.2 (-93.0)
$S^\circ_{(300)}$	59.8 (250.4)	104.1 (435.9)	123.7 (517.9)	122.7 (513.7)	ΔS°	-40.2 (-168.3)	-41.2 (-172.5)
$C^\circ_{p(300)}$	11.1 (46.5)	37.1 (155.3)	50.5 (211.4)	50.4 (211.0)	ΔC°_p	2.3 (9.6)	2.2 (9.2)

Path	log A	E	log k	Conditions	Reference and Experimental Method
1 + 2	7.79	7.0 (29.3)	3.99(399 K)	359-439 K 20 torr	[111] D.G.L.James and G.W.R.Steacie, Proc.Roy.Soc.(London), Ser.A., 244, 297 (1958). Arrhenius parameters relative to $\log k_c = 10.4$ for C_2H_5 combination; $\log (k_{1+2}/k_c^{\frac{1}{2}})(\ell^{\frac{1}{2}}mol^{-\frac{1}{2}}s^{-\frac{1}{2}}) = 2.59 - 7.0/2.3RT$. For details see C_2H_5 + hexene-1.
1 + 2	7.79	7.0 (29.3)	3.23(338 K)	? K 20 torr	[112] A.C.R.Brown and D.G.L.James, Proc.Chem.Soc., 81 (1962). Arrhenius parameters relative to $\log k_c = 10.4$ for C_2H_5 combination; $\log (k_{1+2}/k_c^{\frac{1}{2}})(\ell^{\frac{1}{2}}mol^{-\frac{1}{2}}s^{-\frac{1}{2}}) = 2.59 - 7.0/2.3RT$. Same method as for reference [111] but no experimental details given.

PREFERRED RATE PARAMETERS

Reasonable as given above

COMMENTS

The above measurements refer to the sum of the rate constants for paths 1 and 2, although path 1 will be the favoured route. There are no data for the reverse reactions, but from the preferred value of $\log k_{1+2}$ and the thermodynamic data (path 1), it can be shown that $\log (k_{-1}/s^{-1}) = 14.8 - 30.0/2.3RT$ at 298 K, which is reasonable. See also summary of alkyl additions (page 219) and comments.

ΔH^o_f, ΔH^o and E in kcal mol^{-1} (kJ mol^{-1}); S^o, C^o_p, ΔS^o and ΔC^o_p in cal mol^{-1} K^{-1} (J mol^{-1} K^{-1}); A and k in l mol^{-1} s^{-1}

REACTION: ETHYL RADICAL + 2,3,3-TRIMETHYLBUTENE-1

$CH_3\dot{C}H_2(I) + CH_2{=}C(CH_3)C(CH_3)_2CH_3(II) \xrightarrow{1} C_2H_5CH_2\dot{C}(CH_3)C(CH_3)_2CH_3(III)$

$\xrightarrow{2} \dot{C}H_2C(CH_3)(C_2H_5)C(CH_3)_2CH_3(IV)$

	(I)	(II)	(III)	(IV)		Path 1	Path 2
$\Delta H^\circ_{f(300)}$	26.0 (108.9)	-20.9 (-87.5)	-21.8 (-91.3)	-17.5 (-73.3)	ΔH°	-26.9 (-112.6)	-22.6 (-94.6)
$S^\circ_{(300)}$	59.8 (250.4)	92.7 (388.1)	114.0 (477.3)	111.5 (466.9)	ΔS°	-38.5 (-161.2)	-41.0 (-171.7)
$C^\circ_{p(300)}$	11.1 (46.5)	38.3 (160.4)	50.4 (211.0)	51.2 (214.4)	ΔC°_p	1.0 (4.2)	1.8 (7.5)

Path	log A	E	log k	Conditions	Reference and Experimental Method
1 + 2	6.89	5.6 (23.4)	3.32(343 K)	322-364 K 20 torr	[111] D G.L.James and E.W.R.Steacie, Proc.Roy.Soc.(London) Ser.A., 244, 297 (1958). Arrhenius parameters relative to log k_c = 10.4 for C_2H_5 combination; $\log (k_{1+2}/k_c^{\frac{1}{2}})(\ell^{\frac{1}{2}}mol^{-\frac{1}{2}}s^{-\frac{1}{2}}) = 1.69 - 5.6/2.3RT$. For details see C_2H_5 + hexene-1.

PREFERRED RATE PARAMETERS

Reasonable as given above

COMMENTS

The above measurements refer to the sum of the rate constants for paths 1 and 2, although path 1 will be the favoured route. There are no data for the reverse reactions, but from the preferred value of log k_{1+2} and the thermodynamic data (path 1), it can be shown that $\log (k_{-1}/s^{-1}) = 13.6 - 31.9/2.3RT$ at 298 K, which is reasonable for a radical decomposition. See also summary of alkyl additions (page 219) and comments.

REACTION: ETHYL RADICAL + OCTENE-1

$CH_3\dot{C}H_2(I) + CH_2{=}CH(CH_2)_5CH_3(II) \xrightarrow{1} C_2H_5CH_2\dot{C}H(CH_2)_5CH_3(III)$

$\xrightarrow{2} \dot{C}H_2CH(C_2H_5)(CH_2)_5CH_3(IV)$

	(I)	(II)	(III)	(IV)
$\Delta H^\circ_{f(300)}$	26.0 (108.9)	-19.8 (-82.9)	-17.4 (-72.9)	-15.9 (-66.6)
$S^\circ_{(300)}$	59.8 (250.4)	113.5 (475.2)	133.2 (557.7)	132.1 (553.1)
$C^\circ_{p(300)}$	11.1 (46.5)	42.6 (178.4)	56.0 (234.5)	55.9 (234.1)

	Path 1	Path 2
ΔH°	-23.6 (-98.8)	-22.1 (-92.5)
ΔS°	-40.1 (-167.9)	-41.2 (-172.5)
ΔC°_p	2.3 (9.6)	2.2 (9.2)

Path	log A	E	log k	Conditions	Reference and Experimental Method
1 + 2	7.59	6.7 (28.1)	3.69(375 K)	298-453 K 20 torr	[111] D.G.L.James and E.W.R.Steacie, Proc.Roy.Soc.(London) Ser.A., 244, 297 (1958). Arrhenius parameters relative to log k_c = 10.4 for C_2H_5 combination; $\log(k_{1+2}/k_c^{\frac{1}{2}}) = 2.39 - 6.7/2.3RT$. For details see C_2H_5 + hexene-1.
1 + 2	8.09	7.6 (31.8)	3.74(382 K)	339-425 K ? torr	[113] A.C.R.Brown and D.G.L.James, Can.J.Chem., 43, 1102 (1965). Arrhenius parameters relative to log k_c = 10.4 for C_2H_5 combination; $\log(k_{1+2}/k_c^{\frac{1}{2}})(\ell^{\frac{1}{2}}mol^{-\frac{1}{2}}s^{-\frac{1}{2}}) = 2.89 - 7.6/2.3RT$. For details see C_2H_5 + hexene-1, reference [111].

PREFERRED RATE PARAMETERS

$\log k_{1+2} = 8.09 - 7.6/2.3RT$

COMMENTS

The data of reference [113] are preferred, being more up-to-date results of the original author; the actual rate constants of the two studies are in agreement. Path 1 should be the favoured route for addition. Taking the measured parameters to correspond essentially with k_1, it can be calculated from the thermodynamic data that $\log(k_{-1}/s^{-1}) = 14.5 - 29.7/2.3RT$ at 298 K, which is reasonable for a radical decomposition. See also summary of alkyl radical additions (page 219) and comments.

ΔH^o_f, ΔH^o and E in kcal mol^{-1} (kJ mol^{-1}); S^o, C^o_p, ΔS^o and ΔC^o_p in cal mol^{-1} K^{-1} (J mol^{-1} K^{-1}); A and k in l mol^{-1} s^{-1}

REACTION: ETHYL RADICAL + 2,4,4-TRIMETHYLPENTENE-1

$CH_3\dot{C}H_2(I) + CH_2{=}C(CH_3)CH_2C(CH_3)_3(II) \xrightarrow{1} C_2H_5CH_2\dot{C}(CH_3)CH_2C(CH_3)_3(III)$

$\xrightarrow{2} \dot{C}H_2C(CH_3)(C_2H_5)CH_2C(CH_3)_3(IV)$

	(I)	(II)	(III)	(IV)		Path 1	Path 2
$\Delta H^\circ_{f(300)}$	26.0 (108.9)	-26.7 (-111.8)	-26.8 (-112.2)	-22.5 (-94.2)	ΔH°	-26.1 (-109.3)	-21.8 (-91.3)
$S^\circ_{(300)}$	59.8 (250.4)	102.1 (427.5)	123.4 (516.7)	120.9 (506.2)	ΔS°	-38.5 (-161.2)	-41.0 (-171.7)
$C^\circ_{p(300)}$	11.1 (46.5)	43.8 (183.4)	55.9 (234.1)	56.7 (237.4)	ΔC°_p	1.0 (4.2)	1.8 (7.5)

Path	log A	E	log k	Conditions	Reference and Experimental Method
1 + 2	7.29	5.7 (23.9)	3.59(336 K)	309-364 K 20 torr	[111] D.G.L.James and E.W.R.Steacie, Proc.Roy.Soc.(London) Ser.A., 244, 297 (1958). Arrhenius parameters relative to log k_c = 10.4 for C_2H_5 combination; $\log (k_{1+2}/k_c^{\frac{1}{2}})(\ell^{\frac{1}{2}}mol^{-\frac{1}{2}}s^{-\frac{1}{2}}) = 2.09 - 5.7/2.3RT$. For details see C_2H_5 + hexene-1.

PREFERRED RATE PARAMETERS

Reasonable as given above

COMMENTS

The above measurements refer to the sum of the rate constants for paths 1 and 2, although path 1 will be the favoured route. There are no data for the reverse reactions, but from the preferred value of log k_{1+2} and the thermodynamic data (path 1), it follows that $\log (k_{-1}/s^{-1}) = 14.0 - 31.2/2.3RT$ at 298 K, which is reasonable for a radical decomposition. See also summary of alkyl additions (page 219) and comments.

REACTION: ETHYL RADICAL + ACRYLONITRILE

$CH_3\dot{C}H_2(I) + CH_2{=}CHCN(II) \xrightarrow{1} CH_3CH_2CH_2\dot{C}HCN(III)$

$\xrightarrow{2} \dot{C}H_2CH(C_2H_5)CN(IV)$

	(I)	(II)	(III)	(IV)		Path 1	Path 2
$\Delta H^\circ_{f(300)}$	26.0 (108.9)	43.7 (183.0)	[35 (147)]	[47 (197)]	ΔH°	[-35 (-147)]	[-23 (-96)]
$S^\circ_{(300)}$	59.8 (250.4)	65.6 (274.7)	89.3 (373.9)	90.8 (380.2)	ΔS°	-36.1 (-151.1)	-34.6 (-144.9)
$C^\circ_{p(300)}$	11.1 (46.5)	15.3 (64.1)	26.3 (110.1)	28.7 (120.2)	ΔC°_p	-0.1 (-0.4)	2.3 (9.6)

Path	log A	E	log k	Conditions	Reference and Experimental Method
1 + 2	7.79	3.4 (14.2)	5.80(373 K)	323-454 K ? torr	[114] D.G.L.James and D.McCallum, Can.J.Chem., 43, 633 (1965); D.G.L.James and T.Ogawa, Can.J.Chem., 43, 640 (1965). Arrhenius parameters relative to log k_c = 10.4 for C_2H_5 combination; log $(k_{1+2}/k_c^{\frac{1}{2}})(\ell^{\frac{1}{2}}mol^{-\frac{1}{2}}s^{-\frac{1}{2}})$ = 2.59 - 3.4/2.3RT. See C_2H_5 + hexene-1, reference [111].

PREFERRED RATE PARAMETERS

Reasonable as given above

ΔH^o_f, ΔH^o and E in kcal mol^{-1} (kJ mol^{-1}); S^o, C^o_p, ΔS^o and ΔC^o_p in cal mol^{-1} K^{-1} (J mol^{-1} K^{-1}); A and k in l mol^{-1} s^{-1}

REACTION: ETHYL RADICAL + <u>CIS</u>-CROTONONITRILE

$CH_3\dot{C}H_2(I) + \underline{cis}\text{-}CH_3CH{=}CHCN(II) \xrightarrow{1} CH_3CH(C_2H_5)\dot{C}HCN(III)$

$\xrightarrow{2} CH_3\dot{C}HCH(C_2H_5)CN(IV)$

	(I)	(II)	(III)	(IV)		Path 1	Path 2
$\Delta H^\circ_{f(300)}$	26.0 (108.9)		[26 (109)]	[38 (159)]	ΔH°		
$S^\circ_{(300)}$	59.8 (250.4)		93.7 (392.3)	[96 (402)]	ΔS°		
$C^\circ_{p(300)}$	11.1 (46.5)		31.5 (131.9)	[34.0 (142.4)]	ΔC°_p		

Path	log A	E	log k	Conditions	Reference and Experimental Method
1 + 2	7.19	5.0 (20.9)	4.37(388 K)	323-454 K ? torr	[115] D.G.L.James and T.Ogawa, Can.J.Chem., 43, 640 (1965). Arrhenius parameters relative to $\log k_c = 10.4$ for C_2H_5 combination; $\log (k_{1+2}/k_c^{\frac{1}{2}})(\ell^{\frac{1}{2}}mol^{-\frac{1}{2}}s^{-\frac{1}{2}}) = 1.99 - 5.0/2.3RT$. For details see C_2H_5 + hexene-1, reference [111].

PREFERRED RATE PARAMETERS

Reasonable as given above

REACTION: ETHYL RADICAL + TRANS-CROTONONITRILE

$CH_3\dot{C}H_2(I)$ + trans-$CH_3CH{=}CHCN(II) \xrightarrow{1} CH_3CH(C_2H_5)\dot{C}HCN(III)$

$\xrightarrow{2} CH_3\dot{C}HCH(C_2H_5)CN(IV)$

	(I)	(II)	(III)	(IV)		Path 1	Path 2
$\Delta H^\circ_{f(300)}$	26.0 (108.9)	35.9 (150.3)	[26 (109)]	[38 (159)]	ΔH°	[-36 (-151)]	[-24 (-100)]
$S^\circ_{(300)}$	59.8 (250.4)	71.4 (299.0)	93.7 (392.3)	[96 (402)]	ΔS°	-37.5 (-157.0)	[-35 (-147)]
$C^\circ_{p(300)}$	11.1 (46.5)	19.7 (82.5)	31.5 (131.9)	[34.0 (142.4)]	ΔC°_p	0.7 (2.9)	[3.2 (13.4)]

Path	log A	E	log k	Conditions	Reference and Experimental Method
1 + 2	7.49	5.2 (21.8)	4.56(388 K)	323-454 K ? torr	[115] D.G.L.James and T.Ogawa, Can.J.Chem., 43, 640 (1965). Arrhenius parameters relative to log k_c = 10.4 for C_2H_5 combination; $\log (k_{1+2}/k_c^{\frac{1}{2}})(\ell^{\frac{1}{2}}mol^{-\frac{1}{2}}s^{-\frac{1}{2}}) = 2.29 - 5.2/2.3RT$. For details see C_2H_5 + hexene-1, reference [111].

PREFERRED RATE PARAMETERS

Reasonable as given above

ΔH^o_f, ΔH^o and E in kcal mol^{-1} (kJ mol^{-1}); S^o, C^o_p, ΔS^o and ΔC^o_p in cal mol^{-1} K^{-1} (J mol^{-1} K^{-1}); A and k in l mol^{-1} s^{-1}

REACTION: ETHYL RADICAL + METHACRYLONITRILE

$$CH_3\dot{C}H_2(I) + CH_2{=}C(CH_3)CN(II) \xrightarrow{1} CH_3CH_2CH_2\dot{C}(CH_3)CN(III)$$
$$\xrightarrow{2} \dot{C}H_2C(CH_3)(C_2H_5)CN(IV)$$

	(I)	(II)	(III)	(IV)		Path 1	Path 2
$\Delta H^\circ_{f(300)}$	26.0 (108.9)		[22 (92)]		ΔH°		
$S^\circ_{(300)}$	59.8 (250.4)		94.9 (397.3)	[94 (394)]	ΔS°		
$C^\circ_{p(300)}$	11.1 (46.5)		32.2 (134.8)		ΔC°_p		

Path	log A	E	log k	Conditions	Reference and Experimental Method
1 + 2	8.39	4.6 (19.3)	5.57(356 K)	312-400 K ? torr	[115] D.G.L.James and T.Ogawa, Can.J.Chem., 43, 640 (1965). Arrhenius parameters relative to log k_c = 10.4 for C_2H_5 combination; log $(k_{1+2}/k_c^{\frac{1}{2}})(\ell^{\frac{1}{2}}mol^{-\frac{1}{2}}s^{-\frac{1}{2}})$ = 3.19 - 4.6/2.3RT. For details see C_2H_5 + hexene-1, reference [111].

PREFERRED RATE PARAMETERS

Reasonable as given above

REACTION: ETHYL RADICAL + ALLYL ALCOHOL

$$CH_3\dot{C}H_2(I) + CH_2{=}CHCH_2OH(II) \xrightarrow{1} CH_3CH_2CH_2\dot{C}HCH_2OH(III)$$
$$\xrightarrow{2} \dot{C}H_2CH(C_2H_5)CH_2OH(IV)$$

	(I)	(II)	(III)	(IV)		Path 1	Path 2
$\Delta H^\circ_{f(300)}$	26.0 (108.9)	-29.9 (-125.2)	-28.9 (-121.0)	-27.5 (-115.1)	ΔH°	-25.0 (-104.7)	-23.6 (-98.8)
$S^\circ_{(300)}$	59.8 (250.4)		97.2 (407.0)	96.1 (402.4)	ΔS°		
$C^\circ_{p(300)}$	11.1 (46.5)		31.7 (132.7)	31.5 (131.9)	ΔC°_p		

Path	log A	E	log k	Conditions	Reference and Experimental Method
1 + 2	8.29	7.75 (32.4)	3.70(369 K)	323-415 K ? torr	[116] A.C.R.Brown and D.G.L.James, Can.J.Chem., 40, 796 (1962). Arrhenius parameters relative to log k_c = 10.4 for C_2H_5 combination; $\log (k_{1+2}/k_c^{\frac{1}{2}})(\ell^{\frac{1}{2}}mol^{-\frac{1}{2}}s^{-\frac{1}{2}}) = 3.09 - 7.75/2.3RT$. For details see C_2H_5 + hexene-1, reference [111].

PREFERRED RATE PARAMETERS

Reasonable as given above

ΔH^o_f, ΔH^o and E in kcal mol^{-1} (kJ mol^{-1}); S^o, C^o_p, ΔS^o and ΔC^o_p in cal mol^{-1} K^{-1} (J mol^{-1} K^{-1}); A and k in l mol^{-1} s^{-1}

REACTION: ETHYL RADICAL + VINYL ACETATE

$$CH_3\dot{C}H_2(I) + CH_2{=}CHOCOCH_3(II) \xrightarrow{1} CH_3CH_2CH_2\dot{C}HOCOCH_3(III)$$
$$\xrightarrow{2} \dot{C}H_2CH(C_2H_5)OCOCH_3(IV)$$

	(I)	(II)	(III)	(IV)		Path 1	Path 2
$\Delta H^\circ_{f(300)}$	26.0 (108.9)	-72.4 (-303.1)		-71.0 (-297.3)	ΔH°		-24.6 (-103.0)
$S^\circ_{(300)}$	59.8 (250.4)			109.2 (457.2)	ΔS°		
$C^\circ_{p(300)}$	11.1 (46.5)				ΔC°_p		

Path	log A	E	log k	Conditions	Reference and Experimental Method
1 + 2	7.89	6.9 (28.9)	3.70(360 K)	303-417 K ? torr	[114] D.G.L.James and D.McCallum, Can.J.Chem., 43, 633 (1965). Arrhenius parameters relative to log k_c = 10.4 for C_2H_5 combination; $\log (k_{1+2}/k_c^{\frac{1}{2}})(\ell^{\frac{1}{2}}mol^{-\frac{1}{2}}s^{-\frac{1}{2}}) = 2.69 - 6.9/2.3RT$. For details see C_2H_5 + hexene-1, reference [111].

PREFERRED RATE PARAMETERS

Reasonable as given above

REACTION: ETHYL RADICAL + ALLYL ACETATE

$$CH_3\dot{C}H_2(I) + CH_2{=}CHCH_2OCOCH_3(II) \xrightarrow{1} CH_3CH_2CH_2\dot{C}HCH_2OCOCH_3(III)$$
$$\xrightarrow{2} \dot{C}H_2CH(C_2H_5)CH_2OCOCH_3(IV)$$

	(I)	(II)	(III)	(IV)		Path 1	Path 2
$\Delta H^\circ_{f(300)}$	26.0 (108.9)	-76.8 (-321.6)	-75.8 (-317.4)	-74.4 (-311.5)	ΔH°	-25.0 (-104.7)	-23.6 (-98.8)
$S^\circ_{(300)}$	59.8 (250.4)		119.5 (500.3)	118.4 (495.7)	ΔS°		
$C^\circ_{p(300)}$	11.1 (46.5)				ΔC°_p		

Path	log A	E	log k	Conditions	Reference and Experimental Method
1 + 2	8.39	7.7 (32.2)	3.94(378 K)	308-448 K ? torr	[117] D.G.L.James and G.E.Troughton, J.Polymer Sci. A, 3, 75 (1965). Arrhenius parameters relative to log k_c = 10.4 for C_2H_5 combination; log $(k_{1+2}/k_c^{\frac{1}{2}})(\ell^{\frac{1}{2}}mol^{-\frac{1}{2}}s^{-\frac{1}{2}})$ = 3.19 - 7.7/2.3RT. For details see C_2H_5 + hexene-1, reference [111].

PREFERRED RATE PARAMETERS

Reasonable as given above

ΔH^o_f, ΔH^o and E in kcal mol^{-1} (kJ mol^{-1}); S^o, C^o_p, ΔS^o and ΔC^o_p in cal mol^{-1} K^{-1} (J mol^{-1} K^{-1}); A and k in l mol^{-1} s^{-1}

REACTION: ETHYL RADICAL + VINYL N-BUTYL ETHER

$CH_3\dot{C}H_2(I) + CH_2{=}CHOC_4H_9(II) \xrightarrow{1} CH_3CH_2CH_2\dot{C}HOC_4H_9(III)$

$\xrightarrow{2} \dot{C}H_2CH(C_2H_5)OC_4H_9(IV)$

	(I)	(II)	(III)	(IV)		Path 1	Path 2
$\Delta H^\circ_{f(300)}$	26.0 (108.9)	-44.9 (-188.0)		-37.7 (-157.8)	ΔH°		
$S^\circ_{(300)}$	59.8 (250.4)			123.9 (518.8)	ΔS°		
$C^\circ_{p(300)}$	11.1 (46.5)			48.1 (201.4)	ΔC°_p		

Path	log A	E	log k	Conditions	Reference and Experimental Method
1 + 2	7.39	6.1 (25.5)	3.78(369 K)	303-435 K ? torr	[114] D.G.L.James and D.McCallum, Can.J.Chem., 43, 633 (1965). Arrhenius parameters relative to log k_c = 10.4 for C_2H_5 combination; log $(k_{1+2}/k_c^{\frac{1}{2}})(\ell^{\frac{1}{2}}mol^{-\frac{1}{2}}s^{-\frac{1}{2}})$ = 2.19 - 6.1/2.3RT. For details see C_2H_5 + hexene-1, reference [111].

PREFERRED RATE PARAMETERS

Reasonable as given above

REACTION: ETHYL RADICAL + ALLYL PROPIONATE

$CH_3\dot{C}H_2(I) + CH_2{=}CHCH_2OCOCH_2CH_3(II) \xrightarrow{1} CH_3CH_2CH_2\dot{C}HCH_2OCOC_2H_5(III)$

$\xrightarrow{2} \dot{C}H_2CH(C_2H_5)CH_2OCOCH_2CH_3(IV)$

	(I)	(II)	(III)	(IV)		Path 1	Path 2
$\Delta H^\circ_{f(300)}$	26.0 (108.9)	-81.8 (-342.5)	-80.8 (-338.3)	-79.4 (-332.4)	ΔH°	-25.0 (-104.7)	-23.6 (-98.8)
$S^\circ_{(300)}$	59.8 (250.4)		129.1 (540.5)	128.0 (535.9)	ΔS°		
$C^\circ_{p(300)}$	11.1 (46.5)				ΔC°_p		

Path	log A	E	log k	Conditions	Reference and Experimental Method
1 + 2	8.39	7.7 (32.2)	4.11(393 K)	352-435 K ? torr	[118] D.G.L.James and G.E.Troughton, Trans.Faraday Soc., 62, 120 (1965). Arrhenius parameters relative to log k_c = 10.4 for C_2H_5 combination; log $(k_{1+2}/k_c^{\frac{1}{2}})(\ell^{\frac{1}{2}}mol^{-\frac{1}{2}}s^{-\frac{1}{2}})$ = 3.19 - 7.7/2.3RT. For details see C_2H_5 + hexene-1, reference [111].

PREFERRED RATE PARAMETERS

Reasonable as given above

ΔH^o_f, ΔH^o and E in kcal mol^{-1} (kJ mol^{-1}); S^o, C^o_p, ΔS^o and ΔC^o_p in cal mol^{-1} K^{-1} (J mol^{-1} K^{-1}); A and k in l mol^{-1} s^{-1}

REACTION: ETHYL RADICAL + ALLENE

$$CH_3\dot{C}H_2(I) + CH_2{=}C{=}CH_2(II) \xrightarrow{1} C_2H_5CH_2\dot{C}{=}CH_2(III)$$
$$\xrightarrow{2} \dot{C}H_2C(C_2H_5){=}CH_2(IV)$$

	(I)	(II)	(III)	(IV)
$\Delta H^\circ_{f(300)}$	26.0 (108.9)	45.9 (192.2)	45.0 (188.4)	37.6 (157.4)
$S^\circ_{(300)}$	59.8 (250.4)	58.3 (244.1)	84.8 (355.0)	83.7 (350.5)
$C^\circ_{p(300)}$	11.1 (46.5)	14.2 (59.5)	26.0 (108.9)	25.9 (108.4)
$C^\circ_{p(500)}$	16.3 (68.2)	19.8 (82.9)	39.2 (164.1)	39.4 (165.0)

	Path 1	Path 2
ΔH°	-26.9 (-112.6)	-34.3 (-143.6)
ΔS°	-33.3 (-139.3)	-34.4 (-144.0)
$\Delta C^\circ_{p(300)}$	0.7 (2.9)	0.6 (2.5)
$\Delta C^\circ_{p(500)}$	3.1 (13.0)	3.3 (13.8)
$\langle\Delta C^\circ_p\rangle$	1.9 (8.0)	2.0 (8.4)

Path	log A	E	log k	Conditions	Reference and Experimental Method
1	8.5	9.2 (38.5)	3.74(422 K)	379-465 K 40-100 torr	[104] R.R.Getty, J.A.Kerr and A.F.Trotman-Dickenson, J.Chem.Soc.(A), 979 (1967). Arrhenius parameters relative to log k_c = 10.4 for C_2H_5 combination; log $(k_{1+2}/k_c^{\frac{1}{2}})(\ell^{\frac{1}{2}}mol^{-\frac{1}{2}}s^{-\frac{1}{2}})$ = 3.3 - 9.2/2.3RT. 'Aldehyde method'. For details see CH_3+ $CH_2{=}C{=}CH_2$. It was shown by product analysis that addition occurred exclusively at the terminal C atom in allene (path 1).

PREFERRED RATE PARAMETERS

Tentatively as given above

COMMENTS

See comments on 'aldehyde method' in experimental section of CH_3 + allene system. See also summary of alkyl additions (page 219). It would appear that in relation to the Arrhenius parameters for CH_3 and i-C_3H_7 additions to allene, those for C_2H_5 given above are slight over-estimates, although the rate constants appear to be reasonable.

REACTION: ETHYL RADICAL + 2,3-DIMETHYLBUTADIENE-1,3

$CH_3\dot{C}H_2(I) + CH_2{=}C(CH_3)C(CH_3){=}CH_2(II) \xrightarrow{1} C_2H_5CH_2\dot{C}(CH_3)C(CH_3){=}CH_2(III)$

$\xrightarrow{2} \dot{C}H_2C(CH_3)(C_2H_5)C(CH_3){=}CH_2(IV)$

	(I)	(II)	(III)	(IV)
$\Delta H^\circ_{f(300)}$	26.0 (108.9)	10.1 (42.3)	1.3 (5.4)	18.9 (79.1)
$S^\circ_{(300)}$	59.8 (250.4)	82.4 (345.0)	103.8 (434.6)	[105 (440)]
$C^\circ_{p(300)}$	11.1 (46.5)	31.4 (131.5)	42.8 (179.2)	[43.3 (181.3)]

	Path 1	Path 2
ΔH°	-34.8 (-145.7)	-17.2 (-72.0)
ΔS°	-38.4 (-160.8)	[-37 (-155)]
ΔC°_p	0.3 (1.3)	[0.8 (3.3)]

Path	log A	E	log k	Conditions	Reference and Experimental Method
1 + 2	8.23	4.5 (18.8)	5.32(338 K)	338 K ? torr	[112] A.C.R.Brown and D.G.L.James, Proc.Chem.Soc., 81 (1962). Arrhenius parameters relative to log k_c = 10.4 for C_2H_5 combination; log $(k_{1+2}/k_c^{\frac{1}{2}})(\ell^{\frac{1}{2}}mol^{-\frac{1}{2}}s^{-\frac{1}{2}})$ = 3.03 - 4.5/2.3RT. For details see C_2H_5 + hexene-1, reference [111].
1 + 2	8.19	4.5 (18.8)	5.50(366 K)	318-414 K ? torr	[113] A.C.R.Brown and D.G.L.James, Can.J.Chem., 43, 1102 (1965). Arrhenius parameters relative to log k_c = 10.4 for C_2H_5 combination; log $(k_{1+2}/k_c^{\frac{1}{2}})(\ell^{\frac{1}{2}}mol^{-\frac{1}{2}}s^{-\frac{1}{2}})$ = 2.99 - 4.5/2.3RT. For details see C_2H_5 + hexene-1, reference [111].

PREFERRED RATE PARAMETERS

Reasonable as given above

COMMENTS

Above results refer to the sum of the rate constants for paths 1 and 2, although path 1 should predominate. The Arrhenius parameters are reasonable in relation to the data for CH_3 addition to butadiene-1,3 (see summary of alkyl additions, page 219).

ΔH^o_f, ΔH^o and E in kcal mol^{-1} (kJ mol^{-1}); S^o, C^o_p, ΔS^o and ΔC^o_p in cal mol^{-1} K^{-1} (J mol^{-1} K^{-1}); A and k in l mol^{-1} s^{-1}

REACTION: ETHYL RADICAL + 2,5-DIMETHYLHEXADIENE-2,4

$CH_3\dot{C}H_2(I) + (CH_3)_2C{=}CHCH{=}C(CH_3)_2(II) \xrightarrow{1} (CH_3)_2C(C_2H_5)\dot{C}HCH{=}C(CH_3)_2(III)$

$\xrightarrow{2} (CH_3)_2\dot{C}CH(C_2H_5)CH{=}C(CH_3)_2(IV)$

	(I)	(II)	(III)	(IV)		Path 1	Path 2
$\Delta H^\circ_{f(300)}$	26.0 (109.8)	-6.1 (-25.5)	-9.4 (-39.4)	-0.9 (-3.8)	ΔH°	-29.3 (-122.7)	-20.8 (-87.1)
$S^\circ_{(300)}$	59.8 (250.4)	100.2 (419.5)	117.6 (492.4)	[123 (515)]	ΔS°	-42.4 (-177.5)	[-37 (155)]
$C^\circ_{p(300)}$	11.1 (46.5)	41.9 (175.4)	53.7 (224.8)	[53 (222)]	ΔC°_p	0.7 (2.9)	[0 (0)]

Path	log A	E	log k	Conditions	Reference and Experimental Method
1 + 2	7.79	6.6 (27.6)	3.93(374 K)	328-420 K ? torr	[113] A.C.R.Brown and D.G.L.James, Can.J.Chem., 43, 1102 (1965). Arrhenius parameters relative to log k_c = 10.4 for C_2H_5 combination; $\log (k_{1+2}/k_c^{\frac{1}{2}})(\ell^{\frac{1}{2}}mol^{-\frac{1}{2}}s^{-\frac{1}{2}}) = 2.59 - 6.6/2.3RT$. For details see C_2H_5 + hexene-1, reference [111].

PREFERRED RATE PARAMETERS

Reasonable as given above

REACTION: ETHYL RADICAL + CYCLOHEXA-1,3-DIENE

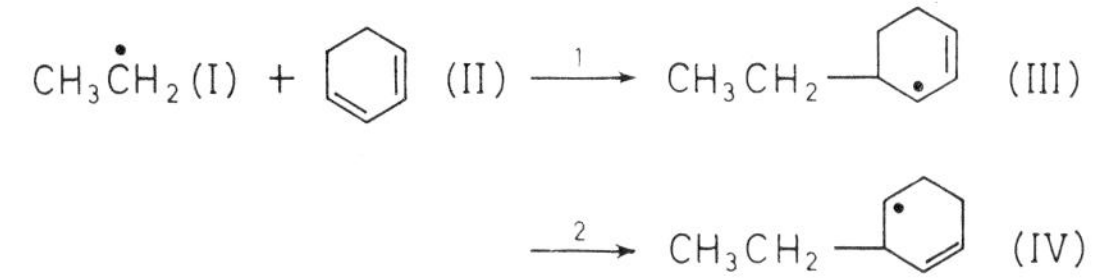

	(I)	(II)	(III)	(IV)		Path 1	Path 2
$\Delta H^\circ_{f(300)}$	26.0 (108.9)	26.0 (108.8)	17.3 (72.4)	29.8 (124.8)	ΔH°	-34.7 (-145.2)	-22.2 (-92.9)
$S^\circ_{(300)}$	59.8 (250.4)		[88.9 (372.2)]	[93.3 (390.6)]	ΔS°		
$C^\circ_{p(300)}$	11.1 (46.5)		[35.5 (148.6)]	[35.7 (149.5)]	ΔC°_p		

Path	log A	E	log k	Conditions	Reference and Experimental Method
1 + 2	8.14	5.4 (22.6)	4.65(338 K)	338 K ? torr	[112] A.C.R.Brown and D.G.L.James, Proc.Chem.Soc., 81 (1962). Arrhenius parameters relative to log k_c = 10.4 for C_2H_5 combination; log $(k_{1+2}/k_c^{\frac{1}{2}})(\ell^{\frac{1}{2}}mol^{-\frac{1}{2}}s^{-\frac{1}{2}})$ = 2.94 - 5.4/2.3RT. For details see C_2H_5 + hexene-1, reference [111].
1 + 2	8.09	5.2 (21.8)	4.90(356 K)	322-389 K ? torr	[113] A.C.R.Brown and D.G.L.James, Can.J.Chem., 43, 1102 (1965). Arrhenius parameters relative to log k_c = 10.4 for C_2H_5 combination; log $(k_{1+2}/k_c^{\frac{1}{2}})(\ell^{\frac{1}{2}}mol^{-\frac{1}{2}}s^{-\frac{1}{2}})$ = 2.89 - 5.2/2.3RT. For details see C_2H_5 + hexene-1, reference [111].

PREFERRED RATE PARAMETERS

Reasonable as given above

COMMENTS

The results refer to the sum of the rate constants for paths 1 and 2, although path 1 should predominate. The above results are reasonable in relation to the data for CH_3 additions to butadienes (see summary of alkyl additions, page 219).

ΔH°_f, ΔH° and E in kcal mol^{-1} (kJ mol^{-1}); S°, C°_p, ΔS° and ΔC°_p in cal mol^{-1} K^{-1} (J mol^{-1} K^{-1}); A and k in l mol^{-1} s^{-1}

<u>REACTION</u>: ETHYL RADICAL + CYCLOHEPTATRIENE

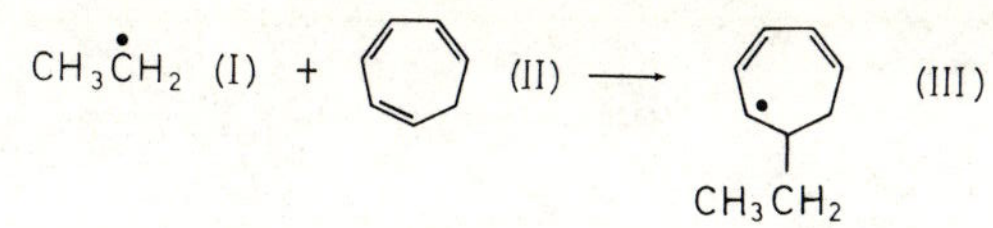

	(I)	(II)	(III)		
$\Delta H^\circ_{f(300)}$	26.0 (108.9)	44.7 (187.2)	[41 (172)]	ΔH°	[-30 (-126)]
$S^\circ_{(300)}$	59.8 (250.4)	[75 (314)]	[53 (222)]	ΔS°	[-81 (-339)]
$C^\circ_{p(300)}$	11.1 (46.5)			ΔC°_p	

log A	E	log k	Conditions	Reference and Experimental Method
8.49	6.4 (26.8)	4.65(364 K)	323-406 K ? torr	[119] A.C.R.Brown and D.G.L.James, <u>Can</u>.<u>J</u>.<u>Chem</u>., <u>43</u>, 1110 (1965). Arrhenius parameters relative to log k_c = 10.4 for C_2H_5 combination; log $(k_1/k_c^{\frac{1}{2}})(\ell^{\frac{1}{2}}mol^{-\frac{1}{2}}s^{-\frac{1}{2}})$ = 3.29 - 6.4/2.3RT. For details see C_2H_5 + hexene-1, reference [111].

<u>PREFERRED RATE PARAMETERS</u>

Reasonable as given above

REACTION: ETHYL RADICAL + CYCLOOCTATETRENE

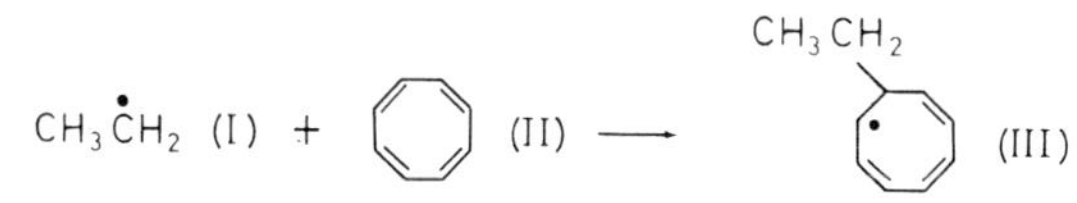

	(I)	(II)	(III)		
$\Delta H^\circ_{f(300)}$	26.0 (108.9)	70.3 (294.3)	70.4 (294.8)	ΔH°	-25.9 (-108.4)
$S^\circ_{(300)}$	59.8 (250.4)			ΔS°	
$C^\circ_{p(300)}$	11.1 (46.5)			ΔC°_p	

log A	E	log k	Conditions	Reference and Experimental Method
7.89	5.6 (23.4)	3.47(358 K)	324-392 K 20 torr	[120] A.C.R.Brown and D.G.L.James, Can.J.Chem., 43, 660 (1965); 43, 1110 (1965). Arrhenius parameters relative to log k_c = 10.4 for C_2H_5 combination; $\log(k_1/k_c^{\frac{1}{2}})(\ell^{\frac{1}{2}}mol^{-\frac{1}{2}}s^{-\frac{1}{2}}) = 2.69 - 5.6/2.3RT$. For details see C_2H_5 + hexene-1, reference [111].

PREFERRED RATE PARAMETERS

Reasonable as given above

ΔH^o_f, ΔH^o and E in kcal mol^{-1} (kJ mol^{-1}); S^o, C^o_p, ΔS^o and ΔC^o_p in cal mol^{-1} K^{-1} (J mol^{-1} K^{-1}); A and k in l mol^{-1} s^{-1}

REACTION: ETHYL RADICAL + BICYCLOHEPTADIENE [BICYCLO [2,2,1] HEPTA 2,5 DIENE]

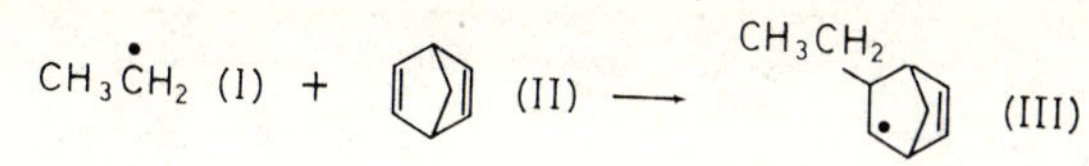

	(I)	(II)	(III)	
$\Delta H^\circ_{f(300)}$	26.0 (108.9)			ΔH°
$S^\circ_{(300)}$	59.8 (250.4)			ΔS°
$C^\circ_{p(300)}$	11.1 (46.5)			ΔC°_p

log A	E	log k	Conditions	Reference and Experimental Method
8.39	7.0 (29.3)	4.19(363 K)	323-404 K ? torr	[119] A.C.R.Brown and D.G.L.James, Can.J.Chem., 43, 1110 (1965). Arrhenius parameters relative to log k_c = 10.4 for C_2H_5 combination; log $(k_1/k_c^{\frac{1}{2}})(\ell^{\frac{1}{2}}mol^{-\frac{1}{2}}s^{-\frac{1}{2}})$ = 3.19 - 7.0/2.3RT. For details see C_2H_5 + hexene-1, reference [111].

PREFERRED RATE PARAMETERS

Reasonable as given above

REACTION: ETHYL RADICAL + STYRENE

$$CH_3\dot{C}H_2(I) + CH_2{=}CHC_6H_5(II) \xrightarrow{1} CH_3CH_2CH_2\dot{C}HC_6H_5(III)$$
$$\xrightarrow{2} \dot{C}H_2CH(C_2H_5)C_6H_5(IV)$$

	(I)	(II)	(III)	(IV)		Path 1	Path 2
$\Delta H^\circ_{f(300)}$	26.0 (108.9)	35.2 (147.4)	26.7 (111.8)	[42 (176)]	ΔH°	-34.5 (-144.5)	[-19 (-80)]
$S^\circ_{(300)}$	59.8 (250.4)	83.8 (350.9)	103.4 (432.9)	[107 (448)]	ΔS°	-40.2 (-168.3)	[-37 (-155)]
$C^\circ_{p(300)}$	11.1 (46.5)	29.4 (123.1)	41.4 (173.3)	[41.4 (173.3)]	ΔC°_p	0.9 (3.8)	[0.9 (3.8)]

Path	log A	E	log k	Conditions	Reference and Experimental Method
1 + 2	7.59	4.1 (17.2)	5.24(382 K)	330-434 K ? torr	[114] D.G.L.James and D.McCallum, Can.J.Chem., 43, 633 (1965). Arrhenius parameters relative to log k_c = 10.4 for C_2H_5 combination; $\log (k_{1+2}/k_c^{1/2})(\ell^{1/2} mol^{-1/2} s^{-1/2}) = 2.39 - 4.1/2.3RT$. For details see C_2H_5 + hexene-1, reference [111].

PREFERRED RATE PARAMETERS

Reasonable as given above

COMMENTS

The estimated enthalpy changes indicate that path 1, giving rise to the resonance stabilized radical, will be the favoured route of addition. It is assumed in reference [114] that there is no addition to the benzene ring, which seems reasonable in relation to the high rate constants for the reaction.

ΔH^o_f, ΔH^o and E in kcal mol^{-1} (kJ mol^{-1}); S^o, C^o_p, ΔS^o and ΔC^o_p in cal mol^{-1} K^{-1} (J mol^{-1} K^{-1}); A and k in l mol^{-1} s^{-1}

REACTION: ETHYL RADICAL + ACETYLENE $CH_3\dot{C}H_2(I) + CH{\equiv}CH(II) \longrightarrow C_2H_5CH{=}\dot{C}H(III)$

	(I)	(II)	(III)		
$\Delta H^\circ_{f(300)}$	26.0 (108.9)	54.2 (226.9)	[56 (234)]	ΔH°	[-24 (-100)]
$S^\circ_{(300)}$	59.8 (250.4)	48.0 (201.0)	74.8 (313.2)	ΔS°	-33.0 (-138.2)
$C^\circ_{p(300)}$	11.1 (46.5)	10.5 (44.0)	20.3 (85.0)	$\Delta C^\circ_{p(300)}$	-1.3 (-5.4)
$C^\circ_{p(500)}$	16.3 (68.2)	13.1 (54.3)	31.0 (129.8)	$\Delta C^\circ_{p(500)}$	1.6 (6.7)
				$\langle\Delta C^\circ_p\rangle$	0.2 (0.8)

log A	E	log k	Conditions	Reference and Experimental Method
7.7	7.0 (29.3)	4.1(423 K)	373-473 K 20-150 torr	[105] J.A.Garcia-Dominguez and A.F.Trotman-Dickenson, J.Chem.Soc., 940 (1962). Arrhenius parameters relative to log k_c = 10.4 for C_2H_5 combination; log $(k_1/k_c^{\frac{1}{2}})(\ell^{\frac{1}{2}}mol^{-\frac{1}{2}}s^{-\frac{1}{2}})$ = 2.5 - 7.0/2.3RT. 'Aldehyde method'. See experimental section of $CH_3 + CH_2{=}C{=}CH_2$ system.

PREFERRED RATE PARAMETERS

Tentatively as given above

COMMENTS

See comments on 'aldehyde method' in experimental section of $CH_3 + CH_2{=}C{=}CH_2$. See also summary of alkyl additions (page 219). In relation to the results for CH_3 added to $CH{\equiv}CH$, the above Arrhenius parameters seem rather low. This is borne out by a comparison of the rate constants.

REACTION: ETHYL RADICAL + HEPTYNE-1

$CH_3\dot{C}H_2(I) + HC{\equiv}C(CH_2)_4CH_3(II) \xrightarrow{1} C_2H_5CH{=}\dot{C}(CH_2)_4CH_3(III)$

$\xrightarrow{2} \dot{C}H{=}C(C_2H_5)(CH_2)_4CH_3(IV)$

	(I)	(II)	(III)	(IV)		Path 1	Path 2
$\Delta H^\circ_{f(300)}$	26.0 (108.9)	24.8 (103.8)	25.6 (107.2)	28.0 (117.2)	ΔH°	-25.2 (-105.5)	-22.8 (-95.5)
$S^\circ_{(300)}$	58.9 (250.4)	97.8 (409.5)	120.0 (502.4)	120.6 (505.0)	ΔS°	-36.6 (-153.2)	-36.1 (-151.2)
$C^\circ_{p(300)}$	11.1 (46.5)	36.0 (150.7)	47.4 (198.5)	48.3 (202.2)	ΔC°_p	0.3 (1.3)	1.2 (5.0)

Path	log A	E	log k	Conditions	Reference and Experimental Method
1 + 2	8.59	8.8 (36.8)	3.50(378 K)	300-455 K 20 torr	[111] D.G.L.James and E.W.R.Steacie, Proc.Roy.Soc.(London) Ser.A., 244, 297 (1958). Arrhenius parameters relative to log k_c = 10.4 for C_2H_5 combination; $\log(k_{1+2}/k_c^{\frac{1}{2}})(\ell^{\frac{1}{2}}mol^{-\frac{1}{2}}s^{-\frac{1}{2}}) = 3.39 - 8.8/2.3RT$. For details see C_2H_5 + hexene-1.

PREFERRED RATE PARAMETERS

Reasonable as given above

COMMENTS

The above measurements refer to the sum of the rate constants for paths 1 and 2, although path 1 will be the favoured route.
See also summary of alkyl additions (page 219) and comments.

ΔH^o_f, ΔH^o and E in kcal mol^{-1} (kJ mol^{-1}); S^o, C^o_p, ΔS^o and ΔC^o_p in cal mol^{-1} K^{-1} (J mol^{-1} K^{-1}); A and k in l mol^{-1} s^{-1}

REACTION: N-PROPYL RADICAL + ETHYLENE

$CH_3CH_2\dot{C}H_2(I) + CH_2{=}CH_2(II) \longrightarrow n\text{-}C_3H_7CH_2\dot{C}H_2(III)$

	(I)	(II)	(III)		
$\Delta H^\circ_{f(300)}$	21.0 (87.9)	12.5 (52.3)	10.9 (45.6)	ΔH°	-22.6 (-94.6)
$S^\circ_{(300)}$	68.5 (286.8)	52.4 (219.4)	87.7 (367.2)	ΔS°	-33.2 (-139.0)
$C^\circ_{p(300)}$	17.1 (71.6)	10.3 (43.1)	28.7 (120.2)	$\Delta C^\circ_{p(300)}$	1.3 (5.4)
$C^\circ_{p(500)}$	25.2 (105.5)	14.9 (62.4)	42.4 (177.5)	$\Delta C^\circ_{p(500)}$	2.3 (9.6)
				$\langle\Delta C^\circ_p\rangle$	1.8 (7.5)

log A	E	log k	Conditions	Reference and Experimental Method
7.29	6.5 (27.2)	3.91(421 K)	375-467 K 70-100 torr	[121] J.A.Kerr and A.F.Trotman-Dickenson, Trans.Faraday Soc., 55, 572 (1959). Arrhenius parameters relative to log k_c = 10.0 for n-C_3H_7 combination; log $(k_1/k_c^{\frac{1}{2}})(\ell^{\frac{1}{2}}mol^{-\frac{1}{2}}s^{-\frac{1}{2}})$ = 2.29 - 6.5/2.3RT. 'Aldehyde method'. For details see $CH_3 + CH_2{=}C{=}CH_2$.
6.88	5.1 (21.4)	4.42(453 K)	403-503 K 100 torr	[92] L.Endrenyi and D.J.LeRoy, J.Phys.Chem., 71, 1334 (1967). Arrhenius parameters relative to log k_c = 10.0 for n-C_3H_7 combination; log $(k_1/k_c^{\frac{1}{2}})(\ell^{\frac{1}{2}}mol^{-\frac{1}{2}}s^{-\frac{1}{2}})$ = 1.88 - 5.1/2.3RT. For details see $CH_3 + CH_2{=}CH_2$. The results were obtained for the secondary reactions of n-C_3H_7 radicals formed in the study of the $CH_3 + CH_2{=}CH_2$ reaction. The rate constants were calculated from the rates of formation of the products of the n-C_5H_{11} radical reactions.

PREFERRED RATE PARAMETERS

Tentatively log k = [7.29 - 6.1/2.3RT]

COMMENTS

As seen from the Arrhenius plot (Fig. 4), there is a discrepancy of about a factor of 2 between the rate constants of references [121] and [92], despite the fact that the Arrhenius parameters do not appear to be too inconsistent. This discrepancy is very difficult to explain and implies a gross experimental error either in one or other or both investigations. On the whole, the radical scavenging method of reference [121] would appear to be the more reliable, but on the other hand it is difficult to explain high rate constants in the type of experiment of reference [92]. Comparison with data for other alkyl radical additions to ethylene (page 219) indicates that the Arrhenius parameters of reference [92] are probably underestimated. At the same time, a comparison of the rate constants at 400 K does not

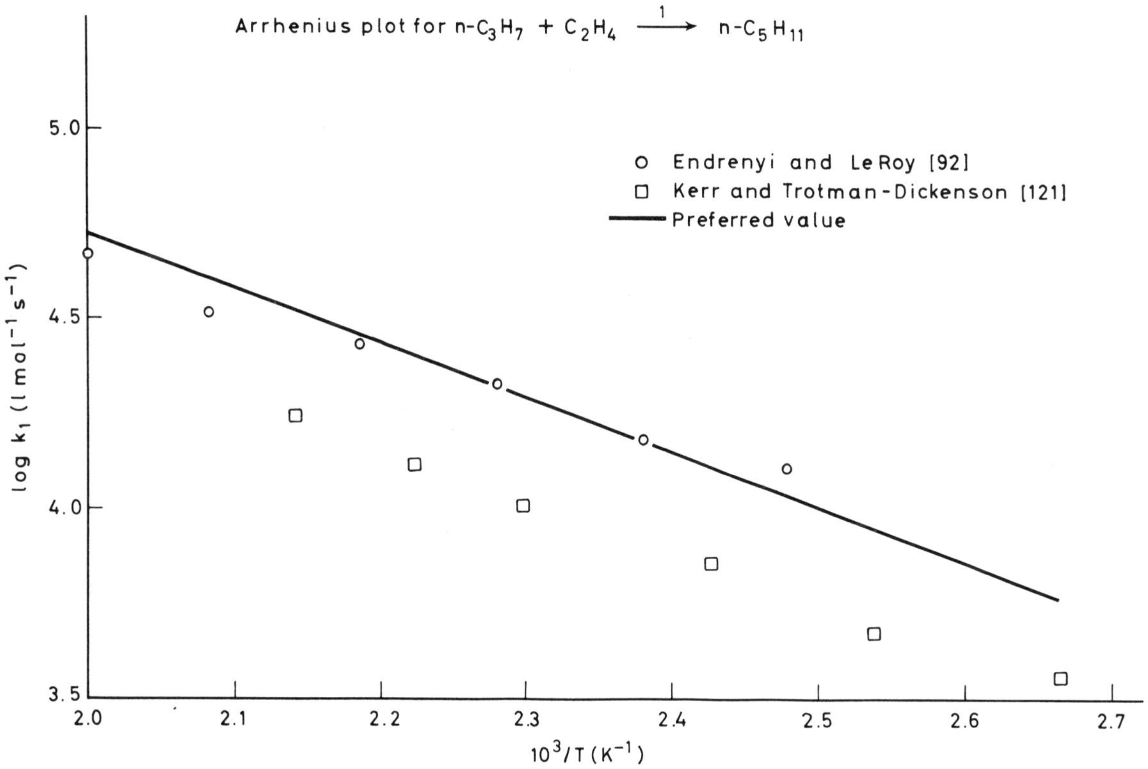

FIGURE 4. Arrhenius plot for n-C_3H_7 + C_2H_4 $\xrightarrow{1}$ n-C_5H_{11}

resolve the difficulty, as that from reference [121] is in agreement with the data for n-C_4H_9 addition to C_2H_4, while that from reference [92] agrees with the data for i-C_3H_7 addition. The above tentative preferred rate constant was obtained by selecting a line which is reasonably consistent with the rate constants of reference [92] and at the same time gives Arrhenius parameters more consistent with those for other alkyl additions to C_2H_4. It is clear that more data is needed on this reaction to resolve the discrepancy. The above tentative Arrhenius parameters and the thermodynamic data yield for the reverse radical decomposition reaction log (k_{-1}/s^{-1}) = 12.7 - 28.1/2.3RT at 298 K. The calculated value of log A_{-1} is considerably less than the lower limit set by transition state theory, log A_{-1} = 13.2 at 298 K, indicating that A_1 is probably underestimated.

REACTION: N-PROPYL RADICAL + BUTENE-1

$CH_3CH_2\dot{C}H_2$(I) + CH_2=$CHCH_2CH_3$(II) $\xrightarrow{1}$ $C_3H_7CH_2\dot{C}HCH_2CH_3$(III)

$\xrightarrow{2}$ $\dot{C}H_2CH(C_3H_7)CH_2CH_3$(IV)

	(I)	(II)	(III)	(IV)		Path 1	Path 2
$\Delta H^o_{f(300)}$	21.0 (87.9)	0.0 (0.0)	-2.5 (-10.5)	-1.1 (-4.6)	ΔH^o	-23.5 (-98.4)	-22.1 (-92.5)
$S^o_{(300)}$	68.5 (286.8)	73.6 (308.0)	104.8 (438.8)	103.8 (434.6)	ΔS^o	-37.3 (-156.2)	-38.3 (-160.4)
$C^o_{p(300)}$	17.1 (71.6)	20.6 (86.2)	39.5 (165.4)	39.4 (165.0)	ΔC^o_p	1.8 (7.5)	1.7 (7.1)

Path	log A	E	log k	Conditions	Reference and Experimental Method
1	-	7.21[a] (30.19)	-	298 K	[100] C.W.Larson, D.C.Tardy and B.S.Rabinovitch, J.Chem.Phys., 49, 299 (1968).
2	-	-	-	-	For details see CH_3 + cis-butene-2 system, reference [98]. Radical (III) generated from H atom addition to trans-heptene-3. The above critical energy is relative to an assumed value of 7.00 kcal mol^{-1} for the critical energy of the CH_3 addition to hexene-1.

[a] critical energy of the reaction

PREFERRED RATE PARAMETERS

Reasonable as given above

REACTION: N-PROPYL RADICAL + CIS-BUTENE-2 $CH_3CH_2\dot{C}H_2(I) + cis\text{-}CH_3CH{=}CHCH_3(II) \longrightarrow CH_3CH(C_3H_7)\dot{C}HCH_3(III)$

	(I)	(II)	(III)		
$\Delta H^\circ_{f(300)}$	21.0 (87.9)	-1.9 (-7.9)	-4.6 (-19.3)	ΔH°	-23.7 (-99.2)
$S^\circ_{(300)}$	68.5 (286.8)	73.6 (308.2)	102.1 (427.5)	ΔS°	-40.0 (-167.5)
$C^\circ_{p(300)}$	17.1 (71.6)	19.4 (81.2)	39.3 (164.5)	ΔC°_p	2.8 (11.7)

log A	E	log k	Conditions	Reference and Experimental Method
-	8.68[a] (36.34)	-	298 K	[100] C.W.Larson, D.C.Tardy and B.S.Rabinovitch, J.Chem.Phys., 49, 299 (1968). For details see CH_3 + cis-butene-2 system, reference [98]. Radical (III) generated from H atom addition to 3-methylhexene-1. The above critical energy is relative to an assumed value of 8.00 kcal mol^{-1} for the critical energy of the CH_3 addition to trans-hexene-2.

[a] critical energy of the reaction

PREFERRED RATE PARAMETERS

Reasonable as given above

ΔH_f°, ΔH° and E in kcal mol^{-1} (kJ mol^{-1}); S°, C_p°, ΔS° and ΔC_p° in cal mol^{-1} K^{-1} (J mol^{-1} K^{-1}); A and k in l mol^{-1} s^{-1}

REACTION: N-PROPYL RADICAL + TRANS-BUTENE-2

$CH_3CH_2\dot{C}H_2$(I) + trans-$CH_3CH{=}CHCH_3$(II) ⟶ $CH_3CH(C_3H_7)\dot{C}HCH_3$(III)

	(I)	(II)	(III)		
$\Delta H^\circ_{f(300)}$	21.0 (87.9)	-2.9 (-12.1)	-4.6 (-19.3)	ΔH°	-22.7 (-95.0)
$S^\circ_{(300)}$	68.5 (286.8)	72.4 (302.8)	102.1 (427.5)	ΔS°	-38.8 (-162.5)
$C^\circ_{p(300)}$	17.1 (71.6)	20.7 (86.6)	39.3 (164.5)	ΔC°_p	1.5 (6.3)

log A	E	log k	Conditions	Reference and Experimental Method
-	8.97[a] (37.56)	-	298 K	[100] C.W.Larson, D.C.Tardy and B.S.Rabinovitch, J.Chem.Phys., 49, 299 (1968). For details see CH_3 + cis-butene-2 system, reference [98]. Radical (III) generated from H atom addition to 3-methylhexene-1. The above critical energy is relative to an assumed value of 8.00 kcal mol^{-1} for the critical energy of the CH_3 addition to trans-hexene-2.

[a] critical energy of the reaction

PREFERRED RATE PARAMETERS

Reasonable as given above

REACTION: N-PROPYL RADICAL + PENTENE-1

$$CH_3CH_2\dot{C}H_2(I) + CH_2{=}CHCH_2CH_2CH_3(II) \xrightarrow{1} C_3H_7CH_2\dot{C}HCH_2CH_2CH_3(III)$$
$$\xrightarrow{2} \dot{C}H_2CH(C_3H_7)CH_2CH_2CH_3(IV)$$

	(I)	(II)	(III)	(IV)
$\Delta H^\circ_{f(300)}$	21.0 (87.9)	-4.9 (-20.5)	-7.4 (-31.0)	-6.0 (-25.1)
$S^\circ_{(300)}$	68.5 (286.8)	83.0 (347.0)	114.2 (478.2)	113.2 (474.0)
$C^\circ_{p(300)}$	17.1 (71.6)	26.1 (109.1)	45.0 (188.4)	44.9 (188.0)

	Path 1	Path 2
ΔH°	-23.5 (-98.4)	-22.1 (-92.5)
ΔS°	-37.3 (-156.2)	-38.3 (-160.4)
ΔC°_p	1.8 (7.5)	1.7 (7.1)

Path	log A	E	log k	Conditions	Reference and Experimental Method
1	-	6.78[a] (28.39)	-	298 K	[100] C.W.Larson, D.C.Tardy and B.S.Rabinovitch, J.Chem.Phys., 49, 299 (1968).
2	-	-	-	-	For details see CH_3 + cis-butene-2 system, reference [98]. Radical (III) generated from H atom addition to trans-octene-4. Above critical energy relative to an assumed value of 6.78 for the critical energy of C_2H_5 addition to hexene-1.

[a] critical energy of the reaction

PREFERRED RATE PARAMETERS

Reasonable as given above

ΔH^o_f, ΔH^o and E in kcal mol^{-1} (kJ mol^{-1}); S^o, C^o_p, ΔS^o and ΔC^o_p in cal mol^{-1} K^{-1} (J mol^{-1} K^{-1}); A and k in l mol^{-1} s^{-1}

REACTION: ISO-PROPYL RADICAL + ETHYLENE $(CH_3)_2\dot{C}H(I) + CH_2{=}CH_2(II) \longrightarrow (CH_3)_2CHCH_2\dot{C}H_2(III)$

	(I)	(II)	(III)		
$\Delta H^\circ_{f(300)}$	17.6 (73.7)	12.5 (52.3)	8.8 (36.8)	ΔH°	-21.3 (-89.2)
$S^\circ_{(300)}$	66.7 (279.3)	52.4 (219.3)	85.0 (355.7)	ΔS°	-34.1 (-142.7)
$C^\circ_{p(300)}$	17.0 (71.2)	10.3 (43.1)	28.4 (118.9)	$\Delta C^\circ_{p(300)}$	1.1 (4.6)
$C^\circ_{p(500)}$	24.9 (104.3)	14.9 (62.3)	42.5 (177.9)	$\Delta C^\circ_{p(500)}$	2.7 (11.3)
				$\langle\Delta C^\circ_p\rangle$	1.9 (8.0)

log A	E	log k	Conditions	Reference and Experimental Method
7.85	6.9 (28.9)	4.21(415 K)	373-457 K 40 torr	[122] J.A.Kerr and A.F.Trotman-Dickenson, Trans.Faraday Soc., 55, 921 (1959). Arrhenius parameters relative to log k_c = 10.0 for iso-C_3H_7 combination; log $(k_1/k_c^{\frac{1}{2}})(\ ^{\frac{1}{2}}mol^{-\frac{1}{2}}s^{-\frac{1}{2}})$ = 2.85 - 6.9/2.3RT [J.A.Kerr, Ph.D.Thesis, Edinburgh, 1959]. 'Aldehyde method'. For details see $CH_3 + CH_2{=}C{=}CH_2$.
7.82	6.93 (29.0)	3.79(376 K)	340-441 K 1-120 torr	[110] K.W.Watkins and L.A.O'Deen, J.Phys.Chem., 73, 4094 (1969). Arrhenius parameters relative to log k_c = 10.0 for iso-C_3H_7 combination; log $(k_1/k_c^{\frac{1}{2}})(\ell^{\frac{1}{2}}mol^{-\frac{1}{2}}s^{-\frac{1}{2}})$ = 2.82 - 6.93/2.3RT. For details see $C_2H_5 + CH_2{=}CH_2$.

PREFERRED RATE PARAMETERS

log k = 7.84 - 6.9/2.3RT

COMMENTS

The agreement between the two sets of measurements of references [122] and [110] is remarkable. The preferred rate constant and the thermodynamic data yield for the reverse radical decomposition reaction, log (k_{-1}/s^{-1}) = 13.4 - 27.6/2.3RT at 298 K, which is reasonable.
See also summary of alkyl radical additions (page 219) and comments.

REACTION: ISO-PROPYL RADICAL + BUTENE-1

$(CH_3)_2\dot{C}H(I) + CH_2{=}CHCH_2CH_3(II) \xrightarrow{1} C_3H_7CH_2\dot{C}HCH_2CH_3(III)$

$\xrightarrow{2} \dot{C}H_2CH(C_3H_7)CH_2CH_3(IV)$

	(I)	(II)	(III)	(IV)
$\Delta H^\circ_{f(300)}$	17.6 (73.7)	0.0 (0.0)	-4.6 (-19.3)	-3.2 (-13.4)
$S^\circ_{(300)}$	66.7 (279.3)	73.6 (308.0)	102.1 (427.5)	101.1 (423.3)
$C^\circ_{p(300)}$	17.0 (71.2)	20.6 (86.2)	39.3 (164.5)	39.1 (163.7)

	Path 1	Path 2
ΔH°	-22.2 (-93.0)	-20.8 (-87.1)
ΔS°	-38.2 (-159.9)	-39.2 (-164.1)
ΔC°_p	1.7 (7.1)	1.5 (6.3)

Path	log A	E	log k	Conditions	Reference and Experimental Method
1	-	6.91[a] (28.93)	-	298 K	[101] C.W.Larson and B.S.Rabinovitch, J.Chem.Phys., 50, 871 (1969). For details see CH_3 + cis-butene-2 system, reference [98]. Radical (III) generated from H atom addition to trans-2-methylhexene-3. The above critical energy is relative to an assumed value of 7.00 kcal mol^{-1} for the critical energy of the CH_3 addition to 4-methylpentene-1.
2	-	-	-	-	

[a] critical energy of the reaction

PREFERRED RATE PARAMETERS

Reasonable as given above

ΔH^o_f, ΔH^o and E in kcal mol^{-1} (kJ mol^{-1}); S^o, C^o_p, ΔS^o and ΔC^o_p in cal mol^{-1} K^{-1} (J mol^{-1} K^{-1}); A and k in l mol^{-1} s^{-1}

REACTION: ISO-PROPYL RADICAL + <u>CIS</u>-BUTENE-2 $(CH_3)_2\dot{C}H(I) + $ <u>cis</u>-$CH_3CH{=}CHCH_3(II) \longrightarrow CH_3CH(C_3H_7)\dot{C}HCH_3(III)$

	(I)	(II)	(III)		
$\Delta H^\circ_{f(300)}$	17.6 (73.7)	-1.9 (-7.9)	-6.7 (-28.1)	ΔH°	-22.4 (-93.8)
$S^\circ_{(300)}$	66.7 (279.3)	73.6 (308.2)	99.4 (416.2)	ΔS°	-40.9 (-171.2)
$C^\circ_{p(300)}$	17.0 (71.2)	19.4 (81.2)	39.0 (163.3)	ΔC°_p	2.6 (10.9)

log A	E	log k	Conditions	Reference and Experimental Method
-	7.85[a] (32.87)	-	298 K	[101] C.W.Larson and B.S.Rabinovitch, J.Chem.Phys., 50, 871 (1969). For details see CH_3 + cis-butene-2 system, reference [98]. Radical (III) generated from H atom addition to 3,4-dimethylpentene-2. The above critical energy is relative to an assumed value of 8.00 kcal mol^{-1} for the critical energy of the CH_3 addition to trans-4-methylpentene-2.

[a] critical energy of the reaction

PREFERRED RATE PARAMETERS

Reasonable as given above

REACTION: ISO-PROPYL RADICAL + TRANS-BUTENE-2 $(CH_3)_2\dot{C}H(I)$ + trans-$CH_3CH{=}CHCH_3(II) \longrightarrow CH_3CH(C_3H_7)\dot{C}HCH_3(III)$

	(I)	(II)	(III)		
$\Delta H^o_{f(300)}$	17.6 (73.7)	-2.9 (-12.1)	-6.7 (-28.1)	ΔH^o	-21.4 (-89.6)
$S^o_{(300)}$	66.7 (279.3)	72.4 (302.8)	99.4 (416.2)	ΔS^o	-39.7 (-166.2)
$C^o_{p(300)}$	17.0 (71.2)	20.7 (86.6)	39.0 (163.3)	ΔC^o_p	1.3 (5.4)

log A	E	log k	Conditions	Reference and Experimental Method
-	8.08[a] (33.83)	-	298 K	[101] C.W.Larson and B.S.Rabinovitch, J.Chem.Phys., 50, 871 (1969). For details see CH_3 + cis-butene-2 system, reference [98]. Radical (III) generated from H atom addition to 3,4-dimethylpentene-2. The above critical energy is relative to an assumed value of 8.00 kcal mol^{-1} for the critical energy of the CH_3 addition to trans-4-methylpentene-2.

[a] critical energy of the reaction

PREFERRED RATE PARAMETERS

Reasonable as given above

ΔH^o_f, ΔH^o and E in kcal mol^{-1} (kJ mol^{-1}); S^o, C^o_p, ΔS^o and ΔC^o_p in cal mol^{-1} K^{-1} (J mol^{-1} K^{-1}); A and k in l mol^{-1} s^{-1}

REACTION: ISO-PROPYL RADICAL + ALLENE

$(CH_3)_2\dot{C}H(I) + CH_2{=}C{=}CH_2(II) \xrightarrow{1} (CH_3)_2CHCH_2\dot{C}{=}CH_2(III)$

$\xrightarrow{2} CH_3CH(CH_3)C(\dot{C}H_2){=}CH_2(IV)$

	(I)	(II)	(III)	(IV)
$\Delta H^\circ_{f(300)}$	17.6 (73.7)	45.9 (192.2)	43.9 (183.8)	18.2 (76.1)
$S^\circ_{(300)}$	66.7 (279.3)	58.3 (244.1)	91.1 (381.4)	[88 (369)]
$C^\circ_{p(300)}$	17.0 (71.2)	14.2 (59.5)	31.3 (131.0)	[31.1 (130.2)]
$C^\circ_{p(500)}$	24.9 (104.3)	19.8 (82.9)	47.6 (199.0)	[47.8 (200.0)]

	Path 1	Path 2
ΔH°	-19.6 (-82.1)	-45.3 (-189.6)
ΔS°	-33.9 (-141.9)	[-37 (-155)]
$\Delta C^\circ_{p(300)}$	0.1 (0.4)	[-0.1 (-0.4)]
$\Delta C^\circ_{p(500)}$	2.9 (12.1)	[3.1 (12.9)]
$\langle \Delta C^\circ_p \rangle$	1.5 (6.3)	1.5 (6.3)

Path	log A	E	log k	Conditions	Reference and Experimental Method
1	7.55	7.2 (30.1)	3.79(419 K)	366-473 K 10-70 torr	[104] R.R.Getty, J.A.Kerr and A.F.Trotman-Dickenson, J.Chem.Soc.(A), 979 (1967). Arrhenius parameters relative to log k_c = 10.0 for iso-C_3H_7 combination; log $(k_{1+2}/k_c^{\frac{1}{2}})(\ell^{\frac{1}{2}}mol^{-\frac{1}{2}}s^{-\frac{1}{2}})$ = 2.55 - 7.2/2.3RT. 'Aldehyde method'. For details see CH_3 + $CH_2{=}C{=}CH_2$. It was shown by product analysis that addition occurred exclusively at the terminal C atom in allene (path 1).

PREFERRED RATE PARAMETERS

Tentatively as given

COMMENTS

See comments on 'aldehyde method' in experimental section of CH_3 + $CH_2{=}C{=}CH_2$ system. It seems that, in relation to the Arrhenius parameters for CH_3 and C_2H_5 additions to allene (see summary of alkyl additions, page 219), those for iso-C_3H_7 given above are slight under-estimates, although the rate constants appear to be reasonable.

REACTION: ISO-PROPYL RADICAL + CYCLOHEXA-1,3-DIENE

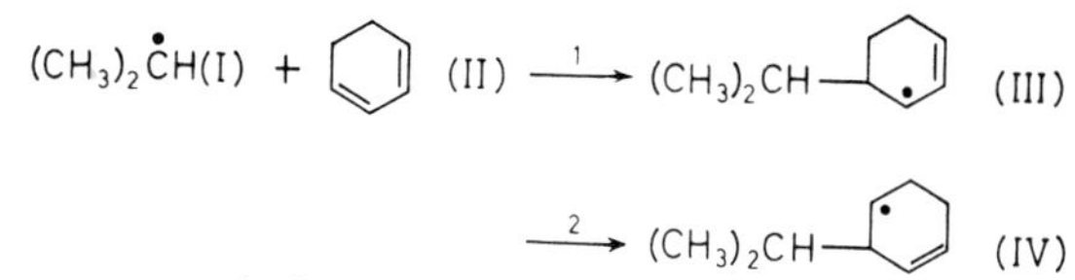

	(I)	(II)	(III)	(IV)
$\Delta H^{\circ}_{f(300)}$	17.6 (73.7)	26.0 (108.9)	10.4 (43.5)	22.8 (95.5)
$S^{\circ}_{(300)}$	66.7 (279.3)		95.0 (397.8)	[100 (419)]
$C^{\circ}_{p(300)}$	17.0 (71.2)		41.1 (172.1)	[40.9 (171.2)]

	Path 1	Path 2
ΔH°	-33.2 (-139.0)	-20.8 (-87.1)
ΔS°		
ΔC°_p		

Path	log A	E	log k	Conditions	Reference and Experimental Method
1 + 2	8.39	5.8 (24.3)	4.87(360 K)	315-406 K ? torr	[123] D.G.L.James and R.D.Suart, J.Phys.Chem., 69, 2362 (1965). Arrhenius parameters relative to log k_c = 10.0 for iso-C_3H_7 combination; log $(k_{1+2}/k_c^{\frac{1}{2}})(\ell^{\frac{1}{2}}mol^{-\frac{1}{2}}s^{-\frac{1}{2}})$ = 3.39 -5.8/2.3RT. Static system. iso-C_3H_7 from photolysis (λ = 313 nm) of diisopropyl ketone. Reaction rates calculated from product analysis using low temperature distillation and gas chromatography.

PREFERRED RATE PARAMETERS

Reasonable as given above

COMMENTS

The results refer to the sum of the rate constants for paths 1 + 2, although path 1 should predominate.

ΔH^{o}_{f}, ΔH^{o} and E in kcal mol^{-1} (kJ mol^{-1}); S^{o}, C^{o}_{p}, ΔS^{o} and ΔC^{o}_{p} in cal mol^{-1} K^{-1} (J mol^{-1} K^{-1}); A and k in l mol^{-1} s^{-1}

REACTION: ISO-PROPYL RADICAL + ACETYLENE

$(CH_3)_2\dot{C}H(I) + CH{\equiv}CH(II) \longrightarrow (CH_3)_2CHCH{=}\dot{C}H(III)$

	(I)	(II)	(III)		
$\Delta H^\circ_{f(300)}$	17.6 (73.7)	54.2 (226.9)	49.1 (205.6)	ΔH°	-22.7 (-95.0)
$S^\circ_{(300)}$	66.7 (279.3)	48.0 (201.0)	[82 (343)]	ΔS°	[-33 (-138)]
$C^\circ_{p(300)}$	17.0 (71.2)	10.5 (44.0)	[25.8 (108.0)]	$\Delta C^\circ_{p(300)}$	[-1.7 (-7.1)]
$C^\circ_{p(500)}$	24.9 (104.3)	13.1 (54.3)	[39.5 (165.4)]	$\Delta C^\circ_{p(500)}$	[1.5 (6.3)]
				$\langle\Delta C^\circ_p\rangle$	[-0.1 (-0.4)]

log A	E	log k	Conditions	Reference and Experimental Method
7.7	6.9 (28.9)	4.49(470 K)	363-577 K 50-100 torr	[105] J.A.Garcia-Dominguez and A.F.Trotman-Dickenson, J.Chem.Soc., 940 (1962). Arrhenius parameters relative to log k_c = 10.0 for iso-C_3H_7 combination; log $(k_1/k_c^{\frac{1}{2}})(\ell^{\frac{1}{2}}mol^{-\frac{1}{2}}s^{-\frac{1}{2}})$ = 2.7 - 6.9/2.3RT. 'Aldehyde method'. For details see $CH_3 + CH_2{=}C{=}CH_2$. Some of the adduct radicals appear to isomerize, $(CH_3)_2CHCH{=}\dot{C}H \longrightarrow CH_3\dot{C}HCH_2CH{=}CH_2$, leading to the formation of the product $CH_3CH_2CH_2CH{=}CH_2$.

PREFERRED RATE PARAMETERS

Tentatively as given above

COMMENTS

See comments on 'aldehyde method' in experimental section of $CH_3 + CH_2{=}C{=}CH_2$. In relation to the data for CH_3 addition to $CH{\equiv}CH$ (see summary of alkyl additions, page 219), the above Arrhenius parameters are slightly low, although the rate constants seem reasonable.

REACTION: ISO-PROPYL RADICAL + PROPYNE

$(CH_3)_2\dot{C}H(I) + CH_3C{\equiv}CH(II) \xrightarrow{1} (CH_3)_2CHC(CH_3){=}\dot{C}H(III)$

$\xrightarrow{2} CH_3\dot{C}{=}CHCH(CH_3)_2(IV)$

	(I)	(II)	(III)	(IV)
$\Delta H^\circ_{f(300)}$	17.6 (73.7)	44.0 (184.2)	40.8 (170.8)	38.4 (160.8)
$S^\circ_{(300)}$	66.7 (279.3)	59.2 (247.9)	[88 (368)]	[90 (377)]
$C^\circ_{p(300)}$	17.0 (71.2)	14.6 (61.1)	[31.9 (133.6)]	[31.0 (129.8)]
$C^\circ_{p(500)}$	24.9 (104.3)	19.7 (82.5)	[48.0 (200.7)]	[47.2 (197.2)]

	Path 1	Path 2
ΔH°	-20.8 (-87.1)	-23.2 (-97.1)
ΔS°	[-38 (-159)]	[-36 (-151)]
$\Delta C^\circ_{p(300)}$	[0.3 (1.3)]	[-0.6 (-2.5)]
$\Delta C^\circ_{p(500)}$	[3.4 (14.2)]	[2.6 (10.9)]
$\langle\Delta C^\circ_p\rangle$	1.9 (8.0)	1.0 (4.2)

Path	log A	E	log k	Conditions	Reference and Experimental Method
2	6.28	5.7 (23.9)	3.17(400 K)	360-439 K 30-70 torr	[106] R.R.Getty, J.A.Kerr and A.F.Trotman-Dickenson, J.Chem.Soc.(A)., 1360 (1967). Results calculated from log $(k_{1+2}/k_d^{\frac{1}{2}})(\ell^{\frac{1}{2}}mol^{-\frac{1}{2}}s^{-\frac{1}{2}})$ = 1.39 - 5.7/2.3RT and log k_d = 9.79 for the reaction $2iso\text{-}C_3H_7 \xrightarrow{d} C_3H_8 + C_3H_6$ based on log k_c = 10.00 and k_d/k_c = 0.6 for iso-C_3H_7 radicals. 'Aldehyde method'. For details see $CH_3 + CH_2{=}C{=}CH_2$. No products were detected from addition at the central C atom in propyne. Some of the adduct radicals from addition at the terminal C atom appear to undergo an isomerization reaction, $CH_3\dot{C}{=}CHCH(CH_3)_2 \longrightarrow CH_3CH{=}CHCH_2\dot{C}HCH_3$, leading to the formation of cis and trans-hexene-2 products. It was not possible to analyse for the iso-propyl dimer and the radical concentrations were monitored by the formation of C_3H_6 in the disproportionation reaction.

PREFERRED RATE PARAMETERS

Above data suspect

COMMENTS

The rate constants and Arrhenius parameters of reference [106] are very low in relation to other data on alkyl additions to propyne and acetylene (see summary of alkyl additions, page 219). Furthermore, the absence of products from addition of iso-C_3H_7 to the central C atom in propyne is completely unexpected, especially since CH_3 and t-C_4H_9 radicals add at both positions.

ΔH°_f, ΔH° and E in kcal mol^{-1} (kJ mol^{-1}); S°, C°_p, ΔS° and ΔC°_p in cal mol^{-1} K^{-1} (J mol^{-1} K^{-1}); A and k in l mol^{-1} s^{-1}

REACTION: N-BUTYL RADICAL + ETHYLENE $CH_3CH_2CH_2\dot{C}H_2(I) + CH_2{=}CH_2(II) \longrightarrow CH_3CH_2CH_2CH_2CH_2\dot{C}H_2(III)$

	(I)	(II)	(III)		
$\Delta H^\circ_{f(300)}$	15.8 (66.1)	12.5 (52.3)	5.9 (24.7)	ΔH°	-22.4 (-93.7)
$S^\circ_{(300)}$	76.9 (327.7)	52.4 (219.3)	97.1 (406.5)	ΔS°	-32.2 (-140.6)
$C^\circ_{p(300)}$	23.2 (97.1)	10.3 (43.1)	34.2 (143.2)	$\Delta C^\circ_{p(300)}$	0.7 (2.9)
$C^\circ_{p(500)}$	34.2 (143.1)	14.9 (62.3)	50.7 (212.1)	$\Delta C^\circ_{p(500)}$	1.8 (7.5)
				$\langle\Delta C^\circ_p\rangle$	1.3 (5.4)

log A	E	log k	Conditions	Reference and Experimental Method
7.60	7.3 (30.6)	3.78(418 K)	334-502 K 15-30 torr	[124] J.A.Kerr and A.F.Trotman-Dickenson, J.Chem.Soc., 1602 (1960). Arrhenius parameters relative to log k_c = 10.0 for n-C_4H_9 combination; $\log (k_1/k_c^{\frac{1}{2}})(\ell^{\frac{1}{2}}mol^{-\frac{1}{2}}s^{-\frac{1}{2}}) = 2.6 - 7.3/2.3RT$. 'Aldehyde method'. For details see $CH_3 + CH_2{=}C{=}CH_2$.
7.37	6.7 (28.1)	3.50(378 K)	352-405 K 6-110 torr	[110] K.W.Watkins and L.A.O'Deen, J.Phys.Chem., 73,4094 (1969). Results calculated from $\log (k_1k_c^{\frac{1}{2}}/k_3) = 2.17 - 6.7/2.3RT$ and log k_c = 10.4 for C_2H_5 combination and log k_3 = 10.2 for $C_2H_5 + n\text{-}C_4H_9 \xrightarrow{3} C_6H_{14}$. For details see $C_2H_5 + CH_2{=}CH_2$. n-C_4H_9 radicals were generated in a study of the addition of C_2H_5 radicals to $CH_2{=}CH_2$.

PREFERRED RATE PARAMETERS

log k = [7.37 - 6.7/2.3RT]

COMMENTS

As seen from the Arrhenius plot (Fig. 5), the rate constants of reference [124] are considerably lower than those of reference [110], although the Arrhenius parameters are not too different. We prefer the rate constants of reference [110], which was a more extensive study, although the temperature range was rather limited. The preferred Arrhenius parameters and the thermodynamic data yield for the reverse radical decomposition $\log (k_{-1}/s^{-1}) = 12.9 - 28.5/2.3RT$ at 298 K. The calculated value of log A_{-1} is significantly below the minimum set by transition state theory, $\log (A_{-1}/s^{-1}) = 13.2$ at 298 K, indicating that A_1 is probably under-estimated. See also summary of alkyl radical additions (page 219) and comments.

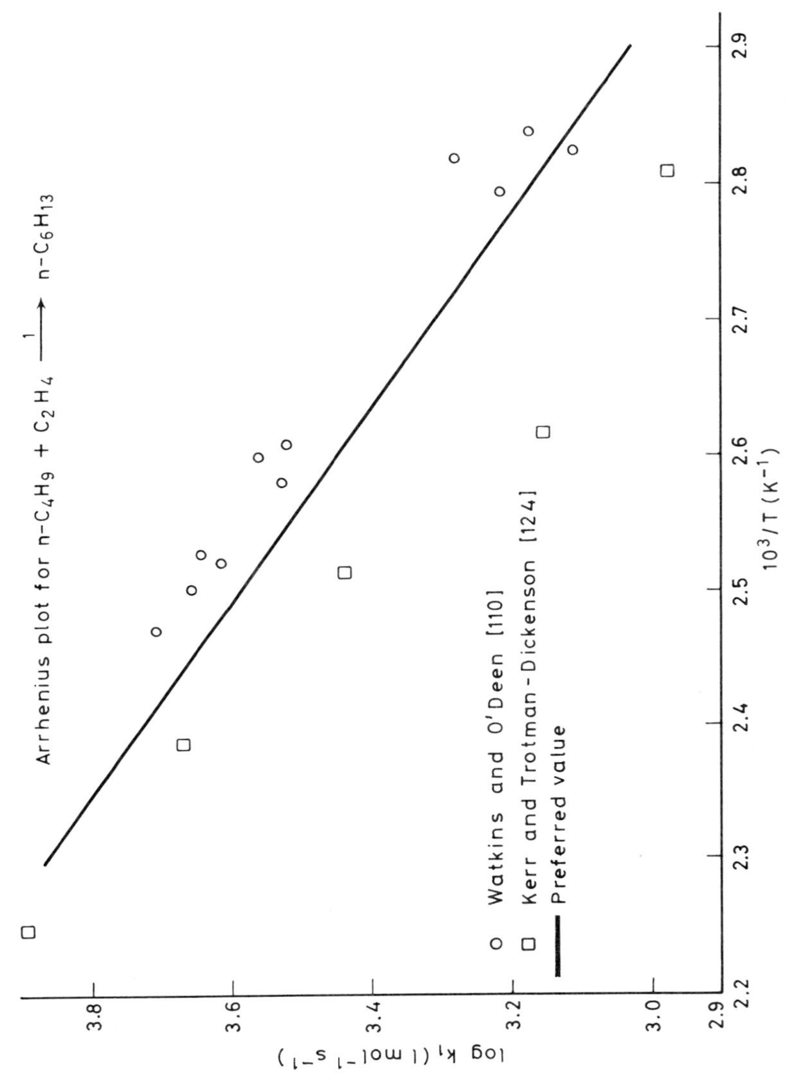

FIGURE 5. Arrhenius plot for $n\text{-}C_4H_9 + C_2H_4 \xrightarrow{1} n\text{-}C_6H_{13}$

REACTION: SEC-BUTYL RADICAL + PROPENE

$CH_3\dot{C}HCH_2CH_3(I) + CH_3CH{=}CH_2(II) \xrightarrow{1} CH_3CH_2CH(CH_3)CH(CH_3)\dot{C}H_2(III)$

$\xrightarrow{2} CH_3CH_2CH(CH_3)CH_2\dot{C}HCH_3(IV)$

	(I)	(II)	(III)	(IV)
$\Delta H^\circ_{f(300)}$	12.3 (51.4)	4.9 (20.5)	-3.2 (-13.4)	-4.6 (-19.3)
$S^\circ_{(300)}$	76.6 (320.4)	63.8 (267.1)	101.1 (423.3)	102.1 (427.5)
$C^\circ_{p(300)}$	23.0 (96.2)	15.3 (64.1)	39.1 (163.7)	39.3 (164.5)
$C^\circ_{p(500)}$	33.9 (141.8)	22.6 (94.6)	59.1 (247.3)	58.7 (245.5)

	Path 1	Path 2
ΔH°	-20.4 (-85.4)	-21.8 (-91.2)
ΔS°	-39.3 (-164.3)	-38.3 (-160.2)
$\Delta C^\circ_{p(300)}$	0.6 (2.5)	1.0 (4.2)
$\Delta C^\circ_{p(500)}$	2.6 (10.8)	2.2 (9.2)
$\langle \Delta C^\circ_p \rangle$	1.6 (6.7)	1.6 (6.7)

Path	log A	E	log k	Conditions	Reference and Experimental Method
1 + 2	∿8.78	∿9.2 (38.5)	∿3.72(397 K)	381-412 K 20 torr	[96] M.Miyoshi and R.K.Brinton, J.Chem.Phys., 36, 3019 (1962). Arrhenius parameters relative to log k_c = 9.9 for sec-C_4H_9 combination; log $(k_{1+2}/k_c^{\frac{1}{2}})(\ell^{\frac{1}{2}}mol^{-\frac{1}{2}}s^{-\frac{1}{2}})$ = 3.83 - 9.2/2.3RT. For details see CH_3 + $CH_3CH{=}CH_2$. sec-C_4H_9 radicals were generated in a study of CH_3 addition to $CH_3CH{=}CH_2$. Some iso-C_4H_9 were also generated in the initial addition step, but since they amount to < 10%, the above results effectively refer to sec-C_4H_9 alone. No attempt appears to have been made to differentiate between paths 1 and 2, but the products were so diverse and difficult to resolve that the rate constants and Arrhenius parameters were considered to be approximate.

PREFERRED RATE PARAMETERS

$\log k_{1+2} = [7.8 - 7.4/2.3RT]$

COMMENTS

The approximate experimental Arrhenius parameters of reference [96] would appear to be over-estimates, since in conjunction with the thermodynamic data (path 2), they lead to log (k_{-2}/s^{-1}) = 15.7 - 30.4/2.3RT for reaction (-2). The preferred tentative parameters were calculated from the measured rate constant at 397 K and an A factor estimated on the basis of the data for other alkyl additions (see page 219).

REACTION: T-BUTYL RADICAL + ETHYLENE $(CH_3)_3\dot{C}(I) + CH_2{=}CH_2(II) \longrightarrow (CH_3)_3CCH_2\dot{C}H_2(III)$

	(I)	(II)	(III)		
$\Delta H^o_{f(300)}$	6.7 (28.1)	12.5 (52.3)	1.1 (4.6)	ΔH^o	-18.1 (-75.8)
$S^o_{(300)}$	74.6 (312.4)	52.4 (219.4)	90.2 (377.7)	ΔS^o	-36.8 (-154.1)
$C^o_{p(300)}$	19.1 (80.0)	10.3 (43.2)	34.4 (144.0)	$\Delta C^o_{p(300)}$	5.0 (20.9)
$C^o_{p(500)}$	30.5 (127.7)	14.9 (62.4)	52.1 (218.1)	$\Delta C^o_{p(500)}$	6.7 (28.1)
				$<\Delta C^o_p>$	5.9 (24.7)

log A	E	log k	Conditions	Reference and Experimental Method
7.45	7.1 (29.7)	4.19(475 K)	300-650 K 30-100 torr	[125] R.N.Birrell and A.F.Trotman-Dickenson, J.Chem.Soc., 4218 (1960). Arrhenius parameters relative to log k_c = 9.5 for t-C_4H_9 combination; $\log (k_1/k_c^{\frac{1}{2}})(\ell^{\frac{1}{2}}mol^{-\frac{1}{2}}s^{-\frac{1}{2}}) = 2.7 - 7.1/2.3RT$. 'Aldehyde method'. For details see $CH_3 + CH_2{=}C{=}CH_2$. It should be noted that only four runs were carried out on this reaction.

PREFERRED RATE PARAMETERS

Reasonable as given above

COMMENTS

See summary of alkyl radical additions (page 219). The preferred Arrhenius parameters and the thermodynamic data yield for the reverse radical decomposition, $\log (k_{-1}/s^{-1}) = 13.7 - 25/2.3RT$ at 298 K, which is reasonable.

ΔH^o_f, ΔH^o and E in kcal mol^{-1} (kJ mol^{-1}); S^o, C^o_p, ΔS^o and ΔC^o_p in cal mol^{-1} K^{-1} (J mol^{-1} K^{-1}); A and k in l mol^{-1} s^{-1}

REACTION: T-BUTYL RADICAL + ACETYLENE $(CH_3)_3\dot{C}(I) + CH{\equiv}CH(II) \longrightarrow (CH_3)_3CCH{=}\dot{C}H(III)$

	(I)	(II)	(III)		
$\Delta H^\circ_{f(300)}$	6.7 (28.1)	54.2 (226.9)	42.2 (176.7)	ΔH°	-18.7 (-78.3)
$S^\circ_{(300)}$	74.6 (312.4)	48.0 (201.0)	[85 (356)]	ΔS°	[-38 (-159)]
$C^\circ_{p(300)}$	19.1 (80.0)	10.5 (44.0)	[31.8 (133.0)]	$\Delta C^\circ_{p(300)}$	[2.2 (9.2)]
$C^\circ_{p(500)}$	30.5 (127.7)	13.1 (54.3)	[49.2 (206.0)]	$\Delta C^\circ_{p(500)}$	[5.6 (23.4)]
				$\langle\Delta C^\circ_p\rangle$	[3.9 (16.2)]

log A	E	log k	Conditions	Reference and Experimental Method
6.81	5.3 (22.2)	4.34(470 K)	363-577 K 50-100 torr	[105] J.A.Garcia-Dominguez and A.F.Trotman-Dickenson, J.Chem.Soc., 940 (1962). Rate constants calculated from $\log (k_1/k_d^{\frac{1}{2}})(\ell^{\frac{1}{2}}mol^{-\frac{1}{2}}s^{-\frac{1}{2}}) = 1.88 - 5.3/2.3RT$ and $\log k_d = 9.86$ assuming $\log k_c = 9.5$ and $k_d/k_c = 2.3$. 'Aldehyde method'. For details see $CH_3 + CH_2{=}C{=}CH_2$. Some of the adduct radicals appear to isomerize, $(CH_3)_3CCH{=}\dot{C}H \longrightarrow (CH_3)_2\dot{C}CH_2CH{=}CH_2$, leading to the formation of the product $(CH_3)_2CHCH_2CH{=}CH_2$. It was not possible to analyse for the t-C_4H_9 radical dimer, and the concentration of radicals was monitored by the rate of formation of the disproportionation product, isobutene.

PREFERRED RATE PARAMETERS

$\log k = [8.0 - 7.7/2.3RT]$

COMMENTS

See comments on 'aldehyde method' in experimental section of $CH_3 + CH_2{=}C{=}CH_2$. A comparison of the above results with other alkyl additions to $CH{\equiv}CH$ (see page 219) indicates that the Arrhenius parameters of reference [105] are too low. The preferred activation energy was derived from the rate constant at 436 K, which seems reasonable, and assuming an A-factor of 10^8 ℓ $mol^{-1}s^{-1}$.

REACTION: T-BUTYL RADICAL + PROPYNE

$$(CH_3)_3\dot{C}(I) + CH_3C{\equiv}CH(II) \xrightarrow{1} (CH_3)_3CC(CH_3){=}\dot{C}H(III)$$
$$\xrightarrow{2} (CH_3)_3CCH{=}\dot{C}CH_3(IV)$$

	(I)	(II)	(III)	(IV)
$\Delta H^\circ_{f(300)}$	6.7 (28.1)	44.0 (184.2)	32.3 (135.2)	31.4 (131.6)
$S^\circ_{(300)}$	74.6 (312.4)	59.2 (247.9)	[95 (398)]	[94 (393)]
$C^\circ_{p(300)}$	19.1 (80.0)	14.6 (61.1)	[38 (159)]	[37 (155)]
$C^\circ_{p(500)}$	30.5 (127.7)	19.7 (82.5)	[58 (243)]	[57 (239)]

	Path 1	Path 2
ΔH°	-18.4 (-77.0)	-19.3 (-80.9)
ΔS°	[-39 (-163)]	[-40 (-167)]
$\Delta C^\circ_{p(300)}$	[4 (17)]	[3 (12)]
$\Delta C^\circ_{p(500)}$	[8 (33)]	[7 (29)]

Path	log A	E	log k	Conditions	Reference and Experimental Method
1 + 2	~5.7	~5.5 (~23.0)	~2.7(400 K)	360-439 K 30-70 torr	[106] R.R.Getty, J.A.Kerr and A.F.Trotman-Dickenson, J.Chem.Soc.(A), 1360 (1967). Rate constants calculated from log $(k_1/k_d^{\frac{1}{2}})(\ell^{\frac{1}{2}}mol^{-\frac{1}{2}}s^{-\frac{1}{2}})$ = 0.80 - 5.5/2.3RT and log k_d = 9.86 assuming log k_c = 9.5 and k_d/k_c = 2.3. 'Aldehyde method'. For details see $CH_3 + CH_2{=}C{=}CH_2$. Some of the adduct radicals appear to isomerize, $(CH_3)_3CCH{=}\dot{C}CH_3 \longrightarrow (CH_3)_2\dot{C}CH_2CH{=}CHCH_3$, leading to the formation of $(CH_3)_2CHCH_2CH{=}CHCH_3$. It was not possible to analyze for the t-butyl dimer, and the concentration of radicals was monitored by the rate of formation of the isobutene from the disproportionation reaction. The rate constants, and hence Arrhenius parameters, were considered to be approximate owing to the large number of possible reaction products, and since the rate constants were dependent upon the ratio of starting materials $[CH_3C{\equiv}CH]/[t\text{-}C_4H_9CHO]$. The above rate constants refer to $[CH_3C{\equiv}CH]/[t\text{-}C_4H_9CHO] \approx 0.5$. It was shown qualitatively that the addition of $t\text{-}C_4H_9$ radicals to propyne occurs predominantly at the terminal C atom (path 2) but no quantitative data were obtained.

PREFERRED RATE PARAMETERS

Above data suspect

COMMENTS

The approximate rate constants and Arrhenius parameters of reference [106] are very low compared with other data on alkyl additions to propyne and acetylene (see summary of alkyl additions, page 219). It is not possible to make a meaningful estimate of the parameters for this reason.

ΔH°_f, ΔH° and E in kcal mol^{-1} (kJ mol^{-1}); S°, C°_p, ΔS° and ΔC°_p in cal mol^{-1} K^{-1} (J mol^{-1} K^{-1}); A and k in l mol^{-1} s^{-1}

REACTION: ISO-PENTYL RADICAL + ETHYLENE $(CH_3)_2CHCH_2\dot{C}H_2(I) + CH_2{=}CH_2(II) \longrightarrow (CH_3)_2CHCH_2CH_2CH_2\dot{C}H_2(III)$

	(I)	(II)	(III)		
$\Delta H^\circ_{f(300)}$	8.8 (36.8)	12.5 (52.3)	-1.1 (-4.6)	ΔH°	-22.4 (-93.8)
$S^\circ_{(300)}$	85.0 (355.9)	52.4 (219.4)	103.8 (434.6)	ΔS°	-33.6 (-140.7)
$C^\circ_{p(300)}$	28.4 (118.9)	10.3 (43.1)	39.4 (165.0)	$\Delta C^\circ_{p(300)}$	0.7 (2.9)
$C^\circ_{p(500)}$	42.5 (177.8)	14.9 (62.4)	59.0 (246.8)	$\Delta C^\circ_{p(500)}$	1.6 (6.7)
				$\langle\Delta C^\circ_p\rangle$	1.2 (5.0)

log A	E	log k	Conditions	Reference and Experimental Method
7.09	6.43 (26.92)	3.36(377 K)	340-413 K 1-110 torr	[110] K.W.Watkins and L.A.O'Deen, J.Phys.Chem., 73, 4094 (1969). Results calculated from $\log(k_1k_c^{\frac{1}{2}}/k_4)(\ell^{\frac{1}{2}}mol^{-\frac{1}{2}}s^{-\frac{1}{2}}) = 2.09 - 6.43/2.3RT$ and $\log k_c = 10.0$ for iso-C_3H_7 combination and $\log k_4 = 10.0$ for iso-C_5H_{11} + iso-$C_3H_7 \longrightarrow C_8H_{18}$. For details see $C_2H_5 + CH_2{=}CH_2$. iso-C_5H_{11} radicals produced in a study of the iso-C_3H_7 addition to $CH_2{=}CH_2$.

PREFERRED RATE PARAMETERS

Tentatively as given above

COMMENTS

See summary of alkyl radical additions (page 219) and comments. The above Arrhenius parameters and the thermodynamic data yield for the reverse radical decomposition, $\log(k_{-1}/s^{-1}) = 12.6 - 28.2/2.3RT$ at 298 K. The calculated value of $\log A_{-1}$ is well below the minimum set by transition state theory, $\log(A_{-1}/s^{-1}) = 13.2$ at 298 K, indicating that A_1 has probably been under-estimated.

TABLE VI. SUMMARY OF KINETIC PARAMETERS* FOR ALKYL RADICAL ADDITIONS

C_1 - C_3 RADICALS

	CH_3			C_2H_5			n-C_3H_7			i-C_3H_7		
Substrate	log A	E	log k	log A	E	log k	log A	E	log k	log A	E	log k
=	8.52	7.7	4.67	8.30	7.5	4.55	[7.3	6.1	4.2]	7.83	6.9	4.38
=—	8.22	7.4	4.52	-	-	-	-	-	-	-	-	-
=——	8.01	7.2	4.41	-	7.3	-	-	7.2	-	-	6.9	-
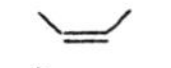	7.64	7.5	3.89	-	8.5	-	-	8.7	-	-	7.9	-
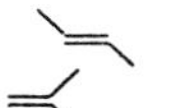	8.15	8.1	4.10	-	8.6	-	-	9.0	-	-	8.1	-
	8.15	6.9	4.70	-	-	-	-	-	-	-	-	-
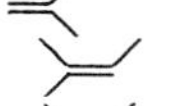	7.13	6.1	4.08	-	-	-	-	-	-	-	-	-
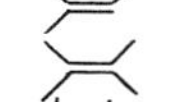	7.01	6.8	3.61	-	-	-	-	-	-	-	-	-
	-	-	-	6.89	5.6	4.09	-	-	-	-	-	-
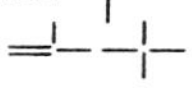	-	-	-	7.29	5.7	4.44	-	-	-	-	-	-
=————	-	-	-	7.59	6.8	4.19	-	-	-	-	-	-
=—————	-	-	-	7.79	7.0	4.29	-	-	-	-	-	-
=——————	-	-	-	7.59	6.7	4.24	-	-	-	-	-	-
CF_2=CF_2	8.05	5.2	5.45	-	-	-	-	-	-	-	-	-
CH_2=CHCℓ	[8.3	6.2	5.2]	-	-	-	-	-	-	-	-	-
==	8.3	8.1	4.25	[8.5	9.2	3.9]	-	-	-	[7.6	7.2	4.0]
=—=	7.91	4.1	5.86	-	-	-	-	-	-	-	-	-
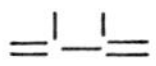	-	-	-	8.23	4.5	5.98	-	-	-	-	-	-
	-	-	-	8.14	5.4	5.44	-	-	-	8.39	5.8	5.49
≡	8.4	7.7	4.55	[7.7	7.0	4.2]	-	-	-	[7.7	6.9	4.3]
≡—	8.7	8.8	4.3	-	-	-	-	-	-	[6.3	5.7	3.4]
≡—————	-	-	-	8.59	8.8	4.19	-	-	-	-	-	-

C_4 & C_5 RADICALS

	n-C_4H_9			s-C_4H_9			t-C_4H_9			i-C_5H_{11}		
Substrate	log A	E	log k	log A	E	log k	log A	E	log k	log A	E	log k
=	7.37	6.7	4.02	-	-	-	7.45	7.1	3.90	[7.09	6.4	3.89]
=—	-	-	-	[7.8	7.4	4.1]	-	-	-	-	-	-
≡	-	-	-	-	-	-	[6.8	5.3	4.2]	-	-	-
≡—	-	-	-	-	-	-	[5.7	5.5	3.0]	-	-	-

*log A and log k (437 K) in ℓ $mol^{-1}s^{-1}$, E in kcal mol^{-1}; values are those 'preferred' in previous tables or, if in parentheses, they are tentative or suspect.

COMMENTS (see over)

COMMENTS

Although the Table is far from complete, one or two interesting features emerge. The Arrhenius parameters for alkyl addition to CH_2=CH_2 all fall within the ranges log A = 7.8 ± 0.7 and E = 7 ± 0.7, and within the experimental errors (log A ⩾ ± 0.5 and E ⩾ ± 1) are essentially equal. A comparison of the rate constants, which are more accurate, indicated that there is a decrease in reactivity in going from CH_3 to i-C_5H_{11}, but the present experimental errors are too large to reveal the small differences in the Arrhenius parameters. It seems likely that this trend will persist for alkyl additions to other unsaturates, i.e. any slight variation in the Arrhenius parameters being in the direction of decreased reactivity as the alkyl radical size increases, and accordingly we have used this idea in evaluating alkyl additions to other substrates. It would be very interesting to have kinetic data on the additions of large alkyl radicals to highly substituted ethylenes.

REACTION: DIFLUOROMETHYLENE RADICAL + TETRAFLUOROETHYLENE - $CF_2(I) + CF_2{=}CF_2(II) \longrightarrow CF_2CF_2CF_2(III)$

	(I)	(II)	(III)	
$\Delta H^\circ_{f(300)}$	-41 (-172)	-157.9 (-661.1)		ΔH°
$S^\circ_{(300)}$	57.5 (241.0)	74.6 (312.0)		ΔS°
$C^\circ_{p(300)}$	9.3 (39.0)	19.4 (81.2)		ΔC°_p

log A	E	log k	Conditions	Reference and Experimental Method
6.81	7.5 (31.4)	2.66(395 K)	294-497 K 0.8-319 torr	[125a] N.Cohen and J.Heicklen, J.Chem.Phys., 43, 871 (1965).

Static system. Hg-photosensitized decomposition of $CF_2{=}CF_2$. At temperatures below 497 K, the only product was perfluorocyclopropane, analyzed by gas chromatography. Quantum yields (Φ) of product measured as a function of $[CF_2{=}CF_2]$ and temperature. Proposed mechanism:

$$Hg^* + C_2F_4 \xrightarrow{I} Hg + C_2F_4^*$$
$$C_2F_4^* + C_2F_4 \xrightarrow{2} 2C_2F_4$$
$$C_2F_4^* \xrightarrow{3} 2CF_2^*$$
$$2CF_2 \xrightarrow{4} C_2F_4$$
$$CF_2 + C_2F_4 \xrightarrow{1} \underline{c}\text{-}C_3F_6$$

It follows that:

$$\Phi(c\text{-}C_3F_6) = [k_1/(I_a k_4^{\frac{1}{2}})][C_2F_4] \quad \text{if } R_4 >> R_1$$

and $$\Phi(c\text{-}C_3F_6) = 2[k_3/(k_3 + k_2[C_2F_4])] \quad \text{if } R_1 >> R_4$$

where R = rate of reaction. $k_1/k_4^{\frac{1}{2}}I_a^{\frac{1}{2}}$ obtained from intercepts of plots of $\log \Phi(c\text{-}C_3F_6)$ versus $\log [C_2F_4]$ and from Arrhenius plot $\log (k_1/k_4^{\frac{1}{2}})$ $(\ell^{\frac{1}{2}}mol^{-\frac{1}{2}}s^{-\frac{1}{2}}) = 2.60 - 6.7/2.3RT$. Above Arrhenius parameters calculated taking $\log k_4 = 8.41 - 1.6/2.3RT$ [F.W.Daby, J.Chem.Phys., 41, 2297 (1964)].

ΔH^o_f, ΔH^o and E in kcal mol^{-1} (kJ mol^{-1}); S^o, C^o_p, ΔS^o and ΔC^o_p in cal mol^{-1} K^{-1} (J mol^{-1} K^{-1}); A and k in l mol^{-1} s^{-1}

PREFERRED RATE PARAMETERS

Reasonable as given above

COMMENTS

The rate of the reverse reaction has been measured, and Benson and O'Neal [NSRDS-NBS 21, 1970] recommend log $(k_{-1}/s^{-1}) = 15.0 - 43.0/2.3RT$. Unfortunately, it is not possible to make an a priori estimate of the thermodynamic properties of perfluorocyclopropane as the corrections for ring strain are unknown, and hence the kinetics of the forward and reverse reactions cannot be compared. Benson and O'Neal suggest an activation energy of 8 kcal mol^{-1} for the addition reaction, which agrees well with the above value. It was proposed that the singlet CF_2 radical is involved in the addition reaction [125a].

REACTION: TRIFLUOROMETHYL RADICAL + ETHYLENE $\dot{C}F_3(I) + CH_2{=}CH_2(II) \longrightarrow CF_3CH_2\dot{C}H_2(III)$

	(I)	(II)	(III)		
$\Delta H^\circ_{f(300)}$	-112.5 (-471.0)	12.5 (52.3)	-127.5 (-533.8)	ΔH°	-27.5 (-115.1)
$S^\circ_{(300)}$	63.8 (267.1)	52.4 (219.4)	80.9 (338.7)	ΔS°	-35.3 (-147.8)
$C^\circ_{p(300)}$	12.2 (51.1)	10.3 (43.1)	24.2 (101.3)	$\Delta C^\circ_{p(300)}$	1.7 (7.1)
$C^\circ_{p(500)}$	15.2 (63.6)	14.9 (62.4)	32.9 (137.7)	$\Delta C^\circ_{p(500)}$	2.8 (11.7)
				$\langle\Delta C^\circ_p\rangle$	2.3 (9.6)

log A	E	log k	Conditions	Reference and Experimental Method
8.21	0.95 (4.0)	7.68(395 K)	338-453 K 700 torr	[126] J.M.Pearson and M.Szwarc, Trans.Faraday Soc., 60, 553 (1964).

Results calculated from log $(A_1/A_3) = 0.37$ and $E_1 - E_3 = -2.35$ kcal mol^{-1}, taking log $k_3 = 7.84 - 3.3/2.3RT$ [J.M.Sangster and J.C.J.Thynne, Int.J. Chem.Kinet., 1, 571 (1969)].

Static system. CF_3 from photolysis of $CF_3N{=}NCF_3$ (full high-pressure Hg arc, Pyrex reaction vessel). Photolysis carried out in presence of alkene and reactive reference hydrocarbon, RH = 2,3-dimethylbutane. Rate constants derived from a mass balance between CF_3 and N_2. In the absence of alkene and with sufficiently high [RH], all CF_3 react by

$$\dot{C}F_3 + RH \xrightarrow{3} CF_3H + \dot{R}$$

and $CF_3H/N_2 = 2$, the products CF_3H and N_2 being analyzed by gas chromatography. In the presence of the alkene, the ratio falls below 2 owing to the reaction

$$\dot{C}F_3 + O\ell \xrightarrow{1} CF_3\dot{O}\ell$$

and the rate constants are calculated from the equation:

ΔH°_f, ΔH° and E in kcal mol^{-1} (kJ mol^{-1}); S°, C°_p, ΔS° and ΔC°_p in cal mol^{-1} K^{-1} (J mol^{-1} K^{-1}); A and k in l mol^{-1} s^{-1}

log A	E	log k	Conditions	Reference and Experimental Method
				$k_1/k_3 = ([RH]/[O\ell])\{(CF_3H/N_2)_o - (CF_3H/N_2)_{expt}\}/(CF_3H/N_2)_{expt}$ in which $(CF_3H/N_2)_o$ and $(CF_3H/N_2)_{expt}$ are the product ratios in experiments without and with alkene added respectively. The method has been criticised on the grounds that, with alkenes containing allylic H atoms, CF_3 attack on the olefin could be an additional source of CF_3H [S.E.Braslavsky, F.Casas and O.Cifuentes, J.Chem.Soc.,(B), 1059 (1970)]. In reference [126], however, $CF_3N{=}NCF_3$(0.2 - 0.6 torr), was photolyzed at 338 and 453 K in presence of 200 torr of isobutene, and neither CF_3H nor C_2F_6 was detected in the products, which indicates that the CF_3 addition reaction is very much more rapid than either the H-abstraction or radical combination under the conditions of the addition studies. For most of the alkene systems it was shown that the CF_3 addition reaction was irreversible by observing k_1 to be independent of total pressure of the system (p_t > 500 torr) ; specific instances of reversible additions are indicated in the experimental sections of the appropriate alkene systems.
8.45	2.78 (11.6)	6.92(395 K)	291-474 K 30-90 torr	[127] J.M.Sangster and J.C.J.Thynne, J.Phys.Chem., 73, 2746 (1969). Results calculated from $\log (k_4/k_1) = 0.26 - 1.41/2.3RT$ and $\log k_4 = 8.81 - 4.19/2.3RT$ [N.L.Arthur and P.Gray, Trans.Faraday Soc., 65, 434 (1969)]. Static system. CF_3 from full Hg arc photolysis of CF_3COCF_3 in presence of alkene and H_2S. With ratios $[H_2S]/[C_2H_4] > 10$, the following simplified reaction scheme accounts for the products: $\dot{C}F_3 + H_2S \xrightarrow{4} CF_3H + H\dot{S}$ $\dot{C}F_3 + CH_2{=}CH_2 \xrightarrow{1} CF_3CH_2\dot{C}H_2$ $CF_3CH_2\dot{C}H_2 + H_2S \xrightarrow{5} CF_3CH_2CH_3 + H\dot{S}$ $CF_3CH_2\dot{C}H_2 + CH_2{=}CH_2 \xrightarrow{6} CF_3CH_2CH_2CH_2\dot{C}H_2$ $CF_3CH_2CH_2CH_2\dot{C}H_2 + H_2S \xrightarrow{7} CF_3CH_2CH_2CH_2CH_3 + H\dot{S}$ Under conditions of high $[H_2S]$ there was no evidence for radical-radical reactions, and the rates of addition of CF_3 and $CF_3CH_2CH_2$ to C_2H_4 were monitored by the rates of formation of $CF_3CH_2CH_3$ and $CF_3(CH_2)_3CH_3$. Results calculated from $\log (k_4/k_1) = 0.26 - 1.41/2.3RT$ and $\log k_4(\ell\ mol^{-1}s^{-1}) = 8.81 - 4.19/2.3RT$ [N.L.Arthur and P.Gray, Trans.Faraday Soc., 65, 434 (1969)]. It should be noted that much lower Arrhenius parameters for reaction (4) have

log A	E	log k	Conditions	Reference and Experimental Method
				also been reported [Ratajczak and Trotman-Dickenson, "Supplementary Tables of Bimolecular Gas Reactions", University of Wales Institute of Science and Technology, Cardiff, 1970].

PREFERRED RATE PARAMETERS

log k = 8.30 - 2.0/2.3RT

COMMENTS

The results of references [126] and [127] are in agreement within the experimental errors, which are large since the errors in each of the results contain a considerable uncertainty from the reference reaction. There is no reason for selecting one set of data in preference to the other, and accordingly the preferred result represents a compromise. In conjunction with the thermodynamic data, the preferred rate constant yields log (k_{-1}/s^{-1}) = 14.2 - 28.9/2.3RT at 298 K for the reverse radical decomposition, $CF_3CH_2\dot{C}H_2 \xrightarrow{-1} \dot{C}F_3 + CH_2{=}CH_2$, which is reasonable in relation to the data for the decomposition of the n-C_3H_7 radical (see $CH_3 + CH_2{=}CH_2$ system).

REACTION: TRIFLUOROMETHYL RADICAL + PROPENE

$$\dot{C}F_3(I) + CH_2{=}CHCH_3(II) \xrightarrow{1} \dot{C}H_2CH(CF_3)CH_3(III)$$
$$\xrightarrow{2} CF_3CH_2\dot{C}HCH_3(IV)$$

	(I)	(II)	(III)	(IV)
$\Delta H^o_{f(300)}$	-112.5 (-471.0)	4.9 (20.5)	-134.6 (-563.6)	-135.5 (-567.3)
$S^o_{(300)}$	63.8 (267.1)	63.8 (267.1)	87.6 (366.8)	88.7 (371.4)
$C^o_{p(300)}$	12.2 (51.1)	15.3 (64.1)	29.4 (123.1)	29.6 (123.9)
$C^o_{p(500)}$	15.2 (63.6)	22.6 (94.7)	41.3 (172.9)	40.9 (171.3)

	Path 1	Path 2
ΔH^o	-27.0 (-113.0)	-28.9 (-121.0)
ΔS^o	-40.0 (-167.5)	-38.9 (-162.9)
$\Delta C^o_{p(300)}$	1.9 (8.0)	2.1 (8.8)
$\Delta C^o_{p(500)}$	3.5 (14.7)	3.1 (12.8)
$\langle \Delta C^o_p \rangle$	2.7 (11.3)	2.6 (10.9)

Path	log A	E	log k	Conditions	Reference and Experimental Method
1 + 2	-	-	7.68	338 K 650 torr	[128] P.S.Dixon and M.Szwarc, Trans.Faraday Soc., 59, 112 (1963). Results calculated for log (k_{1+2}/k_3) = 1.97 at 338 K and log k_3 = 7.84 - 3.3/2.3RT. For details see $CF_3 + CH_2{=}CH_2$, reference [126]. $\dot{C}F_3 + RH \xrightarrow{3} CF_3H + \dot{R}$ $\dot{C}F_3 + O\ell \xrightarrow{1+2} CF_3\dot{O}\ell$

ΔH^o_f, ΔH^o and E in kcal mol^{-1} (kJ mol^{-1}); S^o, C^o_p, ΔS^o and ΔC^o_p in cal mol^{-1} K^{-1} (J mol^{-1} K^{-1}); A and k in l mol^{-1} s^{-1}

Path	log A	E	log k	Conditions	Reference and Experimental Method
1 + 2	8.02	0.54 (2.26)	7.72(395 K)	338-453 K 700 torr	[126] J.M.Pearson and M.Szwarc, Trans.Faraday Soc., 60, 553 (1964) Results calculated from log (A_{1+2}/A_3) = 0.18 and $E_{1+2} - E_3$ = -2.76 kcal mol^{-1}, taking log k_3 = 7.84 - 3.3/2.3RT. For details see $CF_3 + CH_2{=}CH_2$, reference [126].

PREFERRED RATE PARAMETERS

Reasonable as given in reference [126]

COMMENTS

Path 2 will be the favoured mode of addition, but the above method yields only overall rate constants for paths 1 and 2.

REACTION: TRIFLUOROMETHYL RADICAL + CIS-BUTENE-2

$\dot{C}F_3$(I) + cis-$CH_3CH{=}CHCH_3$(II) $\longrightarrow$ $CH_3(CF_3)CH\dot{C}HCH_3$(III)

	(I)	(II)	(III)
$\Delta H^\circ_{f(300)}$	-112.5 (-471.0)	-2.0 (-8.4)	-142.5 (-596.6)
$S^\circ_{(300)}$	63.8 (267.1)	73.6 (308.2)	95.4 (399.4)
$C^\circ_{p(300)}$	12.2 (51.1)	19.4 (81.2)	34.8 (145.7)
$C^\circ_{p(500)}$	15.2 (63.6)	29.6 (123.9)	49.2 (206.0)

ΔH°	-28.0 (-117.2)
ΔS°	-42.0 (-175.9)
$\Delta C^\circ_{p(300)}$	3.2 (13.4)
$\Delta C^\circ_{p(500)}$	4.4 (18.4)
$\langle \Delta C^\circ_p \rangle$	3.8 (15.9)

log A	E	log k	Conditions	Reference and Experimental Method
7.44	-0.15 (-0.63)	7.52(395 K)	338-453 K 700 torr	[129] G.E.Owen, J.M.Pearson and M.Szwarc, Trans.Faraday Soc., 61, 1722 (1965). Results calculated from log (A_1/A_3) = -0.40 and $E_1 - E_3$ = -3.45 kcal mol^{-1}, taking log k_3 = 7.84 - 3.3/2.3RT. Same method as for $CF_3 + CH_2{=}CH_2$ system, reference [126]. $\dot{C}F_3 + RH \longrightarrow CF_3H + \dot{R}$ $\dot{C}F_3 + O\ell \longrightarrow CF_3\dot{O}\ell$

PREFERRED RATE PARAMETERS

log k = 7.5

COMMENTS

The slightly negative E, calculated above, indicates that there are experimental errors in the determination of the Arrhenius parameters for the addition reaction, or the reference reaction, or both. We have assigned the addition reaction $E_1 \approx 0$ kcal mol^{-1}.

REACTION: TRIFLUOROMETHYL RADICAL + TRANS-BUTENE-2. $\dot{C}F_3$(I) + trans-$CH_3CH{=}CHCH_3$(II) ⟶ $CH_3(CF_3)CH\dot{C}HCH_3$(III)

	(I)	(II)	(III)		
$\Delta H^\circ_{f(300)}$	-112.5 (-471.0)	-3.0 (-12.6)	-142.5 (-596.6)	ΔH°	-27.0 (-113.0)
$S^\circ_{(300)}$	63.8 (267.1)	72.4 (303.1)	95.4 (399.4)	ΔS°	-40.8 (-170.8)
$C^\circ_{p(300)}$	12.2 (51.1)	20.7 (86.7)	34.8 (145.7)	$\Delta C^\circ_{p(300)}$	1.9 (8.0)
$C^\circ_{p(500)}$	15.2 (63.6)	30.4 (127.3)	49.2 (206.0)	$\Delta C^\circ_{p(500)}$	3.6 (15.1)
				$\langle\Delta C^\circ_p\rangle$	2.8 (11.7)

log A	E	log k	Conditions	Reference and Experimental Method
7.44	-0.16 (-0.67)	7.53(395 K)	338-453 K 700 torr	[129] G.E.Owen, J.M.Pearson and M.Szwarc, Trans.Faraday Soc., 61, 1722 (1965). Results calculated from log $(A_1/A_3) = -0.40$ and $E_1 - E_3 = -3.46$ kcal mol^{-1}. taking log $k_3 = 7.84 - 3.3/2.3RT$ (see $CF_3 + CH_2{=}CH_2$ system, reference [126]). Same method as for $CF_3 + CH_2{=}CH_2$ system, reference [126]. $\dot{C}F_3 + RH \xrightarrow{3} CF_3H + \dot{R}$ $\dot{C}F_3 + O\ell \xrightarrow{1} CF_3\dot{O}\ell$

PREFERRED RATE PARAMETERS

log k = 7.5

COMMENTS

The slightly negative E, calculated above, indicates that there are experimental errors in the determination of the Arrhenius parameters for the addition reaction or the reference reaction or both. We have assigned the addition reaction $E \approx 0$ kcal mol^{-1}.

ΔH°_f, ΔH° and E in kcal mol^{-1} (kJ mol^{-1}); S°, C°_p, ΔS° and ΔC°_p in cal mol^{-1} K^{-1} (J mol^{-1} K^{-1}); A and k in l mol^{-1} s^{-1}

REACTION: TRIFLUOROMETHYL RADICAL + 2-METHYLPROPENE (ISOBUTENE)

$\dot{C}F_3(I) + CH_2{=}C(CH_3)_2(II) \xrightarrow{1} \dot{C}H_2C(CF_3)(CH_3)_2(III)$

$\xrightarrow{2} CF_3CH_2\dot{C}(CH_3)_2(IV)$

	(I)	(II)	(III)	(IV)		Path 1	Path 2
$\Delta H^\circ_{f(300)}$	-112.5 (-471.0)	-3.6 (-15.1)	-139.5 (-584.1)	-145.5 (-609.2)	ΔH°	-23.4 (-98.0)	-29.4 (-123.1)
$S^\circ_{(300)}$	63.8 (267.1)	71.3 (298.4)	92.8 (388.6)	95.4 (399.4)	ΔS°	-42.3 (-177.1)	-39.7 (-166.2)
$C^\circ_{p(300)}$	12.2 (51.1)	21.6 (90.4)	35.4 (148.2)	34.6 (144.9)	$\Delta C^\circ_{p(300)}$	1.6 (6.7)	0.8 (3.3)
$C^\circ_{p(500)}$	15.2 (63.6)	31.3 (131.1)	50.9 (213.1)	48.9 (204.7)	$\Delta C^\circ_{p(500)}$	4.4 (18.4)	2.4 (10.0)
					$\langle\Delta C^\circ_p\rangle$	3.0 (12.6)	1.6 (6.7)

Path	log A	E	log k	Conditions	Reference and Experimental Method
1 + 2	-	-	8.15	338 K 650 torr	[128] P.S.Dixon and M.Szwarc, Trans.Faraday Soc., 59, 112 (1963). Results calculated from log (k_{1+2}/k_3) = 2.44 at 338 K and log k_3 = 7.84 - 3.3/2.3RT. Same method as for CF_3 + $CH_2{=}CH_2$ system, reference [126]. $\dot{C}F_3 + RH \xrightarrow{3} CF_3H + \dot{R}$ $\dot{C}F_3 + O\ell \xrightarrow{1+2} CF_3\dot{O}\ell$
1 + 2	8.00	-0.22 (-0.9)	8.12(395 K)	338-453 K 700 torr	[126] J.M.Pearson and M.Szwarc, Trans.Faraday Soc., 60, 553 (1964). Results calculated from log (A_{1+2}/A_3) = 0.16 and $E_{1+2} - E_3$ = 3.52 kcal mol^{-1}, taking log k_3 = 7.84 - 3.3/2.3RT. Same method as for [128] above. For details see CF_3 + $CH_2{=}CH_2$.

PREFERRED RATE PARAMETERS

log k_{1+2} = 8.12

COMMENTS

The slightly negative E, calculated above, indicates experimental errors in the Arrhenius parameters for the addition reaction, or the reference reaction, or both. We have assigned the addition reaction E ≃ 0 kcal mol^{-1}. Path 2 will be the favoured mode of addition.

REACTION: TRIFLUOROMETHYL RADICAL + 2,3-DIMETHYLBUTENE-2 (TETRAMETHYLETHYLENE)

$\dot{C}F_3(I) + (CH_3)_2C{=}C(CH_3)_2(II) \longrightarrow (CH_3)_2(CF_3)C\dot{C}(CH_3)_2(III)$

	(I)	(II)	(III)
$\Delta H^\circ_{f(300)}$	-112.5 (-471.0)	-19.6 (-82.1)	-160.1 (-670.3)
$S^\circ_{(300)}$	63.8 (267.1)	87.4 (365.9)	107.3 (449.3)
$C^\circ_{p(300)}$	12.2 (51.1)	33.0 (138.2)	45.9 (192.2)
$C^\circ_{p(500)}$	15.2 (63.6)	47.6 (199.3)	66.8 (279.7)

ΔH°	-28.0 (-117.2)
ΔS°	-43.9 (-183.8)
$\Delta C^\circ_{p(300)}$	0.7 (2.9)
$\Delta C^\circ_{p(500)}$	4.0 (16.7)
$\langle \Delta C^\circ_p \rangle$	2.4 (10.0)

log A	E	log k	Conditions	Reference and Experimental Method
-	-	7.71	338 K 650 torr	[128] P.S.Dixon and M.Szwarc, Trans.Faraday Soc., 59, 112 (1963). Results calculated from log (k_1/k_3) = 2.00, taking log k_3 = 7.84 - 3.3/2.3RT. Same method as for $CF_3 + CH_2{=}CH_2$ system, reference [126]. $\dot{C}F_3 + RH \xrightarrow{3} CF_3H + \dot{R}$ $\dot{C}F_3 + O\ell \xrightarrow{1} CF_3\dot{O}\ell$
7.32	-0.65 (-2.7)	7.68(395 K)	338-453 K 700 torr	[129] G.E.Owen, J.M.Pearson and M.Szwarc, Trans.Faraday Soc., 61, 1722 (1965). Results calculated from log (A_1/A_3) = -0.52 and $E_1 - E_3$ = -3.95 kcal mol^{-1}, taking log k_3 = 7.84 - 3.3/2.3RT. Same method as for reference [128] above. For details see $CF_3 + CH_2{=}CH_2$, reference [126].

PREFERRED RATE PARAMETERS

log k = 7.8

COMMENTS

The slightly negative activation energy calculated above would seem to indicate experimental errors in the Arrhenius parameters of the addition reaction, or the reference reaction, or both. We have assigned the addition reaction $E \simeq 0$ kcal mol^{-1}.

ΔH°_f, ΔH° and E in kcal mol^{-1} (kJ mol^{-1}); S°, C°_p, ΔS° and ΔC°_p in cal mol^{-1} K^{-1} (J mol^{-1} K^{-1}); A and k in l mol^{-1} s^{-1}

REACTION: TRIFLUOROMETHYL RADICAL + STYRENE

$$\dot{C}F_3(I) + CH_2{=}CHC_6H_5(II) \xrightarrow{1} CF_3CH_2\dot{C}HC_6H_5(III)$$
$$\xrightarrow{2} \dot{C}H_2CH(CF_3)C_6H_5(IV)$$

	(I)	(II)	(III)	(IV)		Path 1	Path 2
$\Delta H^\circ_{f(300)}$	-112.5 (-471.0)	35.2 (147.4)	[-120 (-503)]	-101.6 (-425.4)	ΔH°	[-42 (-176)]	-24.3 (-101.7)
$S^\circ_{(300)}$	63.8 (267.1)	83.8 (350.9)	106.0 (443.8)	[109 (457)]	ΔS°	-41.6 (-174.2)	[-38 (-159)]
$C^\circ_{p(300)}$	12.2 (51.1)	29.4 (123.1)	42.4 (177.5)	[42.4 (177.5)]	ΔC°_p	0.8 (3.3)	[0.8 (3.3)]

Path	log A	E	log k	Conditions	Reference and Experimental Method
1 + 2	8.23	-0.15 (-0.63)	8.31(395 K)	338-448 K 700 torr	[126] J.M.Pearson and M.Szwarc, Trans.Faraday Soc., 60, 553 (1964). Results calculated from log (A_{1+2}/A_3) = 0.39 and $E_{1+2} - E_3$ = 3.45 kcal mol^{-1}, and taking log k_3 = 7.84 - 3.3/2.3RT for the reaction. $\dot{C}F_3 + RH \xrightarrow{3} CF_3H + \dot{R}$; $\dot{C}F_3 + O\ell \xrightarrow{1+2} CF_3\dot{O}\ell$. For details see $CF_3 + CH_2{=}CH_2$.

PREFERRED RATE PARAMETERS

log k_{1+2} = 8.3

COMMENTS

The preferred value is based on assigning E ≈ 0 kcal mol^{-1} to the reaction. The addition to the side chain, which will largely involve path 1, is so fast that addition to the ring can be neglected.

REACTION: TRIFLUOROMETHYL RADICAL + α METHYL STYRENE

$$\dot{C}F_3(I) + CH_2{=}C(CH_3)C_6H_5(II) \xrightarrow{1} CF_3CH_2\dot{C}(CH_3)C_6H_5(III)$$
$$\xrightarrow{2} \dot{C}H_2C(CH_3)(CF_3)C_6H_5(IV)$$

	(I)	(II)	(III)	(IV)
$\Delta H^\circ_{f(300)}$	-112.5 (-471.0)	27.0 (113.0)	[-132 (-553)]	-107.8 (-451.4)
$S^\circ_{(300)}$	63.8 (267.1)	91.1 (381.4)	112.4 (470.6)	[115 (481)]
$C^\circ_{p(300)}$	12.2 (51.1)	35.5 (148.6)	48.0 (201.0)	[48.1 (201.4)]

	Path 1	Path 2
ΔH°	[-46 (-193)]	-22.3 (-93.4)
ΔS°	-42.5 (-177.9)	[-40 (-167)]
ΔC°_p	0.3 (1.3)	0.4 (1.7)]

Path	log A	E	log k	Conditions	Reference and Experimental Method
1 + 2	8.17	-0.53 (-2.2)	8.46(395 K)	338-448 K 700 torr	[126] J.M.Pearson and M.Szwarc, Trans.Faraday Soc., 60, 553 (1964). Results calculated from log (A_{1+2}/A_3) = 0.33 and $E_{1+2} - E_3$ = -3.86 kcal mol^{-1}, and taking log k_3 = 7.84 - 3.3/2.3RT for the reaction $\dot{C}F_3 + RH \xrightarrow{3} CF_3H + \dot{R}$; $\dot{C}F_3 + Ol \xrightarrow{1+2} CF_3\dot{O}l$. For details see $CF_3 + CH_2{=}CH_2$.

PREFERRED RATE PARAMETERS

log k_{1+2} = 8.5

COMMENTS

The preferred value is based on assigning E ≈ 0 kcal mol^{-1} to the addition reaction. The addition to the side chain, which will largely involve path 1, is so fast that addition to the ring can be neglected.

ΔH^o_f, ΔH^o and E in kcal mol^{-1} (kJ mol^{-1}); S^o, C^o_p, ΔS^o and ΔC^o_p in cal mol^{-1} K^{-1} (J mol^{-1} K^{-1}); A and k in l mol^{-1} s^{-1}

REACTION: TRIFLUOROMETHYL RADICAL + VINYL FLUORIDE

$$\dot{C}F_3(I) + CH_2{=}CHF(II) \xrightarrow{1} CF_3CH_2\dot{C}HF(III)$$
$$\xrightarrow{2} \dot{C}H_2CHFCF_3(IV)$$

	(I)	(II)	(III)	(IV)		Path 1	Path 2
$\Delta H^\circ_{f(300)}$	-112.5 (-471.0)	-31.6 (-132.3)	[-167 (-700)]	-171.0 (-716.0)	ΔH°	[-23 (-96)]	-26.9 (-112.6)
$S^\circ_{(300)}$	63.8 (267.1)	60.4 (252.9)	86.5 (361.2)	[88 (368)]	ΔS°	-37.7 (-158.7)	[-36 (-151)]
$C^\circ_{p(300)}$	12.2 (51.1)	11.9 (49.8)	26.3 (110.0)		ΔC°_p	2.2 (9.2)	

Path	log A	E	log k	Conditions	Reference and Experimental Method
1 + 2	-	-	6.56	338 K 650 torr	[128] P.S.Dixon and M.Szwarc, Trans.Faraday Soc., 59, 112 (1963). Results calculated from log (k_{1+2}/k_3) = 0.85, taking log k_3 = 7.84 - 3.3/2.3RT. Same method as for $CF_3 + CH_2{=}CH_2$, reference [126] $\dot{C}F_3 + RH \xrightarrow{3} CF_3H + \dot{R}$ $\dot{C}F_3 + O\ell \xrightarrow{1+2} CF_3\dot{O}\ell$ For details see $CF_3 + CH_2{=}CH_2$, reference [126].
1 + 2	8.03	2.28 (9.55)	6.90(395 K)	338-453 K 700 torr	[126] J.M.Pearson and M.Szwarc, Trans.Faraday Soc., 60, 553 (1964). Results calculated from log (A_{1+2}/A_3) = 0.19 and $E_{1+2} - E_3$ = -1.02 kcal mol^{-1}, taking log k_3 = 7.84 - 3.3/2.3RT. Same method as for reference [128] above. For details see CF_3 + $CH_2{=}CH_2$.

PREFERRED RATE PARAMETERS

Reasonable as given above

COMMENTS

The experimental method does not distinguish between addition by path 1 or 2, although Szwarc and co-workers [Trans.Faraday Soc., 61, 1722 (1965)] have argued that addition occurs exclusively by path 2 on the basis of comparisons with the kinetic results for other CF_3 additions. This suggestion is supported by the above thermochemical estimates of ΔH_1 and ΔH_2.

REACTION: TRIFLUOROMETHYL RADICAL + 1,1-DIFLUOROETHYLENE $\dot{C}F_3(I) + CF_2{=}CH_2(II) \xrightarrow{1} CF_3CF_2\dot{C}H_2(III)$

$\xrightarrow{2} CF_3CH_2\dot{C}F_2(IV)$

	(I)	(II)	(III)	(IV)
$\Delta H^o_{f(300)}$	-112.5 (-471.0)	[-71.5 (-299.6)]	[-220 (-921)]	[-216 (-904)]
$S^o_{(300)}$	63.8 (267.1)	64.9 (271.7)	89.3 (373.9)	90.2 (377.7)
$C^o_{p(300)}$	12.2 (51.1)	14.8 (62.0)	28.6 (119.7)	28.1 (117.7)
$C^o_{p(500)}$	15.2 (63.6)	19.5 (81.6)	38.2 (159.9)	

	Path 1	Path 2
ΔH^o	[-36 (-151)]	[-32 (-134)]
ΔS^o	-39.4 (-165.0)	-38.5 (-161.2)
$\Delta C^o_{p(300)}$	1.6 (6.7)	1.1 (4.6)
$\Delta C^o_{p(500)}$	3.5 (14.7)	
$\langle\Delta C^o_p\rangle$	2.6 (10.9)	

Path	log A	E	log k	Conditions	Reference and Experimental Method
2	7.86	4.0 (17.0)	5.15(323 K) 6.66(395 K)	273-373 K 10-70 torr	[130] S.E.Braslavsky, F.Casas and O.Cifuentes, J.Chem.Soc.(B), 1059 (1970).

Static system. CF_3 from photolysis of CF_3I (Pyrex vessel, full Hg-arc) in presence of $CH_2{=}CH_2$ and $CF_2{=}CH_2$. Relative rates of addition from analysis of products $CF_3CH_2CH_2I$ and $CF_3CH_2CF_2I$ by gas chromatography. Reproducible rate constants under conditions of low light intensity (low $[CF_3]$) and values of $[CF_3I]/[olefins] > 2.5$.

Reaction scheme:

$$CF_3I + h\nu \xrightarrow{I} \dot{C}F_3 + \dot{I}$$
$$\dot{C}F_3 + CH_2{=}CH_2 \xrightarrow{1'} CF_3CH_2\dot{C}H_2$$
$$CF_3CH_2\dot{C}H_2 + CF_3I \xrightarrow{3} CF_3CH_2CH_2I + \dot{C}F_3$$
$$\dot{C}F_3 + CF_2{=}CH_2 \xrightarrow{2} CF_3CH_2\dot{C}F_2$$
$$CF_3CH_2\dot{C}F_2 + CF_3I \xrightarrow{4} CF_3CH_2CF_2I + \dot{C}F_3$$

$\log (k_{1'}/k_2) = 0.44 + 2.0/2.3RT$ and above results based on $\log k_{1'}$ ($\ell\ mol^{-1}s^{-1}$) $= 8.3 - 2.0/2.3RT$ (see $CF_3 + CH_2{=}CH_2$ system). The product $CF_3CF_2CH_2I$ was not detected, ruling out addition of CF_3 to $CH_2{=}CF_2$ by path 1, and from the limitations of the analytical method, this set a lower limit of 6 kcal mol^{-1} on the activation energy difference between path 1 and reaction (1') of above scheme.

ΔH^o_f, ΔH^o and E in kcal mol^{-1} (kJ mol^{-1}); S^o, C^o_p, ΔS^o and ΔC^o_p in cal mol^{-1} K^{-1} (J mol^{-1} K^{-1}); A and k in l mol^{-1} s^{-1}

PREFERRED RATE PARAMETERS

Tentatively as given above

COMMENTS

The exclusive addition according to path 2 contradicts the predicted route on the basis of the above thermochemistry. The thermochemical estimates, although tentative, seem reasonable. It is difficult to explain why the above activation energy should be greater than for CF_3 addition to $CF_2{=}CF_2$. At the same time the rate constants of reaction (2) above are very much lower than for CF_3 addition to $CF_2{=}CF_2$, which is equally puzzling.

REACTION: TRIFLUOROMETHYL RADICAL + TETRAFLUOROETHYLENE $\dot{C}F_3(I) + CF_2{=}CF_2(II) \longrightarrow CF_3CF_2\dot{C}F_2(III)$

	(I)	(II)	(III)		
$\Delta H^\circ_{f(300)}$	-112.5 (-471.0)	-157.9 (-661.1)	[-31 (-129)]	ΔH°	[-38 (-159)]
$S^\circ_{(300)}$	63.8 (267.1)	74.6 (312.4)	98.6 (412.8)	ΔS°	-39.8 (-166.6)
$C^\circ_{p(300)}$	12.2 (51.1)	19.4 (81.2)	32.4 (135.7)	ΔC°_p	0.8 (3.3)
$C^\circ_{p(500)}$	15.2 (63.6)	24.0 (100.5)			

log A	E	log k	Conditions	Reference and Experimental Method
7.54	1.4 (5.9)	6.77(395 K)	338-453 K 700 torr	[129] G.E.Owen, J.M.Pearson and M.Szwarc, Trans.Faraday Soc., 61, 1722 (1965). Results calculated from log $(A_1/A_3) = -0.3$ and $E_1 - E_3 = -1.9$ kcal mol^{-1}, taking log $k_3 = 7.84 - 3.3/2.3RT$. Same method as for $CF_3 + CH_2{=}CH_2$, reference [126] $\dot{C}F_3 + RH \xrightarrow{3} CF_3H + \dot{R}$ $\dot{C}F_3 + O\ell \xrightarrow{1} CF_3\dot{O}\ell$

PREFERRED RATE PARAMETERS

Reasonable as given above

REACTION: TRIFLUOROMETHYL RADICAL + 2-FLUOROPROPENE

$$\dot{C}F_3(I) + CH_2{=}CFCH_3(II) \xrightarrow{1} \dot{C}H_2CF(CF_3)CH_3(III)$$
$$\xrightarrow{2} CF_3CH_2\dot{C}FCH_3(IV)$$

	(I)	(II)	(III)	(IV)		Path 1	Path 2
$\Delta H^\circ_{f(300)}$	-112.5 (-471.0)		-176.6 (-739.4)	-176.9 (-740.7)	ΔH°		
$S^\circ_{(300)}$	63.8 (267.1)			[93 (389)]	ΔS°		
$C^\circ_{p(300)}$	12.2 (51.1)				ΔC°_p		

Path	log A	E	log k	Conditions	Reference and Experimental Method
1 + 2	-	-	7.21(338 K)	338 K 650 torr	[128] P.S.Dixon and M.Szwarc, Trans.Faraday Soc., 59, 112 (1963). Results calculated from log (k_{1+2}/k_3) = 1.50, and taking log k_3 = 7.84 - 3.3/2.3RT. Same method as for $CF_3 + CH_2{=}CH_2$ system, reference [126] $\dot{C}F_3 + RH \xrightarrow{3} CF_3H + \dot{R}$ $\dot{C}F_3 + O\ell \xrightarrow{1+2} CF_3\dot{O}\ell$

PREFERRED RATE PARAMETERS

Reasonable as given above

ΔH^o_f, ΔH^o and E in kcal mol^{-1} (kJ mol^{-1}); S^o, C^o_p, ΔS^o and ΔC^o_p in cal mol^{-1} K^{-1} (J mol^{-1} K^{-1}); A and k in l mol^{-1} s^{-1}

REACTION: TRIFLUOROMETHYL RADICAL + 1,1-DICHLORO-2,2-DIFLUOROETHYLENE

$$\dot{C}F_3(I) + CCl_2{=}CF_2(II) \xrightarrow{1} \dot{C}Cl_2CF_2CF_3(III)$$
$$\xrightarrow{2} \dot{C}F_2CCl_2CF_3(IV)$$

	(I)	(II)	(III)	(IV)		Path 1	Path 2
$\Delta H^\circ_{f(300)}$	-112.5 (-471.0)	-79.3 (-332.0)	[-232 (-972)]	[-230 (-963)]	ΔH°	[-40 (-167)]	[-38 (-159)]
$S^\circ_{(300)}$	63.8 (267.1)	79.4 (332.4)	103.0 (431.3)		ΔS°	-40.2 (-168.3)	
$C^\circ_{p(300)}$	12.2 (51.1)	21.1 (88.3)	34.5 (144.5)		ΔC°_p	1.2 (5.0)	

Path	log A	E	log k	Conditions	Reference and Experimental Method
1 + 2	7.14	1.6 (7.0)	6.26(395 K)	338-453 K 700 torr	[129] G.E.Owen, J.M.Pearson and M.Szwarc, Trans.Faraday Soc., 61, 1722 (1965). Results calculated from log (A_{1+2}/A_3) = -0.70 and $E_{1+2} - E_3$ = -1.7 kcal mol^{-1}, taking log k_3 = 7.84 - 3.3/2.3RT. Same method as for $CF_3 + CH_2{=}CH_2$ system, reference [126]. $\dot{C}F_3 + RH \xrightarrow{3} CF_3H + \dot{R}$ $\dot{C}F_3 + Ol \xrightarrow{1+2} CF_3\dot{O}l$

PREFERRED RATE PARAMETERS

Reasonable as given above

COMMENTS

Szwarc and co-workers [129] have argued that argued that addition exclusively by path 1, on the basis of comparisons with the kinetic results for other CF_3 additions, but in view of the results for CF_3 addition to $CF_2{=}CH_2$, this conclusion seems doubtful.

REACTION: TRIFLUOROMETHYL RADICAL + VINYL CHLORIDE

$$\dot{C}F_3(I) + CH_2{=}CHCl(II) \xrightarrow{1} CF_3CH_2\dot{C}HCl(III)$$
$$\xrightarrow{2} \dot{C}H_2CHClCF_3(IV)$$

	(I)	(II)	(III)	(IV)
$\Delta H^o_{f(300)}$	-112.5 (-471.0)	8.6 (36.0)	[-135 (-565)]	[-135 (-565)]
$S^o_{(300)}$	63.8 (267.1)	63.0 (263.6)	88.9 (372.2)	89.1 (373.1)
$C^o_{p(300)}$	12.2 (51.1)	13.0 (54.4)	26.9 (112.6)	27.6 (115.6)
$C^o_{p(500)}$	15.2 (63.6)	17.8 (74.5)	37.0 (154.9)	35.4 (148.2)

	Path 1	Path 2
ΔH^o	[-31 (-128)]	[-31 (-128)]
ΔS^o	-37.9 (-158.7)	-37.7 (-157.8)
$\Delta C^o_{p(300)}$	1.7 (7.1)	2.4 (10.0)
$\Delta C^o_{p(500)}$	4.0 (16.7)	2.4 (10.0)
$\langle\Delta C^o_p\rangle$	2.9 (12.1)	2.4 (10.0)

Path	log A	E	log k	Conditions	Reference and Experimental Method
1 + 2	-	-	7.28	338 K 650 torr	[128] P.S.Dixon and M.Szwarc, Trans.Faraday Soc., 59, 112 (1963). Results calculated from log (k_{1+2}/k_3) = 1.57, taking log k_3 = 7.84 - 3.3/2.3RT. Same method as for $CF_3 + CH_2{=}CH_2$, reference [126]. $\dot{C}F_3 + RH \xrightarrow{3} CF_3H + \dot{R}$ $\dot{C}F_3 + Ol \xrightarrow{1+2} CF_3\dot{O}l$
1 + 2	8.22	1.46 (6.11)	7.41(395 K)	338-453 K 700 torr	[126] J.M.Pearson and M.Szwarc, Trans.Faraday Soc., 60, 553 (1964). Results calculated from log (A_{1+2}/A_3) = 0.38 and $E_{1+2} - E_3$ = -1.84 kcal mol^{-1}, taking log k_3 = 7.84 - 3.3/2.3RT. Same method as for [128] above. For details see $CF_3 + CH_2{=}CH_2$.

PREFERRED RATE PARAMETERS

Reasonable as given above

COMMENTS

The experimental method does not distinguish between addition by path 1 or 2, although Szwarc and co-workers [Trans.Faraday Soc., 61, 1722 (1965)] have argued that addition occurs exclusively by path 2 on the basis of comparisons with kinetic results on other CF_3 additions. The thermochemical estimates given above indicate that $\Delta H^o_1 \approx \Delta H^o_2$, and so it is not possible to distinguish on the basis of these estimates.

ΔH^o_f, ΔH^o and E in kcal mol^{-1} (kJ mol^{-1}); S^o, C^o_p, ΔS^o and ΔC^o_p in cal mol^{-1} K^{-1} (J mol^{-1} K^{-1}); A and k in l mol^{-1} s^{-1}

REACTION: TRIFLUOROMETHYL RADICAL + 1,1-DICHLOROETHYLENE

$\dot{C}F_3(I) + CCl_2{=}CH_2(II) \xrightarrow{1} CF_3CCl_2\dot{C}H_2(III)$

$\xrightarrow{2} CF_3CH_2\dot{C}Cl_2(IV)$

	(I)	(II)	(III)	(IV)
$\Delta H^\circ_{f(300)}$	-112.5 (-471.0)	4.3 (18.0)	[-142 (-595)]	[-139 (-582)]
$S^\circ_{(300)}$	63.8 (267.1)	69.7 (291.8)	[89 (373)]	94.7 (396.5)
$C^\circ_{p(300)}$	12.2 (51.1)	16.5 (69.1)	[27.9 (116.8)]	30.3 (126.9)
$C^\circ_{p(500)}$	15.2 (63.6)	20.8 (87.1)	[36.3 (152.0)]	40.1 (168.0)

	Path 1	Path 2
ΔH°	[-34 (-142)]	[-31 (-130)]
ΔS°	[-45 (-188)]	-38.8 (-162.5)
$\Delta C^\circ_{p(300)}$	[-0.8 (-3.3)]	1.6 (6.7)
$\Delta C^\circ_{p(500)}$	[0.3 (1.3)]	4.1 (17.2)
$\langle\Delta C^\circ_p\rangle$	-0.3 (-1.3)	2.9 (12.1)

Path	log A	E	log k	Conditions	Reference and Experimental Method
1 + 2	7.88	0.5 (2.1)	7.61(395 K)	338-453 K 700 torr	[129] G.E.Owen, J.M.Pearson and M.Szwarc, Trans.Faraday Soc., 61, 1722 (1965). Results calculated from log (A_{1+2}/A_3) = 0.04 and $E_{1+2} - E_3$ = -2.8 kcal mol^{-1}, taking log k_3 = 7.84 - 3.3/2.3RT. Same method as for $CF_3 + CH_2{=}CH_2$ system, reference [126]. $\dot{C}F_3 + RH \xrightarrow{3} CF_3H + \dot{R}$ $\dot{C}F_3 + Ol \xrightarrow{1+2} CF_3\dot{O}l$

PREFERRED RATE PARAMETERS

Reasonable as given above

COMMENTS

The experimental method does not distinguish between addition by path 1 or 2, although Szwarc and co-workers [129] have argued that addition occurs exclusively by path 1 on the basis of comparisons with kinetic results on other CF_3 additions.

The thermochemical estimates of ΔH°_1 and ΔH°_2 given above would appear to be wrong; (they are based on $D(CF_3CCl_2CH_2\text{-}H)$ = 98, and $D(CF_3CH_2CCl_2\text{-}H)$ = 95, and group additivity calculations of the enthalpies of formation of the parent compounds (see Introduction, Section 1.9).

REACTION: TRIFLUOROMETHYL RADICAL + cis-DICHLOROETHYLENE $\dot{C}F_3(I) + cis\text{-}CHCl{=}CHCl(II) \longrightarrow CF_3CHCl\dot{C}HCl(III)$

	(I)		(II)		(III)				
$\Delta H^\circ_{f(300)}$	-112.5	(-471.0)	1.0	(4.2)	[-143	(-599)]	ΔH°	[-32	(-134)]
$S^\circ_{(300)}$	63.8	(267.1)	70.8	(296.4)	97.0	(406.1)	ΔS°	-37.6	(-157.4)
$C^\circ_{p(300)}$	12.2	(51.1)	15.8	(66.2)	30.4	(127.3)	$\Delta C^\circ_{p(300)}$	2.4	(10.0)
$C^\circ_{p(500)}$	15.2	(63.6)	20.6	(86.3)	42.3	(177.1)	$\Delta C^\circ_{p(500)}$	6.5	(27.2)
							$\langle\Delta C^\circ_p\rangle$	4.5	(18.8)

log A	E	log k	Conditions	Reference and Experimental Method
7.62	2.8 (11.7)	6.08(395 K)	338-453 K 700 torr	[129] G.E.Owen, J.M.Pearson and M.Szwarc, Trans.Faraday Soc., 61, 1722 (1965). Results calculated from log $(A_1/A_3) = -0.22$ and $E_1 - E_3 = -0.5$ kcal mol^{-1}, taking log $k_3 = 7.84 - 3.3/2.3RT$. Same method as for $CF_3 + CH_2{=}CH_2$ system, reference [126]. $\dot{C}F_3 + RH \xrightarrow{3} CF_3H + \dot{R}$; $\dot{C}F_3 + Ol \xrightarrow{1} CF_3\dot{O}l$

PREFERRED RATE PARAMETERS

Reasonable as given above

ΔH°_f, ΔH° and E in kcal mol^{-1} (kJ mol^{-1}); S°, C°_p, ΔS° and ΔC°_p in cal mol^{-1} K^{-1} (J mol^{-1} K^{-1}); A and k in l mol^{-1} s^{-1}

REACTION: TRIFLUOROMETHYL RADICAL + TRANS-DICHLOROETHYLENE $\dot{C}F_3$(I) + trans-$CHCl{=}CHCl$(II) ⟶ $CF_3CHCl\dot{C}HCl$(III)

	(I)	(II)	(III)		
$\Delta H^\circ_{f(300)}$	-112.5 (-471.0)	1.2 (5.0)	[-143 (-599)]	ΔH°	[-32 (-134)]
$S^\circ_{(300)}$	63.8 (267.1)	70.8 (296.4)	97.0 (406.1)	ΔS°	-37.6 (-157.4)
$C^\circ_{p(300)}$	12.2 (51.1)	15.8 (66.2)	30.4 (127.3)	$\Delta C^\circ_{p(300)}$	2.4 (10.0)
$C^\circ_{p(500)}$	15.2 (63.6)	20.6 (86.3)	42.3 (177.1)	$\Delta C^\circ_{p(500)}$	6.5 (27.2)
				$\langle\Delta C^\circ_p\rangle$	4.5 (18.8)

log A	E	log k	Conditions	Reference and Experimental Method
7.44	2.0 (8.3)	6.34(395 K)	338-453 K 700 torr	[129] G.E.Owen, J.M.Pearson and M.Szwarc, Trans.Faraday Soc., 61, 1722 (1965). Results calculated from log (A_1/A_3) = -0.40 and $E_1 - E_3$ = -1.3 kcal mol^{-1}, taking log k_3 = 7.84 - 3.3/2.3RT. Same method as for CF_3 + $CH_2{=}CH_2$ system, reference [126].

$$\dot{C}F_3 + RH \xrightarrow{3} CF_3H + \dot{R}$$
$$\dot{C}F_3 + Ol \xrightarrow{1} CF_3\dot{O}l$$

PREFERRED RATE PARAMETERS

Reasonable as given above

REACTION: TRIFLUOROMETHYL RADICAL + TETRACHLOROETHYLENE $\dot{C}F_3(I) + C\ell_2C{=}CC\ell_2(II) \longrightarrow CF_3CC\ell_2\dot{C}C\ell_2(III)$

	(I)	(II)	(III)		
$\Delta H^o_{f(300)}$	-112.5 (-471.0)	-2.7 (-11.3)	[-154 (-645)]	ΔH^o	[-39 (-163)]
$S^o_{(300)}$	63.8 (267.1)	84.2 (352.5)	102.7 (430.0)	ΔS^o	-45.3 (-189.7)
$C^o_{p(300)}$	12.2 (51.1)	22.8 (95.5)	33.6 (140.7)	ΔC^o_p	-1.4 (-5.9)
$C^o_{p(500)}$	15.2 (63.6)	26.6 (111.3)			

log A	E	log k	Conditions	Reference and Experimental Method
7.74	4.1 (17.2)	5.47(395 K)	353-453 K 700 torr	[129] G.E.Owen, J.M.Pearson and M.Szwarc, Trans.Faraday Soc., 61, 1722 (1965). Results calculated from log (A_1/A_3) = -0.10 and $E_1 - E_3$ = 0.8 kcal mol^{-1}, taking log k_3 = 7.84 - 3.3/2.3RT. Same method as for $CF_3 + CH_2{=}CH_2$ system, reference [126]

$$\dot{C}F_3 + RH \xrightarrow{3} CF_3H + \dot{R}$$
$$\dot{C}F_3 + O\ell \xrightarrow{1} CF_3\dot{O}\ell$$

Rate constants at 453 K were not independent of total pressure of the system, and consequently the addition reaction is reversible at this temperature.

PREFERRED RATE PARAMETERS

Reasonable as given above

ΔH^o_f, ΔH^o and E in kcal mol^{-1} (kJ mol^{-1}); S^o, C^o_p, ΔS^o and ΔC^o_p in cal mol^{-1} K^{-1} (J mol^{-1} K^{-1}); A and k in l mol^{-1} s^{-1}

REACTION: TRIFLUOROMETHYL RADICAL + BUTADIENE-1,3

$$\dot{C}F_3(I) + CH_2{=}CHCH{=}CH_2(II) \xrightarrow{1} CF_3CH_2\dot{C}HCH{=}CH_2(III)$$
$$\xrightarrow{2} \dot{C}H_2CH(CF_3)CH{=}CH_2(IV)$$

	(I)	(II)	(III)	(IV)		Path 1	Path 2
$\Delta H^\circ_{f(300)}$	-112.5 (-471.0)	26.1 (109.3)	[-122 (-511)]	-109.2 (-457.2)	ΔH°	[-36 (-151)]	-22.8 (-95.5)
$S^\circ_{(300)}$	63.8 (267.1)	68.0 (284.7)	92.3 (386.5)	[95 (398)]	ΔS°	-39.5 (-165.4)	[-36 (-151)]
$C^\circ_{p(300)}$	12.2 (51.1)	19.1 (80.0)	32.0 (134.0)	[32.1 (134.4)]	ΔC°_p	0.7 (2.9)	[0.8 (3.3)]

Path	log A	E	log k	Conditions	Reference and Experimental Method
1 + 2	-	-	8.60	338 K 650 torr	[128] P.S.Dixon and M.Szwarc, Trans.Faraday Soc., 59, 112 (1963). Results calculated from log (k_{1+2}/k_3) = 2.89, taking log k_3 = 7.84 - 3.3/2.3RT. Same method as for $CF_3 + CH_2{=}CH_2$, reference [126]. $\dot{C}F_3 + RH \xrightarrow{3} CF_3H + \dot{R}$ $\dot{C}F_3 + O\ell \xrightarrow{1+2} CF_3\dot{O}\ell$
1 + 2	8.29	-0.48 (-2.01)	8.56(395 K)	338-453 K 700 torr	[126] J.M.Pearson and M.Szwarc, Trans.Faraday Soc., 60, 553 (1964). Results calculated from log (A_{1+2}/A_3) = 0.45 and $E_{1+2} - E_3$ = -3.78 kcal mol^{-1}, taking log k_3 = 7.84 - 3.3/2.3RT. Same method as for [128] above. For details see $CF_3 + CH_2{=}CH_2$.

PREFERRED RATE PARAMETERS

log k_{1+2} = 8.6

COMMENTS

The slightly negative E, calculated above, indicates experimental errors in the Arrhenius parameters for the addition reaction, or the reference reaction, or both. We have assigned the addition reaction $E_1 \simeq 0$ kcal mol^{-1}. Path 1 will be the favoured route for addition.

REACTION: TRIFLUOROMETHYL RADICAL + ISOPRENE

$$\dot{C}F_3(I) + CH_2{=}C(CH_3)CH{=}CH_2(II) \xrightarrow{1} CF_3CH_2\dot{C}(CH_3)CH{=}CH_2(III)$$
$$\xrightarrow{2} \dot{C}H_2C(CF_3)(CH_3)CH{=}CH_2(IV)$$
$$\xrightarrow{3} CH_2{=}C(CH_3)\dot{C}HCH_2CF_3(V)$$
$$\xrightarrow{4} CH_2{=}C(CH_3)CH(CF_3)\dot{C}H_2(VI)$$

	(I)	(II)	(III)	(IV)	(V)	(VI)
$\Delta H^o_{f(300)}$	-112.5 (-471.0)	18.1 (75.8)	[-132 (-552)]	-116.3 (-486.9)	[-134 (-561)]	-117.5 (-492.0)
$S^o_{(300)}$	63.8 (267.1)	75.2 (314.9)	98.5 (412.4)	[100 (419)]	100.3 (420.0)	[103 (431)]
$C^o_{p(300)}$	12.2 (51.1)	[25.2 (105.5)]	37.7 (157.8)	[38.5 (161.2)]	38.2 (159.9)	[38.2 (159.9)]

	Path 1	Path 2	Path 3	Path 4
ΔH^o	[-38 (-159)]	-21.9 (-91.7)	[-37 (-155)]	-23.1 (-96.7)
ΔS^o	-40.5 (-169.6)	[-39 (-163)]	-38.7 (-162.0)	[-36 (-151)]
ΔC^o_p	[0.3 (12.6)]	[1.1 (4.6)]	[0.8 (3.3)]	[0.8 (3.3)]

Path	log A	E	log k	Conditions	Reference and Experimental Method
1+2+3+4	8.27	-0.75 (-3.14)	8.69(395 K)	338-453 K 700 torr	[126] J.M.Pearson and M.Szwarc, Trans.Faraday Soc., 60, 553 (1964). Results calculated from log (A_a/A_3) = 0.43 and $E_a - E_3$ = -4.05 kcal mol^{-1}, taking log k_3 = 7.84 - 3.3/2.3RT for the reaction $\dot{C}F_3 + RH \xrightarrow{3} CF_3H + \dot{R}$; $\dot{C}F_3 + O\ell \xrightarrow{a} CF_3\dot{O}\ell$. For details see $CF_3 + CH_2{=}CH_2$.

PREFERRED RATE PARAMETERS

log $k_{1+2+3+4}$ = 8.7

COMMENTS

The slightly negative E, calculated above, indicates experimental errors in the Arrhenius parameters for the addition reaction, or the reference reaction, or both. We have assigned the addition reaction E ≈ 0 kcal mol^{-1}. Paths 1 and 3 will be the favoured routes for addition.

ΔH^o_f, ΔH^o and E in kcal mol^{-1} (kJ mol^{-1}); S^o, C^o_p, ΔS^o and ΔC^o_p in cal mol^{-1} K^{-1} (J mol^{-1} K^{-1}); A and k in l mol^{-1} s^{-1}

REACTION: TRIFLUOROMETHYL RADICAL + HEXADIENE-2,4

$\dot{C}F_3(I) + CH_3CH{=}CHCH{=}CHCH_3(II) \xrightarrow{1} CH_3(CF_3)CH\dot{C}HCH{=}CHCH_3(III)$

$\xrightarrow{2} CH_3\dot{C}HCH(CF_3)CH{=}CHCH_3(IV)$

	(I)	(II)	(III)	(IV)
$\Delta H^\circ_{f(300)}$	-112.5 (-471.0)	14.2 (59.5)	[-137 (-575)]	-124.9 (-523.0)
$S^\circ_{(300)}$	63.8 (267.1)	88.3 (369.7)	107.6 (450.5)	111.3 (466.0)
$C^\circ_{p(300)}$	12.2 (51.1)	29.0 (121.4)	42.5 (177.9)	43.1 (180.5)
$C^\circ_{p(500)}$	15.2 (63.6)	42.0 (175.9)	61.2 (256.2)	60.8 (254.6)

	Path 1	Path 2
ΔH°	[-39 (-163)]	-26.6 (-111.4)]
ΔS°	-44.5 (-186.3)	-40.8 (-170.8)
$\Delta C^\circ_{p(300)}$	1.3 (5.4)	1.9 (8.0)
$\Delta C^\circ_{p(500)}$	4.0 (16.7)	3.6 (15.1)
$\langle\Delta C^\circ_p\rangle$	2.7 (11.3)	2.8 (11.7)

Path	log A	E	log k	Conditions	Reference and Experimental Method
1 + 2	7.54	-1.4 (-5.9)	8.31(395 K)	338-453 K 700 torr	[129] G.E.Owen, J.M.Pearson and M.Szwarc, Trans.Faraday Soc., 61, 1722 (1965). Results calculated from log $(A_{1+2}/A_3) = -0.30$ and $E_{1+2} - E_3 = -4.7$ kcal mol^{-1}, taking log $k_3 = 7.84 - 3.3/2.3RT$. Same method as for $CF_3 + CH_2{=}CH_2$ system, reference [126]. $\dot{C}F_3 + RH \xrightarrow{3} CF_3H + \dot{R}$; $\dot{C}F_3 + O\ell \xrightarrow{1+2} CF_3\dot{O}\ell$

PREFERRED RATE PARAMETERS

log $k_{1+2} = 8.3$

COMMENTS

The negative E, calculated above, is unacceptable and indicates experimental errors in the Arrhenius parameters of the addition reaction, or the reference reaction, or both. It seems reasonable to assign $E \approx 0$ kcal mol^{-1} for the addition reaction.

REACTION: TRIFLUOROMETHYL RADICAL + 2,3-DIMETHYLBUTADIENE-1,3

$\dot{C}F_3(I) + CH_2{=}C(CH_3)C(CH_3){=}CH_2(II) \xrightarrow{1} \dot{C}H_2C(CH_3)(CF_3)C(CH_3){=}CH_2(III)$

$\xrightarrow{2} CF_3CH_2\dot{C}(CH_3)C(CH_3){=}CH_2(IV)$

	(I)	(II)	(III)	(IV)
$\Delta H^o_{f(300)}$	-112.5 (-471.0)	10.1 (42.3)	-124.6 (-521.7)	[-141 (-591)]
$S^o_{(300)}$	63.8 (267.1)	82.4 (345.0)	[108 (452)]	[107 (448)]
$C^o_{p(300)}$	12.2 (51.1)	[31.4 (131.5)]	[44.3 (185.5)]	[43.8 (183.4)]

	Path 1	Path 2
ΔH^o	-22.2 (-93.0)	[-38 (-159)]
ΔS^o	[-38 (-159)]	[-40 (-167)]
ΔC^o_p	[0.7 (2.9)	[0.2 (0.8)]

Path	log A	E	log k	Conditions	Reference and Experimental Method
1 + 2	8.22	-1.1 (-4.69)	8.84(395 K)	338-453 K 700 torr	[126] J.M.Pearson and M.Szwarc, Trans.Faraday Soc., 60, 553 (1964). Results calculated from log (A_{1+2}/A_3) = 0.38 and $E_{1+2} - E_3 = -4.42$ kcal mol^{-1}, taking log $k_3 = 7.84 - 3.3/2.3RT$, for the reaction $\dot{C}F_3 + RH \xrightarrow{3} CF_3H + \dot{R}$ $\dot{C}F_3 + Ol \xrightarrow{1+2} CF_3\dot{O}l$ For details see $CF_3 + CH_2{=}CH_2$

PREFERRED RATE PARAMETERS

log k_{1+2} = 8.8

COMMENTS

The negative E, calculated above, indicates experimental errors in the Arrhenius parameters for the addition reaction, or the reference reaction, or both. We have assigned the addition reaction E ≃ 0 kcal mol^{-1}. Path 2 will be the favoured route for addition.

ΔH^o_f, ΔH^o and E in kcal mol^{-1} (kJ mol^{-1}); S^o, C^o_p, ΔS^o and ΔC^o_p in cal mol^{-1} K^{-1} (J mol^{-1} K^{-1}); A and k in l mol^{-1} s^{-1}

REACTION: TRIFLUOROMETHYL RADICAL + CYCLOPENTENE

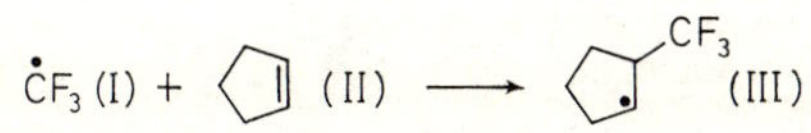

	(I)	(II)	(III)		
$\Delta H^\circ_{f(300)}$	-112.5 (-471.0)	8.6 (36.0)	-131.4 (-550.2)	ΔH°	-27.5 (-115.1)
$S^\circ_{(300)}$	63.8 (267.1)	70.8 (296.4)	94.5 (395.7)	ΔS°	-40.1 (-167.9)
$C^\circ_{p(300)}$	12.2 (51.1)	18.1 (75.8)	31.4 (131.5)	$\Delta C^\circ_{p(300)}$	1.1 (4.6)
$C^\circ_{p(500)}$	15.2 (63.6)	31.6 (132.3)	49.7 (208.1)	$\Delta C^\circ_{p(500)}$	2.9 (12.1)
				$\langle\Delta C^\circ_p\rangle$	2.0 (8.4)

log A	E	log k	Conditions	Reference and Experimental Method
-	-	7.41(338 K)	338 K 650 torr	[128] P.S.Dixon and M.Szwarc, Trans.Faraday Soc., 59, 112 (1963). Results calculated from log k_1/k_3 = 1.70 and log k_3 = 7.84 - 3.3/2.3RT. Same method as for CF_3 + CH_2=CH_2 system, reference [126]. $\dot{C}F_3 + RH \xrightarrow{3} CF_3H + \dot{R}$ $\dot{C}F_3 + O\ell \xrightarrow{1} CF_3\dot{O}\ell$
7.92	0.8 (3.3)	7.48(395 K)	338-453 K 700 torr	[129] G.E.Owen, J.M.Pearson and M.Szwarc, Trans.Faraday Soc., 61, 1722 (1965). Results calculated from log (A_1/A_3) = 0.08 and $E_1 - E_3$ = -2.50 kcal mol^{-1}, and log k_3 = 7.84 - 3.3/2.3RT. Same method as for [128] above. For details see CF_3 + CH_2=CH_2, reference [126].

PREFERRED RATE PARAMETERS

Reasonable as given in reference [129]

REACTION: TRIFLUOROMETHYL RADICAL + CYCLOHEXENE

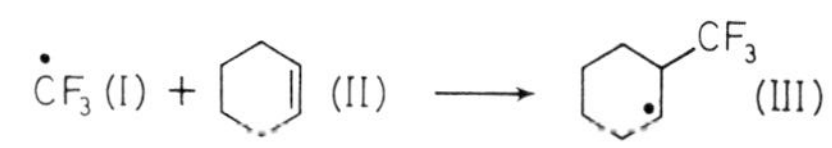

	(I)	(II)	(III)
$\Delta H^o_{f(300)}$	-112.5 (-471.0)	-0.8 (-3.3)	-142.7 (-597.5)
$S^o_{(300)}$	63.8 (267.1)	74.3 (311.1)	95.5 (399.9)
$C^o_{p(300)}$	12.2 (51.1)	25.3 (105.9)	38.0 (159.1)
$C^o_{p(500)}$	15.2 (63.6)	42.8 (179.2)	60.1 (251.6)

ΔH^o	-29.4	(-123.1)
ΔS^o	-42.6	(-178.4)
$\Delta C^o_{p(300)}$	0.5	(2.1)
$\Delta C^o_{p(500)}$	2.1	(8.8)
$\langle\Delta C^o_p\rangle$	1.3	(5.4)

log A	E	log k	Conditions	Reference and Experimental Method
7.92	1.2 (5.0)	7.26(395 K)	353-453 K 700 torr	[129] G.E.Owen, J.M.Pearson and M.Szwarc, Trans.Faraday Soc., 61, 1722 (1965). Results calculated from log (A_1/A_3) = 0.08 and $E_1 - E_3$ = -2.10 kcal mol^{-1}, taking log k_3 = 7.84 - 3.3/2.3RT. Same method as for $CF_3 + CH_2{=}CH_2$ system, reference [126] $\dot{C}F_3 + RH \xrightarrow{3} CF_3H + \dot{R}$ $\dot{C}F_3 + O\ell \xrightarrow{1} CF_3\dot{O}\ell$

PREFERRED RATE PARAMETERS

Reasonable as given above

ΔH^o_f, ΔH^o and E in kcal mol^{-1} (kJ mol^{-1}); S^o, C^o_p, ΔS^o and ΔC^o_p in cal mol^{-1} K^{-1} (J mol^{-1} K^{-1}); A and k in l mol^{-1} s^{-1}

REACTION: TRIFLUOROMETHYL RADICAL + BENZENE

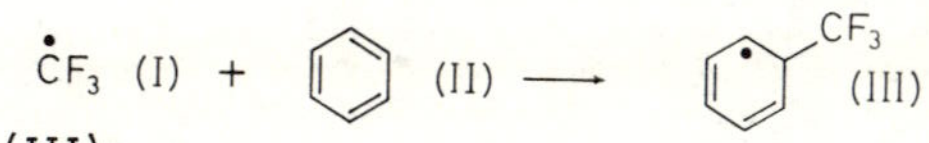

	(I)	(II)	(III)
$\Delta H^\circ_{f(300)}$	-112.5 (-471.0)	19.8 (82.9)	-121.5 (-508.7)
$S^\circ_{(300)}$	63.8 (267.1)	64.5 (270.1)	[81 (339)]
$C^\circ_{p(300)}$	12.2 (51.1)	19.4 (81.2)	[35.8 (149.9)]
$C^\circ_{p(500)}$	15.2 (63.6)	32.8 (137.3)	[54.0 (226.1)]

ΔH°	-28.8 (-120.6)
ΔS°	[-48 (-201)]
$\Delta C^\circ_{p(300)}$	[4.2 (17.6)]
$\Delta C^\circ_{p(500)}$	[6.0 (25.1)]
$<\Delta C^\circ_p>$	[5.1 (21.4)]

log A	E	log k	Conditions	Reference and Experimental Method
9.16	5.44 (22.78)	5.63(338 K)	296-359 K 35-50 torr	[131] S.W.Charles and E.Whittle, Trans.Faraday Soc., 56, 794 (1960).

Arrhenius parameters relative to log k_c = 10.36 for the recombination of CF_3 radicals; log $(A_1/A_c^{\frac{1}{2}})(\ell^{\frac{1}{2}}mol^{-\frac{1}{2}}s^{-\frac{1}{2}})$ = 3.98 and $E_1 - \frac{1}{2}E_c$ = 5.44. Static system. CF_3 from photolysis (λ > 286 nm) of CF_3COCF_3. In the absence of a substrate, the mechanism can be written

$$CF_3COCF_3 + h\nu \xrightarrow{I} 2\,\dot{C}F_3 + CO$$
$$2\,\dot{C}F_3 \xrightarrow{4} C_2F_6$$

and CO/C_2F_6 = 1. In the presence of an aromatic compound the following reactions must be included

$$\dot{C}F_3 + A \xrightarrow{3} CF_3H + (A - H)$$
$$\dot{C}F_3 + A \xrightarrow{1} CF_3\dot{A}$$

The rate constant for the addition reaction can be calculated from

$$k_1/k_4^{\frac{1}{2}} = R_{Ad}/R_{C_2F_6}^{\frac{1}{2}}[ArH]$$

where $R_{Ad} = 2\,R_{CO} - 2\,R_{C_2F_6} - 2\,R_{CF_3H}$. The products were analyzed by low-temperature distillation and gas chromatography. Szwarc [Trans. Faraday Soc., 60, 553 (1964)] has criticised this method on the basis that the reaction

$$\dot{C}F_3 + CF_3\dot{O}\ell \longrightarrow CF_3O\ell CF_3$$

which was not originally considered by Whittle and co-workers, could lead to an error in the rate constants of a factor of 2 and a consequent error of 1.5 kcal mol^{-1} in E.

log A	E	log k	Conditions	Reference and Experimental Method
8.46	4.3 (18.0)	5.66(338 K)	296-375 K 25-160 torr	[132] J.L.Holmes and K.O.Kutschke, <u>Trans.Faraday Soc.</u>, <u>58</u>, 333 (1962). Arrhenius parameters calculated from log $(k_5/k_1) = -0.31 - 0.4/2.3RT$ and taking $k_5 = 8.15 - 4.7/2.3RT$ for the reaction $\dot{C}F_3 + iso\text{-}C_4H_{10} \xrightarrow{5} CF_3H + t\text{-}\dot{C}_4H_9$ [A.F.Trotman-Dickenson and G.S.Milne, NSRDS-NBS 9, 1967]. Static system. CF_3 from photolysis (λ = 300-400 nm) of CF_3COCF_3 in the presence of an aromatic compound plus a large excess of iso-C_4H_{10}. The simplified mechanism was assumed to be: $CF_3COCF_3 + h\nu \xrightarrow{I} 2\,\dot{C}F_3 + CO$ $2\,\dot{C}F_3 \xrightarrow{4} C_2F_6$ $\dot{C}F_3 + A \xrightarrow{3} CF_3H + (A - H)$ $\dot{C}F_3 + A \xrightarrow{1} CF_3\dot{A}$ $\dot{C}F_3 + iso\text{-}C_4H_{10} \xrightarrow{5} CF_3H + t\text{-}\dot{C}_4H_9$ and the function $r = R_{CF_3H}/(2\,R_{CO} - 2\,R_{C_2F_6} - R_{CF_3H})$ is given by $r = k_3/k_1 + (k_5/k_1)[iso\text{-}C_4H_{10}]/[A]$. The products were analyzed by low-temperature distillation and gas chromatography. Szwarc [<u>Trans.Faraday Soc.</u>, <u>60</u>, 553 (1964)] has criticised the method on the grounds that the CF_3 radicals may also react with other radicals in the system.
-	-	5.62(338 K)	338 K 650 torr	[128] P.S.Dixon and M.Szwarc, <u>Trans.Faraday Soc.</u>, <u>59</u>, 112 (1963). Results calculated from log $(k_1/k_3) = -0.09$ at 338 K, taking log $k_3 = 7.84 - 3.3/2.3RT$. Same method as for $CF_3 + CH_2{=}CH_2$, reference [126] $\dot{C}F_3 + RH \xrightarrow{3} CF_3H + \dot{R}$ $\dot{C}F_3 + O\ell \xrightarrow{1} CF_3\dot{O}\ell$
7.76	3.0 (12.6)	6.02(338 K)	338-373 K 700 torr	[129] G.E.Owen, J.M.Pearson and M.Szwarc, <u>Trans.Faraday Soc.</u>, <u>61</u>, 1722 (1965). Results calculated from log $(A_1/A_3) = -0.08$ and $E_1 - E_3 = -0.3$ kcal mol^{-1}, taking log $k_3 = 7.84 - 3.3/2.3RT$. Same method as for $CF_3 + CH_2{=}CH_2$ system, reference [126]. $\dot{C}F_3 + RH \xrightarrow{3} CF_3H + \dot{R}$ $\dot{C}F_3 + O\ell \xrightarrow{1} CF_3\dot{O}\ell$ At temperatures of 403 and 423 K, the rate constants for reaction (1) were

ΔH_f^o, ΔH^o and E in kcal mol^{-1} (kJ mol^{-1}); S^o, C_p^o, ΔS^o and ΔC_p^o in cal mol^{-1} K^{-1} (J mol^{-1} K^{-1}); A and k in l mol^{-1} s^{-1}

log A	E	log k	Conditions	Reference and Experimental Method
				not independent of pressure, indicating the reversibility of the addition reaction at these temperatures; (for further studies of the reversibility of the addition, see G.A.Chamberlain and E.Whittle, Trans. Faraday Soc., 67, 2077 (1971)).

PREFERRED RATE PARAMETERS

log k = 7.76 - 3.0/2.3RT

COMMENTS

The rate constants in all of the above references are in reasonable agreement. The Arrhenius parameters of reference [129] are preferred on the grounds that the experimental method seems to be the most reliable.

REACTION: TRIFLUOROMETHYL RADICAL + TOLUENE

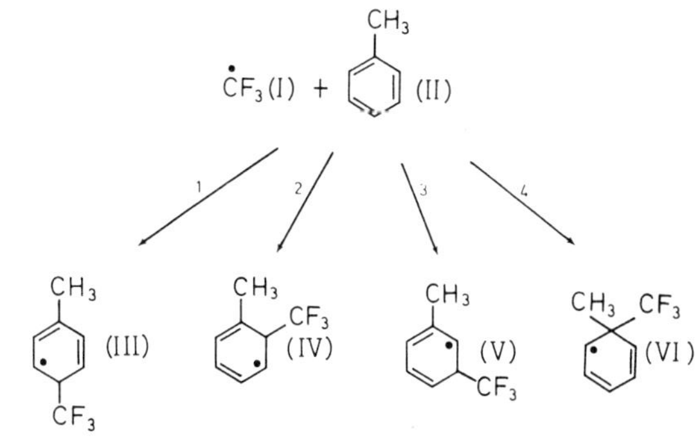

	(I)	(II)	(III)	(IV)	(V)	(VI)
$\Delta H^\circ_{f(300)}$	-112.5 (-471.0)	12.0 (50.2)	-130.3 (-545.6)	-128.8 (-539.1)	-129.2 (-541.0)	-132.5 (-554.8)
$S^\circ_{(300)}$	63.8 (267.1)	76.4 (319.9)	89.7 (375.6)	[88 (368)]	89.7 (375.6)	108.9 (456.0)
$C^\circ_{p(300)}$	12.2 (51.1)	24.8 (103.8)	41.1 (172.1)	[42.6 (178.4)]	41.5 (173.8)	41.7 (174.6)

	Path 1	Path 2	Path 3	Path 4
ΔH°	-29.8 (-124.8)	-28.3 (-118.5)	-28.7 (-120.1)	-32.0 (-133.9)
ΔS°	-50.5 (-211.4)	[-52 (-218)]	-50.5 (-211.4)	-31.3 (-131.1)
ΔC°_p	4.1 (17.2)	[5.6 (23.4)]	4.5 (18.8)	4.7 (19.7)

Path	log A	E	log k	Conditions	Reference and Experimental Method
1+2+3+4	8.97	4.90 (20.52)	5.80(338 K)	296-359 K 35-50 torr	[131] S.W.Charles and E.Whittle, Trans.Faraday Soc., 56, 794 (1960). Arrhenius parameters relative to log k_c = 10.36 for the combination of CF_3 radicals; log $(A_a/A_c^{\frac{1}{2}})(\ell^{\frac{1}{2}}mol^{-\frac{1}{2}}s^{-\frac{1}{2}})$ = 3.79 and $E_a - \frac{1}{2}E_c$ = 4.90 kcal mol^{-1}. For details see CF_3 + benzene.
1+2+3+4	8.22	3.5 (14.91)	5.99(338 K)	302-381 K 30-130 torr	[132] J.L.Holmes and K.O.Kutschke, Trans.Faraday Soc., 58, 333 (1962). Arrhenius parameters calculated from log (k_5/k_1) = -0.07 - 1.2/2.3RT and taking k_5 = 8.15 - 4.7/2.3RT for the reaction $\dot{C}F_3$ + iso-C_4H_{10} $\xrightarrow{5}$ CF_3H + t-$\dot{C}_4H_9$ [A.F.Trotman-Dickenson and G.S.Milne, NSRDS-NBS 9, 1967].

PREFERRED RATE PARAMETERS

log $k_{1+2+3+4}$ = 8.60 - 4.2/2.3RT

ΔH^o_f, ΔH^o and E in kcal mol^{-1} (kJ mol^{-1}); S^o, C^o_p, ΔS^o and ΔC^o_p in cal mol^{-1} K^{-1} (J mol^{-1} K^{-1}); A and k in l mol^{-1} s^{-1}

COMMENTS

There is reasonable agreement between the rate constants of references [131] and [132]. The preferred Arrhenius parameters are the average of the two determinations. There is no information on the products of the reaction, and consequently the site of addition is unknown.

REACTION: TRIFLUOROMETHYL RADICAL + o-XYLENE

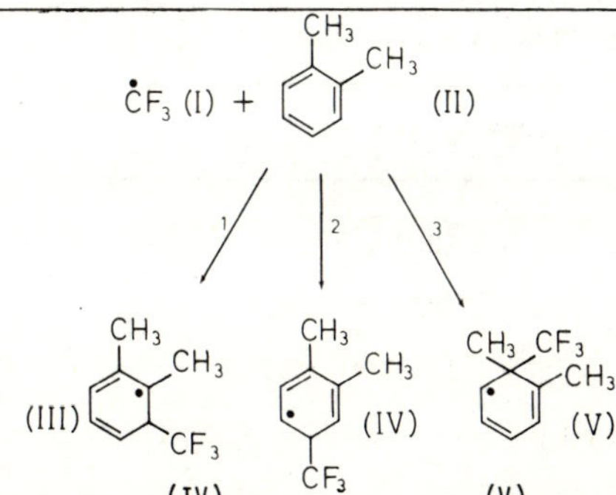

	(I)	(II)	(III)	(IV)	(V)
$\Delta H^\circ_{f(300)}$	-112.5 (-471.0)	4.8 (20.1)	-138.4 (-579.5)	-136.7 (-572.4)	-136.1 (-569.9)
$S^\circ_{(300)}$	63.8 (267.1)	84.0 (351.7)	[99 (415)]	[97 (406)]	[95 (397)]
$C^\circ_{p(300)}$	12.2 (63.6)	31.3 (131.1)	[51 (214)]	[48 (201)]	[49 (205)]

	Path 1	Path 2	Path 3
ΔH°	-30.7 (-128.5)	-29.0 (-121.4)	-28.4 (-118.9)
ΔS°	[-49 (-205)]	[-51 (-214)]	[-53 (-222)]
ΔC°_p	[7.5 (31.4)]	[4.5 (18.8)]	[5.5 (23.0)]

Path	log A	E	log k	Conditions	Reference and Experimental Method
1+2+3	9.03	4.35 (18.21)	6.22(338 K)	310-341 K 30 torr	[131] S.W.Charles and E.Whittle, Trans.Faraday Soc., 56, 794 (1960). Arrhenius parameters relative to $\log k_c = 10.36$ for the combination of CF_3 radicals; $\log (A_a/A_c^{\frac{1}{2}})(\ell^{\frac{1}{2}}mol^{-\frac{1}{2}}s^{-\frac{1}{2}}) = 3.85$ and $E_a - \frac{1}{2}E_c = 2.9$. For details see CF_3 + benzene system.

PREFERRED RATE PARAMETERS

Reasonable as given above

COMMENTS

There is no information on the products of the reaction and consequently it is not possible to identify which path predominates, if any.

ΔH_f^o, ΔH^o and E in kcal mol^{-1} (kJ mol^{-1}); S^o, C_p^o, ΔS^o and ΔC_p^o in cal mol^{-1} K^{-1} (J mol^{-1} K^{-1}); A and k in l mol^{-1} s^{-1}

REACTION: TRIFLUOROMETHYL RADICAL + FLUOROBENZENE

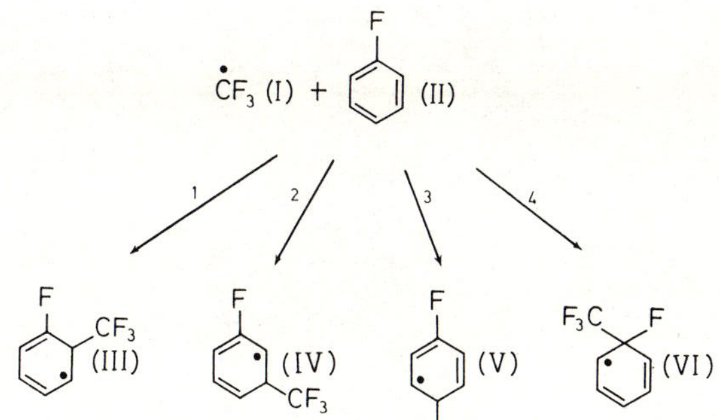

	(I)	(II)	(III)	(IV)	(V)	(VI)
$\Delta H^{\circ}_{f(300)}$	-112.5 (-471.0)	-26.3 (-110.1)	-167.6 (-701.7)	-167.6 (-701.7)	-167.6 (-701.7)	
$S^{\circ}_{(300)}$	63.8 (267.1)	73.8 (309.0)	[85 (356)]	[85 (356)]	[85 (356)]	
$C^{\circ}_{p(300)}$	12.2 (51.1)	22.5 (94.2)	[38.9 (162.9)]	[38.9 (162.9)]	[38.9 (162.9)]	

	Paths 1,2 and 3	Path 4
ΔH°	-28.8 (-120.6)	
ΔS°	[-52 (-217)]	
ΔC°_{p}	[4.2 (17.6)]	

Path	log A	E	log k	Conditions	Reference and Experimental Method
1+2+3+4	7.77	4.0 (16.7)	5.19(338 K)	302-398 K 30 torr	[133] S.W.Charles, J.T.Pearson and E.Whittle, Trans.Faraday Soc., 59,1156 (1963). Arrhenius parameters relative to log k_c = 10.36 for combination of CF_3 radicals; log $(A_a/A_c^{\frac{1}{2}})(\ell^{\frac{1}{2}}mol^{-\frac{1}{2}}s^{-\frac{1}{2}})$ = 2.59 and $E_a - \frac{1}{2}E_c$ = 4.0 kcal mol^{-1}. For details see CF_3 + benzene system, reference [131].

PREFERRED RATE PARAMETERS

Reasonable as given above

COMMENTS

There is no information on the products of the reaction and consequently it is not possible to say which path, if any, will be favoured.

REACTION: TRIFLUOROMETHYL RADICAL + o-DIFLUOROBENZENE

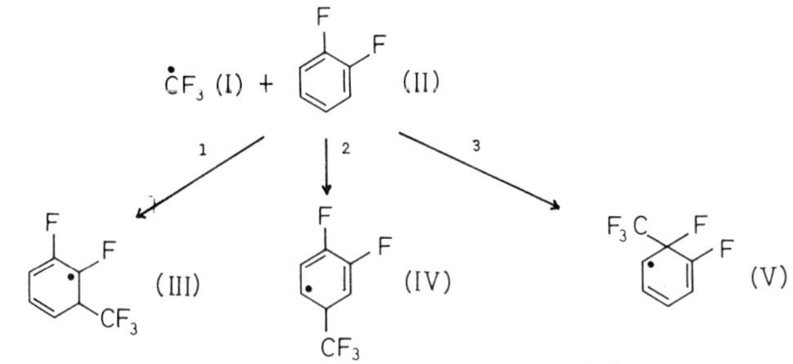

	(I)	(II)	(III)	(IV)	(V)
$\Delta H^\circ_{f(300)}$	-112.5 (-471.0)	-67.4 (-282.2)		-208.7 (-873.8)	
$S^\circ_{(300)}$	63.8 (267.1)	78.3 (327.8)		[90 (377)]	
$C^\circ_{p(300)}$	12.2 (51.1)	25.6 (107.2)		[42.0 (175.9)]	

	Paths 1 and 3	Path 2
ΔH°		-28.8 (-120.6)
ΔS°		[-52 (-217)]
ΔC°_p		[4.8 (17.6)]

Path	log A	E	log k	Conditions	Reference and Experimental Method
1+2+3	8.06	4.7 (19.7)	5.02(338 K)	300-400 K 30 torr	[133] S.W.Charles, J.T.Pearson and E.Whittle, Trans.Faraday Soc., 59, 1156 (1963). Arrhenius parameters relative to log k_c = 10.36 for combination of CF_3 radicals; log $(A_a/A_c^{\frac{1}{2}})(\ell^{\frac{1}{2}}mol^{-\frac{1}{2}}s^{-\frac{1}{2}})$ = 2.88 and $E_a - \frac{1}{2}E_c$ = 4.7 kcal mol^{-1}. For details see CF_3 + benzene system, reference [131].

PREFERRED RATE PARAMETERS

Reasonable as given above

COMMENTS

There is no information on the products of the reaction and consequently it is not possible to say which path, if any, will be favoured.

ΔH^o_f, ΔH^o and E in kcal mol^{-1} (kJ mol^{-1}); S^o, C^o_p, ΔS^o and ΔC^o_p in cal mol^{-1} K^{-1} (J mol^{-1} K^{-1}); A and k in l mol^{-1} s^{-1}

REACTION: TRIFLUOROMETHYL RADICAL + M-DIFLUOROBENZENE

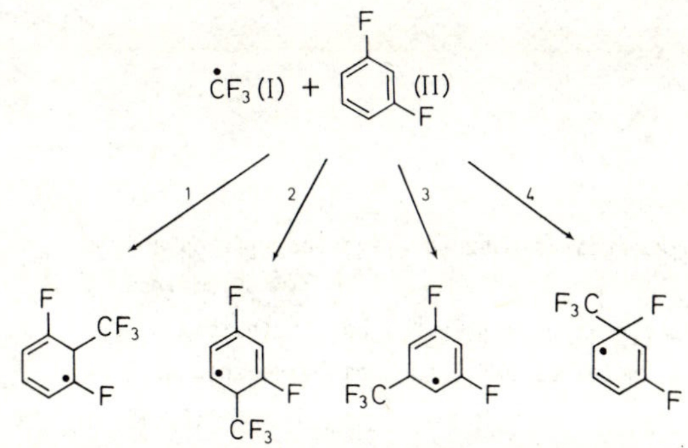

(III) (IV) (V) (VI)

	(I)	(II)	(III)	(IV)	(V)	(VI)
$\Delta H^\circ_{f(300)}$	-112.5 (-471.0)	-72.4 (-303.1)		-213.7 (-894.8)	-213.7 (-894.8)	
$S^\circ_{(300)}$	63.8 (267.1)	78.3 (327.8)		[90 (377)]	[90 (377)]	
$C^\circ_{p(300)}$	12.2 (51.1)	25.6 (107.2)		[42.0 (175.9)]	[42.0 (175.9)]	

	Paths 1 and 4	Paths 2 and 3
ΔH°		-28.8 (-120.6)
ΔS°		[-52 (-217)]
ΔC°_p		[4.8 (17.6)]

Path	log A	E	log k	Conditions	Reference and Experimental Method
1+2+3+4	8.06	4.9 (20.5)	4.89(338 K)	300-400 K 30 torr	[133] S.W.Charles, J.T.Pearson and E.Whittle, Trans.Faraday Soc., 59, 1156 (1963) Arrhenius parameters relative to log k_c = 10.36 for combination of CF_3 radicals; log $(A_a/A_c^{\frac{1}{2}})(\ell^{\frac{1}{2}}mol^{-\frac{1}{2}}s^{-\frac{1}{2}})$ = 2.88 and $E_a - \frac{1}{2}E_c$ = 4.9 kcal mol^{-1}. For details see CF_3 + benzene system, reference [131].

PREFERRED RATE PARAMETERS

Reasonable as given above

COMMENTS

There is no information on the products of the reaction, and consequently it is not possible to say which path, if any, will be favoured.

REACTION: TRIFLUOROMETHYL RADICAL + p-DIFLUOROBENZENE

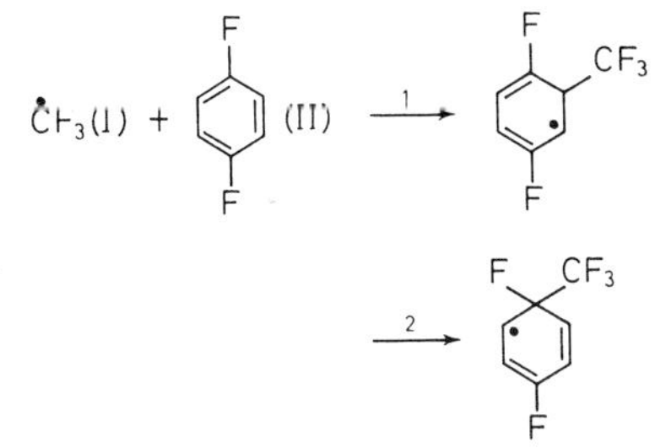

	(I)	(II)	(III)	(IV)
$\Delta H^o_{f(300)}$	-112.5 (-471.0)	-72.4 (-303.1)	-213.7 (-894.8)	
$S^o_{(300)}$	63.8 (267.1)	78.3 (327.8)	[90 (377)]	
$C^o_{p(300)}$	12.2 (51.1)	25.6 (162.5)	[42.0 (175.9)]	

	Path 1	Path 2
ΔH^o	-28.8 (-120.6)	
ΔS^o	[-52 (-217)]	
ΔC^o_p	[4.8 (17.6)]	

Path	log A	E	log k	Conditions	Reference and Experimental Method
1 + 2	7.72	4.1 (17.2)	5.07(338 K)	300-400 K 30 torr	[133] S.W.Charles, J.T.Pearson and E.Whittle, Trans.Faraday Soc.; 59, 1156 (1963). Arrhenius parameters relative to log k_C = 10.36 for combination of CF_3 radicals; $\log(A_{1+2}/A_C^{\frac{1}{2}})(\ell^{\frac{1}{2}}mol^{-\frac{1}{2}}s^{-\frac{1}{2}}) = 2.54$ and $E_{1+2} - \frac{1}{2}E_C$ = 4.1 kcal mol^{-1}. For details see CF_3 + benzene system, reference [131].

PREFERRED RATE PARAMETERS

Reasonable as given above

ΔH^o_f, ΔH^o and E in kcal mol^{-1} (kJ mol^{-1}); S^o, C^o_p, ΔS^o and ΔC^o_p in cal mol^{-1} K^{-1} (J mol^{-1} K^{-1}); A and k in l mol^{-1} s^{-1}

REACTION: TRIFLUOROMETHYL RADICAL + 1,2,4-TRIFLUOROBENZENE

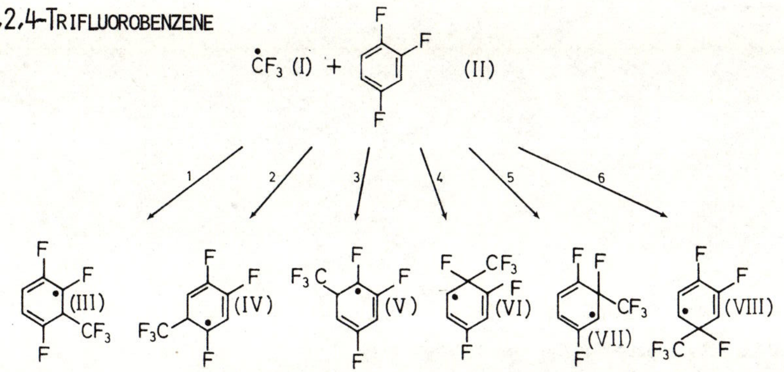

	(I)	(II)	(III)	(IV) and (V)	(VI) - (VIII)		Paths 2 and 3
$\Delta H^\circ_{f(300)}$	-112.5 (-471.0)	-113.5 (-475.2)		-254.8 (-1066.8)		ΔH°	-28.8 (-120.6)
$S^\circ_{(300)}$	63.8 (267.1)	82.9 (342.9)		[94 (394)]		ΔS°	[-52 (-219)]
$C^\circ_{p(300)}$	12.2 (51.1)	28.6 (119.7)		[45.0 (188.4)]		$\Delta C^\circ_{p(300)}$	[4.2 (17.6)]
$C^\circ_{p(500)}$	15.2 (63.6)	41.9 (175.4)		[63.1 (264.2)]		$\Delta C^\circ_{p(500)}$	[6.0 (25.1)]
						$<\Delta C^\circ_p>$	[5.1 (21.4)]

Path	log A	E	log k	Conditions	Reference and Experimental Method
1+2+3+ 4+5+6.	7.91	5.0 (20.9)	4.68(338 K)	300-400 K 30 torr	[133] S.W.Charles, J.T.Pearson and E.Whittle, Trans.Faraday Soc., 59, 1156 (1963). Arrhenius parameters relative to log k_c = 10.36 for combination of CF_3 radicals; $\log(A_a/A_c^{\frac{1}{2}})(\ell^{\frac{1}{2}}mol^{-\frac{1}{2}}s^{-\frac{1}{2}})$ = 2.73 and $E_a - \frac{1}{2}E_c$ = 5.0 kcal mol^{-1}. For details see CF_3 + benzene system, reference [131].

PREFERRED RATE PARAMETERS

Reasonable as given above

COMMENTS

No information on products of reaction.

REACTION: TRIFLUOROMETHYL RADICAL + 1,3,5-TRIFLUOROBENZENE

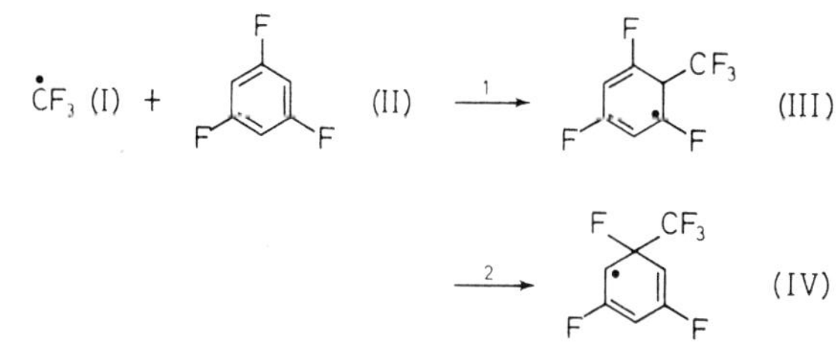

	(I)	(II)	(III)	(IV)		Path 1	Path 2
$\Delta H^o_{f(300)}$	-112.5 (-471.0)	-118.5 (-496.2)			ΔH^o		
$S^o_{(300)}$	63.8 (267.1)	82.9 (347.1)			ΔS^o		
$C^o_{p(300)}$	12.2 (51.1)	28.6 (119.7)			ΔC^o_p		

Path	log A	E	log k	Conditions	Reference and Experimental Method
1 + 2	7.36	4.5 (18.8)	4.45(338 K)	300-400 K 30 torr	[133] S.W.Charles, J.T.Pearson and E.Whittle, Trans.Faraday Soc., 59, 1156 (1963). Arrhenius parameters relative to log k_c = 10.36 for combination of CF_3 radicals; $\log(A_a/A_c^{\frac{1}{2}})(\ell^{\frac{1}{2}}mol^{-\frac{1}{2}}s^{-\frac{1}{2}})$ = 2.18 and $E_a - \frac{1}{2}E_c$ = 4.5 kcal mol^{-1}. For details see CF_3 + benzene system, reference [131].

PREFERRED RATE PARAMETERS

Reasonable as given above

COMMENTS

No information on products of the reaction.

ΔH^o_f, ΔH^o and E in kcal mol^{-1} (kJ mol^{-1}); S^o, C^o_p, ΔS^o and ΔC^o_p in cal mol^{-1} K^{-1} (J mol^{-1} K^{-1}); A and k in l mol^{-1} s^{-1}

REACTION: TRIFLUOROMETHYL RADICAL + 1,2,3,4-TETRAFLUOROBENZENE

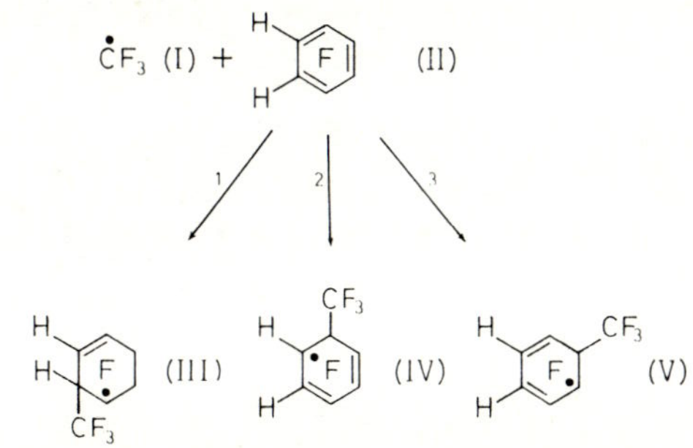

	(I)	(II)	(III)	(IV)	(V)		Path 1	Path 2	Path 3
$\Delta H^\circ_{f(300)}$	-112.5 (-471.0)	-149.6 (-626.4)				ΔH°			
$S^\circ_{(300)}$	63.8 (267.1)	87.5 (366.4)				ΔS°			
$C^\circ_{p(300)}$	12.2 (51.1)	31.7 (132.7)				ΔC°_p			

Path	log A	E	log k	Conditions	Reference and Experimental Method
1+2+3	8.31	5.8 (24.3)	4.56(338 K)	300-400 K 30 torr	[133] S.W.Charles, J.T.Pearson and E.Whittle, Trans.Faraday Soc., 59, 1156 (1963). Arrhenius parameters relative to log k_c = 10.36 for combination of CF_3 radicals; log $(A_a/A_c^{\frac{1}{2}})(\ell^{\frac{1}{2}}mol^{-\frac{1}{2}}s^{-\frac{1}{2}})$ = 3.13 and $E_a - \frac{1}{2}E_c$ = 5.8 kcal mol^{-1}. For details see CF_3 + benzene system, reference [131].

PREFERRED RATE PARAMETERS

Reasonable as given above

COMMENTS

No information on products of the reaction.

REACTION: TRIFLUOROMETHYL RADICAL + 1,2,4,5-TETRAFLUOROBENZENE

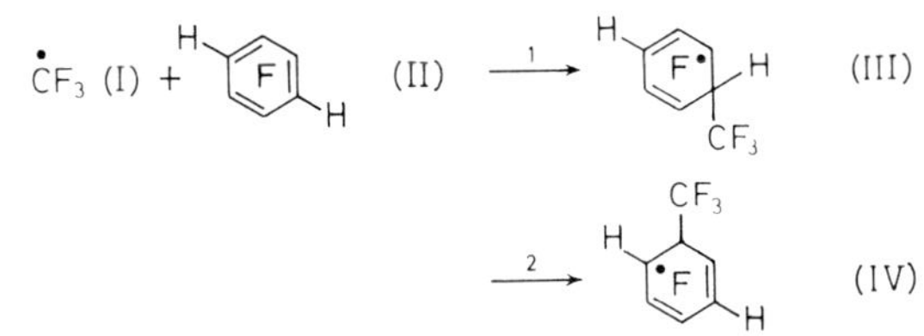

	(I)	(II)	(III)	(IV)		Path 1	Path 2
$\Delta H^\circ_{f(300)}$	-112.5 (-471.0)	-154.6 (-647.3)			ΔH°		
$S^\circ_{(300)}$	63.8 (267.1)	87.5 (366.8)			ΔS°		
$C^\circ_{p(300)}$	12.2 (51.1)	31.7 (132.7)			$\Delta C^\circ_{p(300)}$		
$C^\circ_{p(500)}$	15.2 (63.6)	44.9 (188.0)			$\Delta C^\circ_{p(500)}$		

Path	log A	E	log k	Conditions	Reference and Experimental Method
1 + 2	7.19	4.1 (17.2)	4.54(338 K)	300-400 K 30 torr	[133] S.W.Charles, J.T.Pearson and E.Whittle, Trans.Faraday Soc., 59, 1156 (1963). Arrhenius parameters relative to log k_c = 10.36 for combination of CF_3 radicals; log $(A_{1+2}/A_c^{\frac{1}{2}})(\ell^{\frac{1}{2}}mol^{-\frac{1}{2}}s^{-\frac{1}{2}})$ = 2.01 and $E_{1+2} - \frac{1}{2}E_c$ = 4.1 kcal mol^{-1}. For details see CF_3 + benzene system, reference [131].

PREFERRED RATE PARAMETERS

Reasonable as given above

COMMENTS

No information on products of the reaction.

ΔH°_f, ΔH° and E in kcal mol^{-1} (kJ mol^{-1}); S°, C°_p, ΔS° and ΔC°_p in cal mol^{-1} K^{-1} (J mol^{-1} K^{-1}); A and k in l mol^{-1} s^{-1}

REACTION: TRIFLUOROMETHYL RADICAL + PENTAFLUOROBENZENE

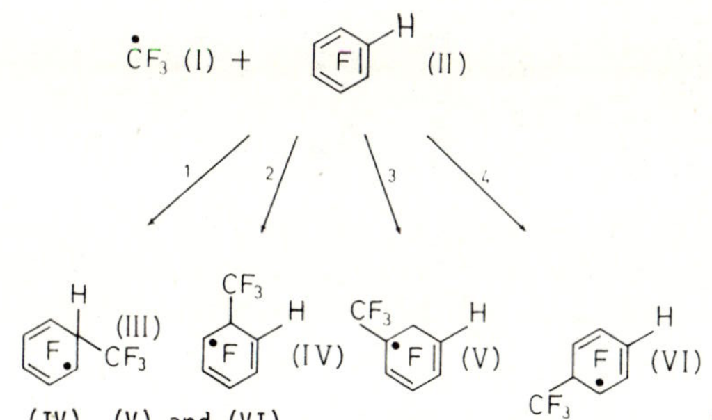

	(I)	(II)	(III), (IV), (V) and (VI)	Paths 1, 2, 3 and 4
$\Delta H^\circ_{f(300)}$	-112.5 (-471.0)	-190.7 (-798.5)		ΔH°
$S^\circ_{(300)}$	63.8 (267.1)	92.0 (385.2)		ΔS°
$C^\circ_{p(300)}$	12.2 (51.1)	34.7 (145.3)		$\Delta C^\circ_{p(300)}$
$C^\circ_{p(500)}$	15.2 (63.6)	48.0 (201.0)		$\Delta C^\circ_{p(500)}$

Path	log A	E	log k	Conditions	Reference and Experimental Method
1+2+3+4	7.19	4.1 (17.2)	4.54(338 K)	300-400 K 30 torr	[133] S.W.Charles, J.T.Pearson and E.Whittle, Trans.Faraday Soc., 59, 1156 (1963). Arrhenius parameters relative to log k_c = 10.36 for combination of CF_3 radicals; log $(A_a/A_c^{\frac{1}{2}})(\ell^{\frac{1}{2}}mol^{-\frac{1}{2}}s^{-\frac{1}{2}})$ = 2.01 and $E_a - \frac{1}{2}E_c$ = 4.1 kcal mol^{-1}. For details see CF_3 + benzene system, reference [131].

PREFERRED RATE PARAMETERS

Reasonable as given above

COMMENTS

No information on the products of the reaction.

REACTION: TRIFLUOROMETHYL RADICAL + HEXAFLUOROBENZENE

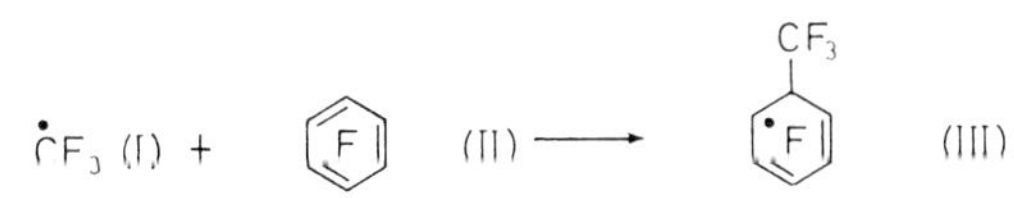

	(I)	(II)	(III)	
$\Delta H^\circ_{f(300)}$	-112.5 (-471.0)	-226.8 (-949.6)		ΔH°
$S^\circ_{(300)}$	63.8 (267.1)	96.6 (404.5)		ΔS°
$C^\circ_{p(300)}$	12.2 (51.1)	37.8 (158.3)		ΔC°_p

log A	E	log k	Conditions	Reference and Experimental Method
8.01	5.8 (24.3)	4.26(338 K)	300-400 K 35 torr	[133] S.W.Charles, J.T.Pearson and E.Whittle, <u>Trans.Faraday Soc.</u>, <u>59</u>, 1156 (1963). Arrhenius parameters relative to log k_C = 10.36 for combination of CF_3 radicals; log $(A_1/A_C^{\frac{1}{2}})(\ell^{\frac{1}{2}}mol^{-\frac{1}{2}}s^{-\frac{1}{2}})$ = 2.83 and $E_1 - \frac{1}{2}E_C$ = 5.8 kcal mol^{-1}. For details see CF_3 + benzene system, reference [131].
7.70	5.1 (21.4)	4.88(395 K)	358-453 K 700 torr	[129] G.E.Owen, J.M.Pearson and M.Szwarc, <u>Trans.Faraday Soc.</u>, <u>61</u>, 1722 (1965). Results calculated from log (A_1/A_3) = -0.14 and $E_1 - E_3$ = 1.8 kcal mol^{-1}, taking log k_3 = 7.84 - 3.3/2.3RT. Same method as for CF_3 + $CH_2{=}CH_2$ system, reference [126]. $\dot{C}F_3 + RH \xrightarrow{3} CF_3H + \dot{R}$ $\dot{C}F_3 + O\ell \xrightarrow{1} CF_3\dot{O}\ell$

<u>PREFERRED RATE PARAMETERS</u>

log k = 7.86 - 5.5/2.3RT

<u>COMMENTS</u>

The agreement between the two sets of data is quite good.

The preferred parameters are the averages of the two determinations.

ΔH°_f, ΔH° and E in kcal mol^{-1} (kJ mol^{-1}); S°, C°_p, ΔS° and ΔC°_p in cal mol^{-1} K^{-1} (J mol^{-1} K^{-1}); A and k in l mol^{-1} s^{-1}

REACTION: TRIFLUOROMETHYL RADICAL + PENTAFLUOROCHLOROBENZENE

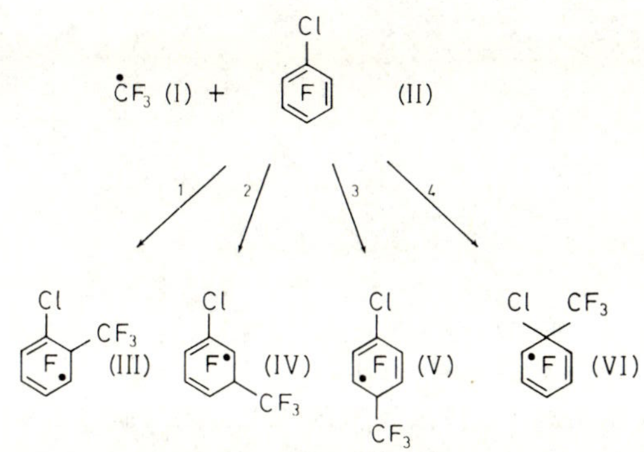

	(I)	(II)	(III), (IV), (V) and (VI)		Paths 1, 2, 3 and 4
$\Delta H^\circ_{f(300)}$	-112.5 (-471.0)	-193.8 (-811.4)		ΔH°	
$S^\circ_{(300)}$	63.8 (267.1)	99.4 (416.2)		ΔS°	
$C^\circ_{p(300)}$	12.2 (51.1)	38.9 (162.9)		ΔC°_p	

Path	log A	E	log k	Conditions	Reference and Experimental Method
1+2+3+4	7.83	5.2 (21.8)	4.49(338 K)	295-384 K 30 torr	[134] R.D.Giles and E.Whittle, Trans.Faraday Soc., 62, 128 (1966). Arrhenius parameters relative to log k_c = 10.36 for combination of CF_3 radicals; log $(A_a/A_c^{\frac{1}{2}})(\ell^{\frac{1}{2}}mol^{-\frac{1}{2}}s^{-\frac{1}{2}})$ = 2.65 and $E_a - \frac{1}{2}E_c$ = 5.2 kcal mol^{-1}. For details see CF_3 + benzene system, reference [131].

PREFERRED RATE PARAMETERS

Reasonable as given above

COMMENTS

No information on the products of the reaction.

REACTION: TRIFLUOROMETHYL RADICAL + PENTAFLUOROBROMOBENZENE

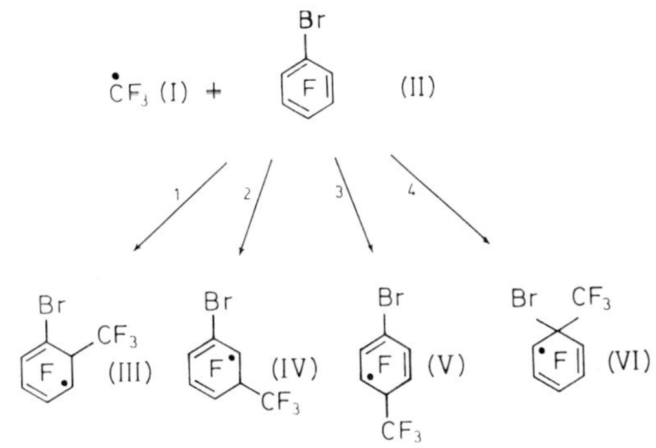

	(I)	(II)	(III)	(IV) (V) (VI)	Paths 1, 2, 3 and 4
$\Delta H^\circ_{f(300)}$	-112.5 (-471.0)	-183.3 (-767.5)			ΔH°
$S^\circ_{(300)}$	63.8 (267.1)	102.1 (427.5)			ΔS°
$C^\circ_{p(300)}$	12.2 (51.1)	39.3 (164.5)			ΔC°_p

Path	log A	E	log k	Conditions	Reference and Experimental Method
1+2+3+4	7.87	5.2 (21.8)	4.55(338 K)	312-380 K 30 torr	[133] S.W.Charles, J.T.Pearson and E.Whittle, Trans.Faraday Soc., 59, 1156 (1963). Arrhenius parameters relative to log k_c = 10.36 for combination of CF_3 radicals; $\log (A_a/A_c^{\frac{1}{2}})(\ell^{\frac{1}{2}}mol^{-\frac{1}{2}}s^{-\frac{1}{2}}) = 2.69$ and $E_a - \frac{1}{2}E_c$ = 5.2 kcal mol^{-1}. For details see CF_3 + benzene system, reference [131].
1+2+3+4	7.91	5.3 (22.2)	4.58(338 K)	312-380 K 30 torr	[134] R.D.Giles and E.Whittle, Trans.Faraday Soc., 62, 128 (1966). Arrhenius parameters relative to log k_c = 10.36 for combination of CF_3 radicals; $\log (A_a/A_c^{\frac{1}{2}})(\ell^{\frac{1}{2}}mol^{-\frac{1}{2}}s^{-\frac{1}{2}}) = 2.73$ and $E_a - \frac{1}{2}E_c$ = 5.3 kcal mol^{-1}. As for reference [133] above.

PREFERRED RATE PARAMETERS

Reasonable as given above

COMMENTS

No information on the products of the reaction.

ΔH^o_f, ΔH^o and E in kcal mol^{-1} (kJ mol^{-1}); S^o, C^o_p, ΔS^o and ΔC^o_p in cal mol^{-1} K^{-1} (J mol^{-1} K^{-1}); A and k in l mol^{-1} s^{-1}

REACTION: TRIFLUOROMETHYL RADICAL + CHLOROBENZENE

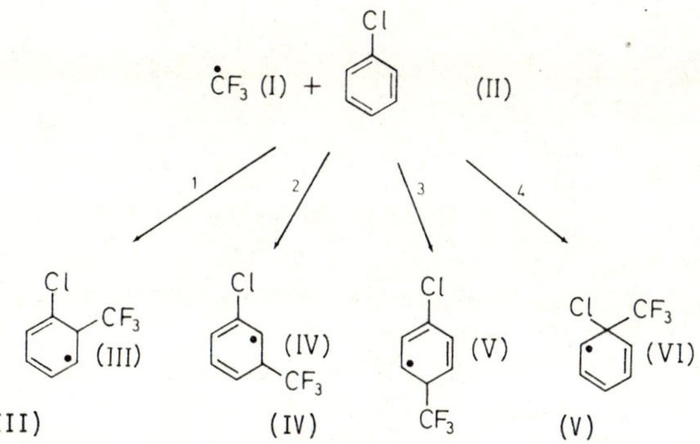

	(I)	(II)	(III)	(IV)	(V)	(VI)
$\Delta H^\circ_{f(300)}$	-112.5 (-471.0)	12.7 (53.2)	-128.0 (-535.9)	-128.6 (-538.4)	-128.6 (-538.4)	
$S^\circ_{(300)}$	63.8 (267.1)	76.6 (320.7)	[88 (369)]	[88 (369)]	[88 (369)]	
$C^\circ_{p(300)}$	12.2 (51.1)	23.6 (98.8)	[40.0 (167.5)]	[40.0 (167.5)]	[40.0 (167.5)]	

	Path 1	Paths 2 and 3
ΔH°	-28.2 (-118.1)	-28.8 (-120.6)
ΔS°	[-52 (-217)]	[-52 (-217)]
ΔC°_p	[4.2 (17.6)]	[4.2 (17.6)]

Path	log A	E	log k	Conditions	Reference and Experimental Method
1+2+3+4	8.28	4.4 (18.4)	5.43(338 K)	293-391 K 30 torr	[134] R.D.Giles and E.Whittle, Trans.Faraday Soc., 62, 128 (1966). Arrhenius parameters relative to log k_c = 10.36 for combination of CF_3 radicals; log $(A_a/A_c^{\frac{1}{2}})(\ell^{\frac{1}{2}}mol^{-\frac{1}{2}}s^{-\frac{1}{2}})$ = 3.10 and $E_a - \frac{1}{2}E_c$ = 4.4 kcal mol^{-1}. For details see CF_3 + benzene system, reference [131].

PREFERRED RATE PARAMETERS

Reasonable as given above

COMMENTS

No information on products of the reaction.

REACTION: TRIFLUOROMETHYL RADICAL + BROMOBENZENE

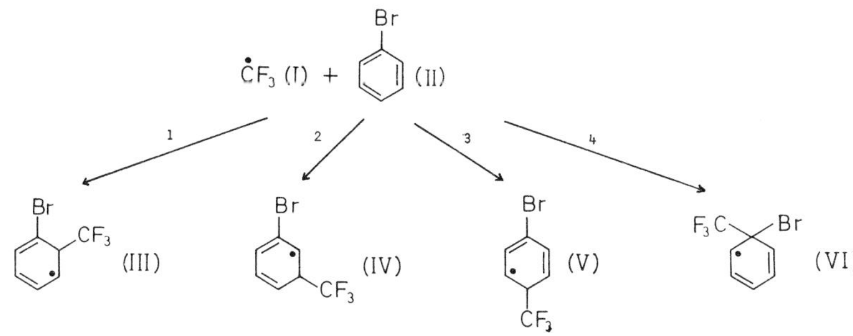

	(I)	(II)	(III)	(IV)	(V)	(VI)
$\Delta H^\circ_{f(300)}$	-112.5 (-471.0)	27.2 (113.9)	-113.5 (-475.2)	-114.1 (-477.7)	-114.1 (-477.7)	
$S^\circ_{(300)}$	63.8 (267.1)	79.3 (332.0)	[91 (381)]	[91 (381)]	[91 (381)]	
$C^\circ_{p(300)}$	12.2 (51.1)	24.0 (100.5)	[40.4 (169.2)]	[40.4 (169.2)]	[40.4 (169.2)]	

	Path 1	Paths 2 and 3	Path 4
ΔH°	-28.2 (-118.1)	-28.8 (-120.6)	
ΔS°	[-52 (-217)]	[-52 (-217)]	
ΔC°_p	[4.2 (17.6)]	[4.2 (17.6)]	

Path	log A	E	log k	Conditions	Reference and Experimental Method
1+2+3+4	9.89	6.7 (28.1)	5.66(338 K)	300-400 K 30 torr	[133] S.W.Charles, J.T.Pearson and E.Whittle, Trans.Faraday Soc., 59, 1156 (1963). Arrhenius parameters relative to log k_c = 10.36 for combination of CF_3 radicals; log $(A_a/A_c^{\frac{1}{2}})(\ell^{\frac{1}{2}}mol^{-\frac{1}{2}}s^{-\frac{1}{2}})$ = 4.71 and $E_a - \frac{1}{2}E_c$ = 6.7 kcal mol^{-1}. For details see CF_3 + benzene system, reference [131].
1+2+3+4	8.79	4.8 (20.1)	5.69(338 K)	300-400 K 30 torr	[134] R.D.Giles and E.Whittle, Trans.Faraday Soc., 62, 128 (1966). Arrhenius parameters relative to log k_c = 10.36 for combination of CF_3 radicals; log $(A_a/A_c^{\frac{1}{2}})(\ell^{\frac{1}{2}}mol^{-\frac{1}{2}}s^{-\frac{1}{2}})$ = 3.61 and $E_a - \frac{1}{2}E_c$ = 4.8 kcal mol^{-1}. As for reference [133] above.

ΔH^o_f, ΔH^o and E in kcal mol^{-1} (kJ mol^{-1}); S^o, C^o_p, ΔS^o and ΔC^o_p in cal mol^{-1} K^{-1} (J mol^{-1} K^{-1}); A and k in l mol^{-1} s^{-1}

PREFERRED RATE PARAMETERS

log k = 8.79 - 4.8/2.3RT

COMMENTS

No information on products.

REACTION: TRIFLUOROMETHYL RADICAL + BENZOTRIFLUORIDE

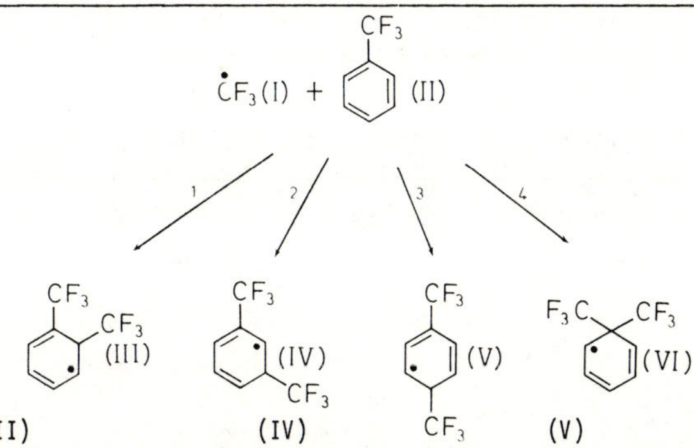

	(I)	(II)	(III)	(IV)	(V)	(VI)
$\Delta H^\circ_{f(300)}$	-112.5 (-471.0)	-136.4 (-571.1)	-281.4 (-1178.2)	-282.0 (-1180.7)	-282.0 (-1180.7)	-276.1 (-1156.0)
$S^\circ_{(300)}$	63.8 (267.1)	90.3 (378.1)	[100 (420)]	[102 (427)]	[102 (427)]	[97 (406)]
$C^\circ_{p(300)}$	12.2 (51.1)	31.6 (132.3)	[48.9 (204.7)]	[47.8 (200.1)]	[47.8 (200.1)]	[48.0 (201.0)]

	Path 1	Path 2 and 3	Path 4
ΔH°	-32.5 (-136.1)	-33.1 (-138.6)	-27.2 (-113.9)
ΔS°	[-54 (-226)]	[-52 (-217)]	[-57 (-239)]
ΔC°_p	[5.1 (21.5)]	[4.0 (16.7)]	[4.2 (17.6)]

Path	log A	E	log k	Conditions	Reference and Experimental Method
1+2+3+4	8.82	5.8 (24.3)	5.07(338 K)	300-400 K 30 torr	[133] S.W.Charles, J.T.Pearson and E.Whittle, Trans.Faraday Soc., 59, 1156 (1963). Arrhenius parameters relative to log k_c = 10.36 for combination of

Path	log A	E	log k	Conditions	Reference and Experimental Method
					CF_3 radicals; $\log (A_a/A_c^{\frac{1}{2}})(\ell^{\frac{1}{2}}mol^{-\frac{1}{2}}s^{-\frac{1}{2}}) = 3.64$ and $E_a - \frac{1}{2}E_c = 5.8$ kcal mol^{-1}. For details see CF_3 + benzene system, reference [131].

PREFERRED RATE PARAMETERS

Reasonable as given above

COMMENTS

There is no information on the products of the reaction, and consequently it is not possible to say which path, if any, will be favoured.

ΔH_f^o, ΔH^o and E in kcal mol^{-1} (kJ mol^{-1}); S^o, C_p^o, ΔS^o and ΔC_p^o in cal mol^{-1} K^{-1} (J mol^{-1} K^{-1}); A and k in l mol^{-1} s^{-1}

REACTION: TRIFLUOROMETHYL RADICAL + PENTAFLUOROTOLUENE

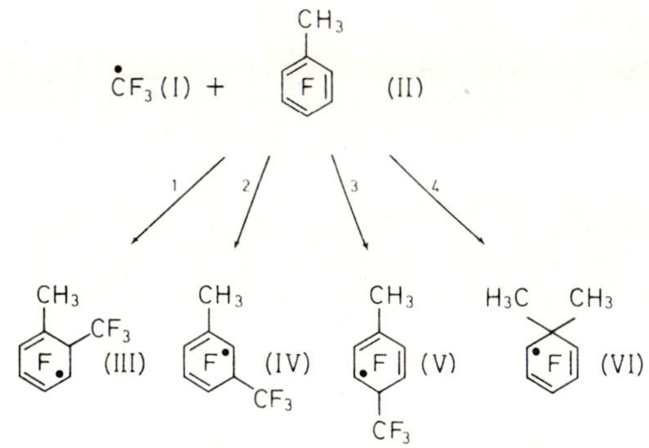

	(I)	(II)	(III) - (VI)		Paths 1, 2, 3 and 4
$\Delta H^\circ_{f(300)}$	-112.5 (-471.0)	-198.6 (-831.5)		ΔH°	
$S^\circ_{(300)}$	63. 8 (267.1)	101.0 · (422.9)		ΔS°	
$C^\circ_{p(300)}$	12.2 (51.1)	40.4 (169.2)		ΔC°_p	

Path	log A	E	log k	Conditions	Reference and Experimental Method
1+2+3+4	6.95	3.7 (15.5)	4.56(338 K)	300-400 K 40 torr	[133] S.W.Charles, J.T.Pearson and E.Whittle, Trans.Faraday Soc., 59, 1156 (1963). Arrhenius parameters relative to log k_c = 10.36 for the combination of CF_3 radicals; log $(A_a/A_c^{\frac{1}{2}})(\ell^{\frac{1}{2}}mol^{-\frac{1}{2}}s^{-\frac{1}{2}})$ = 1.77 and $E_a - \frac{1}{2}E_c$ = 3.7 kcal mol^{-1}. For details see CF_3 + benzene system, reference [131].

PREFERRED RATE PARAMETERS

Reasonable as given above

COMMENTS

No information on the products of the reaction.

REACTION: TRIFLUOROMETHYL RADICAL + BENZYL CHLORIDE

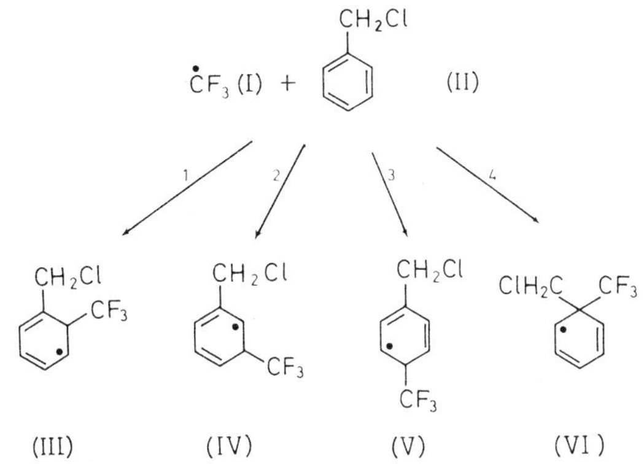

	(I)	(II)	(III) - (VI)		Paths 1, 2, 3 and 4
$\Delta H^\circ_{f(300)}$	-112.5 (-471.0)			ΔH°	
$S^\circ_{(300)}$	63.8 (267.1)			ΔS°	
$C^\circ_{p(300)}$	12.2 (51.1)			ΔC°_p	

Path	log A	E	log k	Conditions	Reference and Experimental Method
1+2+3+4	7.72	3.5 (14.7)	5.79(395 K)	340-424 K 40 torr	[134] R.D.Giles and E.Whittle, Trans.Faraday Soc., 62, 128 (1966). Arrhenius parameters relative to log k_c = 10.36 for combination of CF_3 radicals; log $(A_a/A_c^{\frac{1}{2}})(\ell^{\frac{1}{2}} mol^{-\frac{1}{2}} s^{-\frac{1}{2}})$ = 2.54 and $E_a - \frac{1}{2}E_c$ = 3.5 kcal mol^{-1}. For details see CF_3 + benzene system, reference [131].

PREFERRED RATE PARAMETERS

Reasonable as given above

COMMENTS

No information on products of the reaction.

ΔH^o_f, ΔH^o and E in kcal mol^{-1} (kJ mol^{-1}); S^o, C^o_p, ΔS^o and ΔC^o_p in cal mol^{-1} K^{-1} (J mol^{-1} K^{-1}); A and k in l mol^{-1} s^{-1}

REACTION: TRIFLUOROMETHYL RADICAL + BENZOTRICHLORIDE

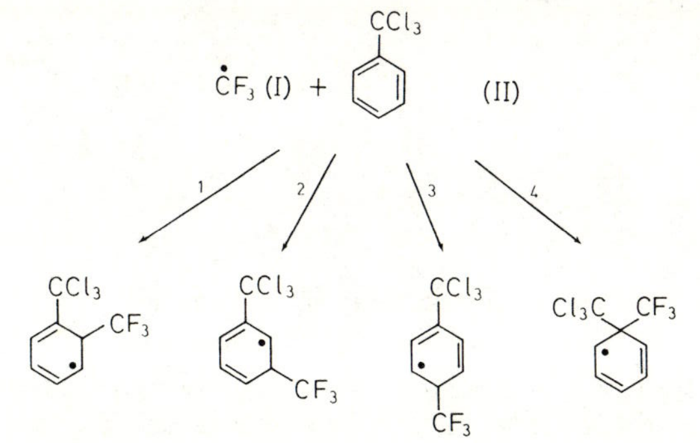

	(I)	(II)	(III) - (VI)		Paths 1, 2, 3 and 4
$\Delta H^\circ_{f(300)}$	-112.5 (-471.0)			ΔH°	
$S^\circ_{(300)}$	63.8 (267.1)			ΔS°	
$C^\circ_{p(300)}$	12.2 (51.1)			ΔC°_p	

Path	log A	E	log k	Conditions	Reference and Experimental Method
1+2+3+4	8.44	5.0 (20.9)	5.74(395 K)	377-444 K 40 torr	[134] R.D.Giles and E.Whittle, Trans.Faraday Soc., 62, 128 (1966). Arrhenius parameters relative to log k_c = 10.36 for combination of CF_3 radicals; log $(A_a/A_c^{\frac{1}{2}})(\ell^{\frac{1}{2}}mol^{-\frac{1}{2}}s^{-\frac{1}{2}})$ = 3.26 and $E_a - \frac{1}{2}E_c$ = 5.0 kcal mol^{-1}. For details see CF_3 + benzene system, reference [131].

PREFERRED RATE PARAMETERS

Reasonable as given above

COMMENTS

No information on products of the reaction.

REACTION: TRIFLUOROMETHYL RADICAL + PERFLUOROTOLUENE

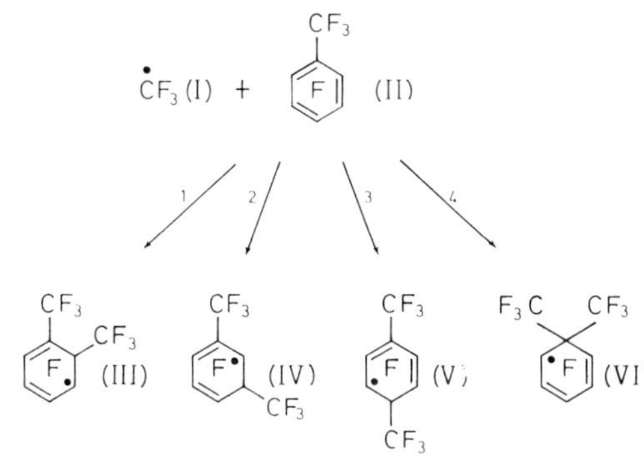

	(I)	(II)	(III) - (VI)		Paths 1, 2, 3 and 4
$\Delta H^{\circ}_{f(300)}$	-112.5 (-471.0)	-351.2 (-1470.5)		ΔH°	
$S^{\circ}_{(300)}$	63.8 (267.1)	113.4 (474.8)		ΔS°	
$C^{\circ}_{p(300)}$	12.2 (51.1)	46.7 (195.5)		ΔC°_{p}	

Path	log A	E	log k	Conditions	Reference and Experimental Method
1+2+3+4	7.29	4.6 (19.3)	4.32(338 K)	300-400 K 40 torr	[133] S.W.Charles, J.T.Pearson and E.Whittle, Trans.Faraday Soc., 59, 1156 (1963). Arrhenius parameters relative to log k_c = 10.36 for combination of CF_3 radicals; log $(A_a/A_c^{\frac{1}{2}})(\ell^{\frac{1}{2}}mol^{-\frac{1}{2}}s^{-\frac{1}{2}})$ = 2.11 and $E_a - \frac{1}{2}E_c$ = 4.6 kcal mol^{-1}. For details see CF_3 + benzene system, reference [131].

PREFERRED RATE PARAMETERS

Reasonable as given above

COMMENTS

No information on the products of the reaction.

ΔH^{o}_{f}, ΔH^{o} and E in kcal mol^{-1} (kJ mol^{-1}); S^{o}, C^{o}_{p}, ΔS^{o} and ΔC^{o}_{p} in cal mol^{-1} K^{-1} (J mol^{-1} K^{-1}); A and k in l mol^{-1} s^{-1}

REACTION: TRIFLUOROMETHYL RADICAL + PERFLUORO-P-XYLENE

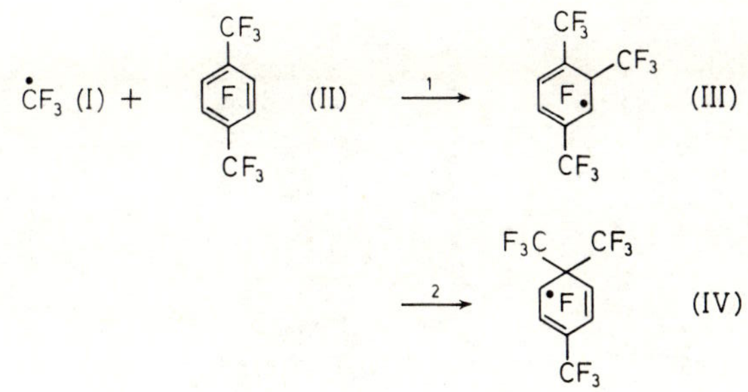

	(I)	(II)	(III)	(IV)		Paths 1 and 2
$\Delta H^\circ_{f(300)}$	-112.5 (-471.0)	-475.6 (-1991.3)			ΔH°	
$S^\circ_{(300)}$	63.8 (267.1)	134.6 (563.6)			ΔS°	
$C^\circ_{p(300)}$	12.2 (51.1)	55.5 (232.4)			ΔC°_p	

Path	log A	E	log k	Conditions	Reference and Experimental Method
1 + 2	6.41	4.2 (17.6)	3.69(338 K)	300-400 K 40 torr	[133] S.W.Charles, J.T.Pearson and E.Whittle, Trans.Faraday Soc., 59, 1156 (1963). Arrhenius parameters relative to log k_C = 10.36 for combination of CF_3 radicals; log $(A_{1+2}/A_C^{\frac{1}{2}})(\ell^{\frac{1}{2}}mol^{-\frac{1}{2}}s^{-\frac{1}{2}})$ = 1.23 and $E_{1+2} - \frac{1}{2}E_C$ = 4.2 kcal mol^{-1}. For details see CF_3 + benzene system, reference [131].

PREFERRED RATE PARAMETERS

Reasonable as given above

COMMENTS

No information on the products of the reaction.

REACTION: TRIFLUOROMETHYL RADICAL + ANISOLE

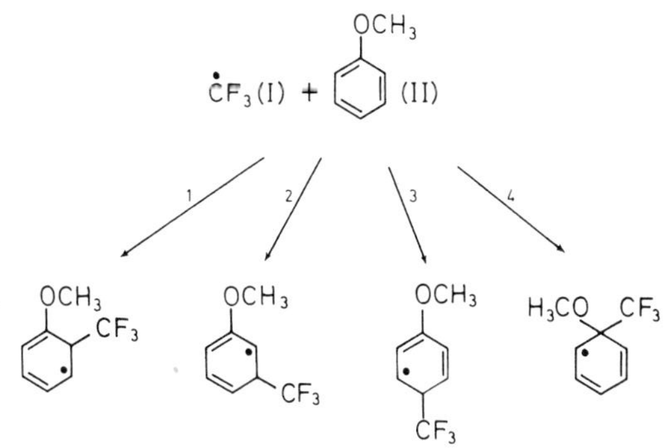

(III) (IV) (V) (VI)

	(I)	(II)	(III) - (V)	(VI)		Paths 1, 2 and 3
$\Delta H^\circ_{f(300)}$	-112.5 (-471.0)	-18.0 (-75.3)	-159.3 (-667.0)		ΔH°	-28.8 (-120.6)
$S^\circ_{(300)}$	63.8 (267.1)				ΔS°	
$C^\circ_{p(300)}$	12.2 (51.1)				ΔC°_p	

Path	log A	E	log k	Conditions	Reference and Experimental Method
1+2+3+4	-	-	5.90	338 K 30 torr	[133] S.W.Charles, J.T.Pearson and E.Whittle, Trans.Faraday Soc., 59, 1156 (1963). Rate constant relative to log k_c = 10.36 for combination of CF_3 radicals; $(k_a/k_c^{\frac{1}{2}})(\ell^{\frac{1}{2}}mol^{-\frac{1}{2}}s^{-\frac{1}{2}})$ = 5.35 For details see CF_3 + benzene system, reference [131].

PREFERRED RATE PARAMETERS

Reasonable as given above

COMMENTS

No information on products of reaction.

ΔH^o_f, ΔH^o and E in kcal mol^{-1} (kJ mol^{-1}); S^o, C^o_p, ΔS^o and ΔC^o_p in cal mol^{-1} K^{-1} (J mol^{-1} K^{-1}); A and k in l mol^{-1} s^{-1}

REACTION: TRIFLUOROMETHYL RADICAL + PYRIDINE

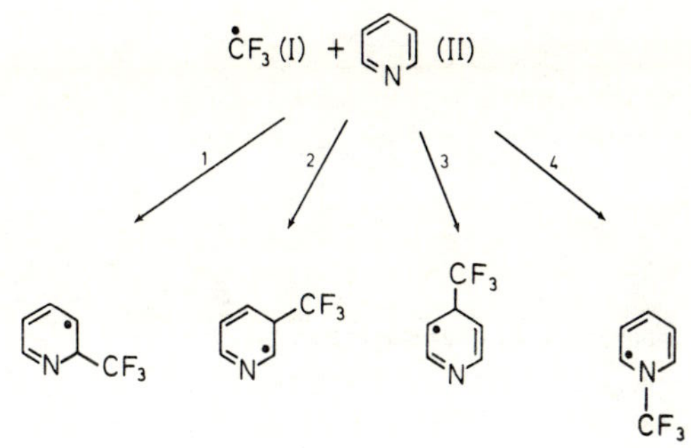

	(I)	(II)	(III)	(IV)	(V)	(VI)
$\Delta H^\circ_{f(300)}$	-112.5 (-471.0)	33.2 (139.0)	-113.8 (-476.4)		-110.5 (-462.4)	
$S^\circ_{(300)}$	63.8 (267.1)	67.7 (283.5)				
$C^\circ_{p(300)}$	12.2 (51.1)	18.8 (78.7)				

	Path 1	Path 2	Path 3	Path 4
ΔH°	-34.5 (-144.4)		-31.2 (-130.6)	
ΔS°				
ΔC°_p				

Path	log A	E	log k	Conditions	Reference and Experimental Method
1+2+3+4	∿7.49	3.0 (12.6)	∿5.89(393 K)	368-393 K 700 torr	[129] G.E.Owen, J.M.Pearson and M.Szwarc, Trans.Faraday Soc., 65, 1722 (1965). Results calculated from log $(A_a/A_3) \simeq -0.35$ and $E_a - E_3 = -0.3$ kcal mol^{-1}, taking log $k_3 = 7.84 - 3.3/2.3RT$. Same method as for $CF_3 + CH_2{=}CH_2$, reference [126] $\dot{C}F_3 + RH \xrightarrow{3} CF_3H + \dot{R}$ $\dot{C}F_3 + O\ell \xrightarrow{a} CF_3\dot{O}\ell$ At 453 K the rate constants of reaction (a) were not independent of pressure indicating that the addition reaction is reversible at this temperature.

PREFERRED RATE PARAMETERS

Reasonable as given above

COMMENTS

No information on products of the reaction.

REACTION: TRIFLUOROMETHYL RADICAL + ACETYLENE $\dot{C}F_3(I) + CH{\equiv}CH(II) \longrightarrow CF_3CH{=}\dot{C}H(III)$

	(I)	(II)	(III)		
$\Delta H^\circ_{f(300)}$	-112.5 (-471.0)	54.2 (226.9)	-87.6 (-366.8)	ΔH°	-29.3 (-122.7)
$S^\circ_{(300)}$	63.8 (267.1)	48.0 (201.0)	77.2 (323.2)	ΔS°	-34.6 (-144.9)
$C^\circ_{p(300)}$	12.2 (51.1)	10.5 (44.0)	21.8 (91.3)	$\Delta C^\circ_{p(300)}$	-0.9 (-3.8)
$C^\circ_{p(500)}$	15.2 (63.6)	13.1 (54.8)	29.5 (123.5)	$\Delta C^\circ_{p(500)}$	1.2 (5.0)
				$\langle \Delta C^\circ_p \rangle$	0.2 (0.8)

log A	E	log k	Conditions	Reference and Experimental Method
8.74	2.8 (11.7)	7.20(395 K)	338-453 K 700 torr	[129] G.E.Owen, J.M.Pearson and M.Szwarc, Trans.Faraday Soc., 61, 1722 (1965). Results calculated from log $(A_1/A_3) = 0.90$ and $E_1 - E_3 = -0.5$ kcal mol^{-1}, taking log $k_3 = 7.84 - 3.3/2.3RT$ (see $CF_3 + CH_2{=}CH_2$ system, reference [126]). Same method as for $CF_3 + CH_2{=}CH_2$ system, reference [126] $\dot{C}F_3 + RH \xrightarrow{3} CF_3H + \dot{R}$ $\dot{C}F_3 + O\ell \xrightarrow{1} CF_3\dot{O}\ell$

PREFERRED RATE PARAMETERS

Reasonable as given above

ΔH°_f, ΔH° and E in kcal mol^{-1} (kJ mol^{-1}); S°, C°_p, ΔS° and ΔC°_p in cal mol^{-1} K^{-1} (J mol^{-1} K^{-1}); A and k in l mol^{-1} s^{-1}

REACTION: TRIFLUOROMETHYL RADICAL + PROPYNE

$\dot{C}F_3(I) + CH\equiv CCH_3(II) \xrightarrow{1} \dot{C}H=C(CF_3)CH_3(III)$

$\xrightarrow{2} CF_3CH=\dot{C}CH_3(IV)$

	(I)	(II)	(III)	(IV)
$\Delta H^\circ_{f(300)}$	-112.5 (-471.0)	44.4 (185.9)	[-96 (-402)]	[-95 (-398)]
$S^\circ_{(300)}$	63.8 (267.1)	61.5 (257.5)	84.7 (354.6)	85.9 (359.7)
$C^\circ_{p(300)}$	12.2 (51.1)	14.6 (61.1)	28.0 (117.2)	27.1 (113.5)
$C^\circ_{p(500)}$	15.2 (63.6)	19.7 (82.5)	38.3 (160.4)	37.3 (156.2)

	Path 1	Path 2
ΔH°	[-28 (-117)]	[-27 (-113)]
ΔS°	-40.6 (-169.9)	-39.4 (-165.0)
$\Delta C^\circ_{p(300)}$	1.2 (5.0)	0.3 (1.3)
$\Delta C^\circ_{p(500)}$	3.4 (14.2)	2.4 (10.0)
$\langle\Delta C^\circ_p\rangle$	2.3 (9.6)	1.4 (5.9)

Path	log A	E	log k	Conditions	Reference and Experimental Method
1 + 2	8.72	2.2 (9.2)	7.51(395 K)	338-453 K 700 torr	[129] G.E.Owen, J.M.Pearson and M.Szwarc, Trans.Faraday Soc., 61, 1722 (1965). Results calculated from log (A_{1+2}/A_3) = 0.88 and $E_{1+2} - E_3$ = -1.1 kcal mol^{-1}, taking log k_3 = 7.84 - 3.3/2.3RT. Same method as for $CF_3 + CH_2=CH_2$ system, reference [126]. $\dot{C}F_3 + RH \xrightarrow{3} CF_3H + \dot{R}$; $\dot{C}F_3 + O\ell \xrightarrow{1+2} CF_3\dot{O}\ell$

PREFERRED RATE PARAMETERS

Reasonable as given above

REACTION: TRIFLUOROMETHYL RADICAL + BUTYNE-2 $\dot{C}F_3(I) + CH_3C{\equiv}CCH_3(II) \longrightarrow CH_3(CF_3)C{=}\dot{C}CH_3(III)$

	(I)	(II)	(III)		
$\Delta H^\circ_{f(300)}$	-112.5 (-471.0)	34.9 (146.1)	[-104 (-435)]	ΔH°	[-26 (-109)]
$S^\circ_{(300)}$	63.8 (267.1)	69.1 (289.3)	93.4 (391.1)	ΔS°	-39.5 (-165.4)
$C^\circ_{p(300)}$	12.2 (51.1)	18.6 (77.9)	33.3 (139.4)	$\Delta C^\circ_{p(300)}$	2.5 (10.5)
$C^\circ_{p(500)}$	15.2 (63.6)	26.4 (110.5)	46.0 (192.6)	$\Delta C^\circ_{p(500)}$	4.4 (18.4)
				$\langle\Delta C^\circ_p\rangle$	3.5 (14.7)

log A	E	log k	Conditions	Reference and Experimental Method
8.29	2.1 (8.8)	7.13(395 K)	338-453 K 700 torr	[129] G.E.Owen, J.M.Pearson and M.Szwarc, Trans.Faraday Soc., 61, 1722 (1965). Results calculated from log $(A_1/A_3) = 0.45$ and $E_1 - E_3 = -1.2$ kcal mol^{-1}, taking log $k_1 = 7.84 - 3.3/2.3RT$. Same method as for $CF_3 + CH_2{=}CH_2$ system, reference [126]. $\dot{C}F_3 + RH \xrightarrow{3} CF_3H + \dot{R}$ $\dot{C}F_3 + O\ell \xrightarrow{1} CF_3\dot{O}\ell$

PREFERRED RATE PARAMETERS

Reasonable as given above

ΔH°_f, ΔH° and E in kcal mol^{-1} (kJ mol^{-1}); S°, C°_p, ΔS° and ΔC°_p in cal mol^{-1} K^{-1} (J mol^{-1} K^{-1}); A and k in l mol^{-1} s^{-1}

REACTION: TRIFLUOROMETHYL RADICAL + PHENYLACETYLENE (ETHYNYL BENZENE)

$\dot{C}F_3(I) + CH{\equiv}CC_6H_5(II) \xrightarrow{1} \dot{C}H{=}C(CF_3)C_6H_5(III)$

$\xrightarrow{2} CF_3CH{=}\dot{C}C_6H_5(IV)$

	(I)	(II)	(III)	(IV)
$\Delta H^o_{f(300)}$	-112.5 (-471.0)	[78.1 (327.0)]	[-65 (-272)]	[-77 (-322)]
$S^o_{(300)}$	63.8 (267.1)	81.0 (339.1)	[104.6 (438.0)]	106.3 (445.1)
$C^o_{p(300)}$	12.2 (51.1)	27.6 (115.6)	[41.9 (175.4)]	41.1 (172.1)
$C^o_{p(500)}$	15.2 (63.6)	41.7 (174.6)	[61.5 (257.5)]	59.7 (250.0)

	Path 1	Path 2
ΔH^o	[-31 (-130)]	[-42 (-176)]
ΔS^o	[-40 (-167)]	-38.5 (-161.2)
$\Delta C^o_{p(300)}$	[2.1 (8.8)]	1.3 (5.4)
$\Delta C^o_{p(500)}$	[4.6 (19.3)]	2.8 (11.7)
$<\Delta C^o_p>$	3.4 (14.2)]	2.1 (8.8)

Path	log A	E	log k	Conditions	Reference and Experimental Method
1 + 2	8.47	1.1 (4.6)	7.86(395 K)	358-453 K 700 torr	[129] G.E.Owen, J.M.Pearson and M.Szwarc, Trans.Faraday Soc., 61, 1722 (1965). Results calculated from log (A_{1+2}/A_3) = 0.63 and $E_{1+2} - E_3$ = -2.2 kcal mol^{-1}, taking log k_3 = 7.84 - 3.3/2.3RT. Same method as for $CF_3 + CH_2{=}CH_2$ system, reference [126]. $\dot{C}F_3 + RH \xrightarrow{3} CF_3H + \dot{R}$ $\dot{C}F_3 + O\ell \xrightarrow{1+2} CF_3\dot{O}\ell$

PREFERRED RATE PARAMETERS

Reasonable as given above

COMMENTS

Path 1 will be the favoured route for addition. It is assumed that addition to the triple bond is very much faster than addition to the aromatic ring.

REACTION: FLUOROMETHYL RADICAL + ETHYLENE $\dot{C}H_2F(I) + CH_2{=}CH_2(II) \longrightarrow CH_2FCH_2\dot{C}H_2(III)$

	(I)	(II)	(III)		
$\Delta H^\circ_{f(300)}$	[-6 (-25)]	12.5 (52.3)	-20.9 (-87.5)	ΔH°	[-27 (-113)]
$S^\circ_{(300)}$	54.8 (229.4)	52.4 (219.4)	76.0 (318.2)	ΔS°	-31.2 (-130.6)
$C^\circ_{p(300)}$	8.6 (36.0)	10.3 (43.1)	19.6 (82.1)	ΔC°_p	0.7 (2.9)

log A	E	log k	Conditions	Reference and Experimental Method
7.59	4.3 (18.0)	5.10(378 K)	290-465 K 20-160 torr	[135] J.M.Sangster and J.C.J.Thynne, Trans.Faraday Soc., 65, 2110 (1969).

Static system. CH_2F radicals from the photolysis of $(CH_2F)_2CO$ (full Hg arc) in the presence of $CH_2{=}CH_2$ and H_2S as a radical scavenger. Principal reactions:

$$(CH_2F)_2CO + h\nu \xrightarrow{I} 2\dot{C}H_2F + CO$$
$$\dot{C}H_2F + CH_2{=}CH_2 \xrightarrow{1} CH_2FCH_2\dot{C}H_2$$
$$\dot{C}H_2F + H_2S \xrightarrow{2} CH_3F + H\dot{S}$$
$$CH_2FCH_2\dot{C}H_2 + H_2S \xrightarrow{3} CH_2FCH_2CH_3 + H\dot{S}$$

The products CH_3F and $n\text{-}C_3H_7F$ were analyzed by gas chromatography and rate constant ratios calculated from

$$k_2/k_1 = R_{CF_3H}\,[C_2H_4]/R_{C_3H_7F}\,[H_2S]$$

The mechanism was established by variation of the conditions; hence

$$\log\,(k_2/k_1) = 0.91 + 1.3/2.3RT$$

and the above results are calculated from $\log k_2 = 8.50 - 3.0/2.3RT$, which is an estimate based on the known parameters for CH_3 and CF_3 attack on H_2S.

PREFERRED RATE PARAMETERS

Reasonable as given above

COMMENTS

Comparison of the rate constants or activation energies for CH_3, CH_2F and CF_3 addition to $CH_2{=}CH_2$ indicates that CH_2F is intermediate in reactivity between CH_3 and CF_3, which seems reasonable.
From the above preferred parameters and the thermodynamic data, it can be calculated that for the radical decomposition reaction, $CH_2FCH_2\dot{C}H_2 \longrightarrow \dot{C}H_2F + CH_2{=}CH_2$, $\log\,(k_{-1}/s^{-1}) = (12.6 - 31/2.3RT)$ which would imply that the A-factor for the CH_2F addition reaction has been slightly underestimated.

ΔH°_f, ΔH° and E in kcal mol^{-1} (kJ mol^{-1}); S°, C°_p, ΔS° and ΔC°_p in cal mol^{-1} K^{-1} (J mol^{-1} K^{-1}); A and k in l mol^{-1} s^{-1}

REACTION: 3,3,3-TRIFLUOROPROPYL-1 RADICAL + ETHYLENE $CF_3CH_2\dot{C}H_2(I) + CH_2{=}CH_2(II) \longrightarrow CF_3CH_2CH_2CH_2\dot{C}H_2(III)$

	(I)	(II)	(III)		
$\Delta H^\circ_{f(300)}$	-127.5 (-533.8)	12.5 (52.4)	-137.4 (-575.3)	ΔH°	-22.4 (-93.8)
$S^\circ_{(300)}$	83.1 (347.9)	52.4 (219.4)	102.0 (427.1)	ΔS°	-33.5 (-140.3)
$C^\circ_{p(300)}$	24.2 (101.3)	10.3 (43.1)	35.2 (147.4)	ΔC°_p	0.7 (2.9)

log A	E	log k	Conditions	Reference and Experimental Method
-	-	∿ 6.6(437 K)	291-473 K 30-90 torr	[127] J.M.Sangster and J.C.J.Thynne, J.Phys.Chem., 73, 2746 (1969). Rate constant relative to the assumed value log k_3 = 9.7, where k_3 refers to $CF_3CH_2\dot{C}H_2 + H_2S \xrightarrow{3} CF_3CH_2CH_3 + \dot{S}H$ and k_1/k_3 = ∿ 0.9 $CF_3CH_2CH_2$ radicals produced by $\dot{C}F_3 + CH_2{=}CH_2 \longrightarrow CF_3CH_2\dot{C}H_2$ reaction and ratio of rate constants based on analyses of $CF_3CH_2CH_2CH_2CH_3$ and $CF_3CH_2CH_3$. For details see $CF_3 + CH_2{=}CH_2$.

PREFERRED RATE PARAMETERS

Tentatively as given above

COMMENTS

The approximate rate constant derived above is considerably higher than for n-C_3H_7 addition to $CH_2{=}CH_2$ (log k = 4.2 at 437 K) and more akin to that for CF_3 addition to $CH_2{=}CH_2$ (log k = 7.3 at 437 K). It is surprising that the presence of the F atoms so far removed from the site of the odd electron should have such a marked effect, but it seems unlikely that the experimental method would yield a rate constant grossly in error.

REACTION: N-HEPTAFLUOROPROPYL RADICAL + ETHYLENE $CF_3CF_2\dot{C}F_2(I) + CH_2{=}CH_2(II) \longrightarrow CF_3CF_2CF_2CH_2\dot{C}H_2(III)$

	(I)	(II)	(III)		
$\Delta H^\circ_{f(300)}$	[-306 (-1281)]	12.5 (52.3)	-321.5 (-1346.1)	ΔH°	[-28 (-117)]
$S^\circ_{(300)}$	98.6 (412.8)	52.4 (219.4)	116.5 (487.8)	ΔS°	-34.5 (-144.5)
$C^\circ_{p(300)}$	32.0 (134.0)	10.3 (43.1)	44.0 (184.2)	ΔC°_p	1.7 (7.1)

log A	E	log k	Conditions	Reference and Experimental Method
[7.5]	[2.0] [(8.4)]	[6.4(393 K)]	313-473 K 40-120 torr	[136] J.Gibb, M.J.Peters, J.M.Tedder, J.C.Walton and K.D.R.Winton, Chem.Comm., 978 (1970). The reference is a preliminary communication with very little experimental detail. $n\text{-}C_3F_7$ was produced from the photolysis of $n\text{-}C_3F_7I$ and the rates of addition to a fluoroethylene measured relative to ethylene, apparently by analysis of the products C_3F_7OlI in experiments with pairs of olefins by a method analogous to that for CCl_3 additions (see $CCl_3 + CH_2{=}CH_2$ system, references [140] and [141]). The results were placed on an absolute basis by assuming the above Arrhenius parameters for $n\text{-}C_3F_7$ addition to $CH_2{=}CH_2$, based on the data for CF_3 addition to $CH_2{=}CH_2$.

PREFERRED RATE PARAMETERS

log k = [8.3 - 2.0/2.3RT]

COMMENTS

The preferred tentative parameters are based on the assumption that $A_{C_3F_7} = A_{CF_3}$ and $E_{C_3F_7} = E_{CF_3}$ (see summary of alkyl radical additions, p.219).

ΔH°_f, ΔH° and E in kcal mol^{-1} (kJ mol^{-1}); S°, C°_p, ΔS° and ΔC°_p in cal mol^{-1} K^{-1} (J mol^{-1} K^{-1}); A and k in l mol^{-1} s^{-1}

REACTION: n-HEPTAFLUOROPROPYL RADICAL + VINYL FLUORIDE

$CF_3CF_2\dot{C}F_2(I) + CH_2{=}CHF(II) \xrightarrow{1} CF_3CF_2CF_2CH_2\dot{C}HF(III)$

$\xrightarrow{2} CF_3CF_2CF_2CHF\dot{C}H_2(IV)$

	(I)	(II)	(III)	(IV)		Path 1	Path 2
$\Delta H^\circ_{f(300)}$	[-306 (-1281)]	-31.6 (-132.3)	[-361 (-1510)]	-365.0 (-1528.3)	ΔH°	[-24 (-101)]	[-28 (-117)]
$S^\circ_{(300)}$	98.6 (412.8)	60.4 (252.9)	122.1 (511.2)	121.1 (507.0)	ΔS°	-36.9 (-154.5)	-37.9 (-158.7)
$C^\circ_{p(300)}$	32.0 (134.0)	11.9 (49.8)	46.0 (192.6)		ΔC°_p	2.1 (8.8)	

Path	log A	E	log k	Conditions	Reference and Experimental Method
1	7.1	3.1 (13.0)	5.4(393 K)	313-473 K 40-120 torr	[136] J.Gibb, M.J.Peters, J.M.Tedder, J.C.Walton and K.D.R.Winton, Chem. Comm., 978 (1970). For details see $n\text{-}C_3F_7 + CH_2{=}CH_2$.
2	6.9	5.4 (22.6)	3.9(393 K)	313-473 K 40-120 torr	

PREFERRED RATE PARAMETERS

$\log k_1 = [7.9 - 3.1/2.3RT]$

$\log k_2 = [7.7 - 5.4/2.3RT]$

COMMENTS

The preferred tentative data were calculated from the relative results $E_1 - E_{C_2H_4} = 1.1$ kcal mol^{-1}, $\log (A_1/A_{C_2H_4}) = -0.4$, and $E_2 - E_{C_2H_4} = 3.4$, $\log (A_2/A_{C_2H_4}) = -0.6$, and assuming $\log k = 8.3 - 2.0/2.3RT$ for the reaction $n\text{-}C_3F_7 + C_2H_4 \longrightarrow C_3F_7CH_2CH_2$.

REACTION: N-HEPTAFLUOROPROPYL RADICAL + 1,1-DIFLUOROETHYLENE

$CF_3CF_2\dot{C}F_2(I) + CF_2{=}CH_2(II) \xrightarrow{1} CF_3CF_2CF_2CF_2\dot{C}H_2(III)$

$\xrightarrow{2} CF_3CF_2CF_2CH_2\dot{C}F_2(IV)$

	(I)		(II)		(III)		(IV)			Path 1		Path 2	
$\Delta H^\circ_{f(300)}$	[-306	(-1281)]	[-71.5	(-299.4)]	-413.6	(-1731.7)	[-410	(-1717)]	ΔH°	[-36	(-151)]	[-33	(-138)]
$S^\circ_{(300)}$	98.6	(412.8)	64.9	(271.7)	124.9	(523.0)	125.8	(526.7)	ΔS°	-38.6	(-161.6)	-37.7	(-157.8)
$C^\circ_{p(300)}$	32.0	(134.0)	14.8	(62.0)	48.4	(202.7)	47.9	(200.6)	ΔC°_p	1.6	(6.7)	1.1	(4.6)

Path	log A	E	log k	Conditions	Reference and Experimental Method
1	6.0	9.2 (38.5)	0.9(393 K)	313-473 K 40-120 torr	[136] J.Gibb, M.J.Peters, J.M.Tedder, J.C.Walton and K.D.R.Winton, Chem. Comm., 978 (1970).
2	7.2	3.9 (16.3)	5.0(393 K)	313-473 K 40-120 torr	For details see $n\text{-}C_3F_7 + CH_2{=}CH_2$.

PREFERRED RATE PARAMETERS

$\log k_1 = [6.8 - 9.2/2.3RT]$

$\log k_2 = [8.0 - 3.9/2.3RT]$

COMMENTS

The preferred tentative data were calculated from the relative results $E_1 - E_{C_2H_4} = 7.2$ kcal mol^{-1}, $\log(A_1/A_{C_2H_4}) = -1.5$ and $E_2 - E_{C_2H_4} = 1.9$ kcal mol^{-1}, $\log(A_2/A_{C_2H_4}) = -0.3$, and assuming $\log k = 8.3 - 2.0/2.3RT$ for the reaction $n\text{-}C_3F_7 + C_2H_4 \longrightarrow C_3F_7CH_2CH_2$. The A-factor for reaction (2) seems rather low.

ΔH°_f, ΔH° and E in kcal mol^{-1} (kJ mol^{-1}); S°, C°_p, ΔS° and ΔC°_p in cal mol^{-1} K^{-1} (J mol^{-1} K^{-1}); A and k in l mol^{-1} s^{-1}

REACTION: n-HEPTAFLUOROPROPYL RADICAL + TRIFLUOROETHYLENE

$CF_3CF_2\dot{C}F_2(I) + CF_2{=}CFH(II) \xrightarrow{1} CF_3CF_2CF_2CF_2\dot{C}FH(III)$

$\xrightarrow{2} CF_3CF_2CF_2CFH\dot{C}F_2(IV)$

	(I)	(II)	(III)	(IV)		Path 1	Path 2
$\Delta H^\circ_{f(300)}$	[-306 (-1281)]	-117.3 (-491.9)	-453.5 (-1898.8)	[-453 (-1897)]	ΔH°	[-30 (-126)]	[-30 (-126)]
$S^\circ_{(300)}$	98.6 (412.8)	70.1 (293.5)	130.5 (546.4)	[130.4 (546.0)]	ΔS°	-38.2 (-159.9)	[-38.3 (-160.4)]
$C^\circ_{p(300)}$	32.0 (134.0)	16.5 (69.1)	50.0 (209.4)		ΔC°_p	1.5 (6.3)	

Path	log A	E	log k	Conditions	Reference and Experimental Method
1	[7.8]	5.8 (24.3)	4.6(393 K)	313-473 K 40-120 torr	[136] J.Gibb, M.J.Peters, J.M.Tedder, J.C.Walton and K.D.R.Winton, Chem. Comm., 978 (1970).
2	7.8	4.1 (17.2)	5.5(393 K)	313-473 K 40-120 torr	For details see $n\text{-}C_3F_7 + CH_2{=}CH_2$.

PREFERRED RATE PARAMETERS

$\log k_1 = [8.6 - 5.8/2.3RT]$

$\log k_2 = [8.6 - 4.1/2.3RT]$

COMMENTS

The preferred tentative data were calculated from the relative results $E_1 - E_{C_2H_4} = 3.8$ kcal mol^{-1}, $\log (A_1/A_{C_2H_4}) = 0.3$, and $E_2 - E_{C_2H_4} = 2.1$ kcal mol^{-1}, $\log (A_2/A_{C_2H_4}) = 0.3$, and assuming $\log k = 8.3 - 2.0/2.3RT$ for the reaction $n\text{-}C_3F_7 + C_2H_4 \longrightarrow C_3F_7CH_2CH_2$.

REACTION: CHLORODIFLUOROMETHYL RADICAL + BENZENE

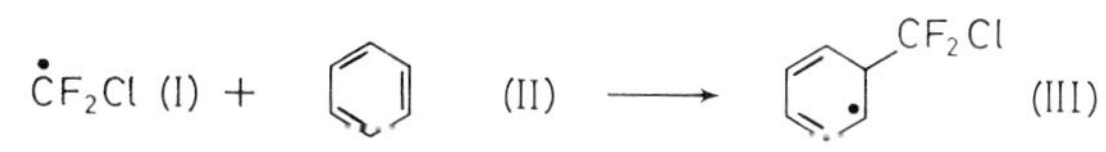

	(I)	(II)	(III)	
$\Delta H^\circ_{f(300)}$	-63.7 (-266.6)	19.8 (82.9)		ΔH°
$S^\circ_{(300)}$	67.8 (283.9)	64.5 (270.1)		ΔS°
$C^\circ_{p(300)}$	13.0 (54.4)	19.7 (82.5)		ΔC°_p

log A	E	log k	Conditions	Reference and Experimental Method
8.66	5.3 (22.2)	5.57(375 K)	298-347 K 3 torr	[137] J.R.Majer, D.Phillips and J.C.Robb, Trans.Faraday Soc., 61, 110 (1965).

Static system. $CF_2C\ell$ from photolysis (λ = 313 nm) of $(CF_2C\ell)_2CO$. Products CO and $(CF_2C\ell)_2$ analyzed by gas chromatography in identical runs with and without added unsaturate. The rate of formation of the adduct radical was then calculated from a mass balance of $CF_2C\ell$ radicals, assuming the basic mechanism for addition to be:

$$\dot{C}F_2C\ell + O\ell \xrightarrow{1} CF_2C\ell\dot{O}\ell$$
$$CF_2C\ell\dot{O}\ell + \dot{C}F_2C\ell \xrightarrow{3} CF_2C\ell O\ell CF_2C\ell$$
$$2CF_2C\ell\dot{O}\ell \xrightarrow{4} (CF_2C\ell O\ell)_2$$
$$2\dot{C}F_2C\ell \xrightarrow{c} (CF_2C\ell)_2$$

The rate constant ratio was calculated from $k_1/k_c^{\frac{1}{2}} = R_{Adduct}/R^{\frac{1}{2}}_{(CF_2C\ell)_2}[O\ell]$. At temperatures > 373 K the calculated rates of adduct formation started to decrease and this was ascribed to the onset of the adduct radical decomposition reaction (-1). The rate constants are relative to log k_c = 10.12 for $CF_2C\ell$ radicals [J.R.Majer, C.Olavesen and J.C.Robb, Trans.Faraday Soc., 65, 2988 (1969)].

This experimental method is extremely indirect and depends upon a mechanism for which no kinetic evidence was obtained. The method also suffers from the lack of a radical trap for the adduct radicals.

PREFERRED RATE PARAMETERS

Reasonable as given above

COMMENTS

In relation to the data on CF_3 addition to benzene the above results seem reasonable, although the determination of the rate constants involved a very indirect method.

ΔH°_f, ΔH° and E in kcal mol^{-1} (kJ mol^{-1}); S°, C°_p, ΔS° and ΔC°_p in cal mol^{-1} K^{-1} (J mol^{-1} K^{-1}); A and k in l mol^{-1} s^{-1}

REACTION: CHLORODIFLUOROMETHYL RADICAL + 1-HYDROHEPTAFLUOROCYCLOPENTENE

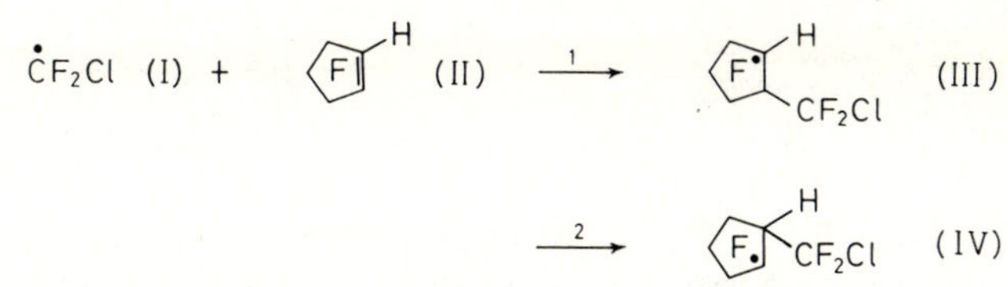

	(I)	(II)	(III)	(IV)		Path 1	Path 2
$\Delta H^\circ_{f(300)}$	-63.7 (-266.6)				ΔH°		
$S^\circ_{(300)}$	67.8 (283.9)				ΔS°		
$C^\circ_{p(300)}$	13.0 (54.4)				ΔC°_p		

Path	log A	E	log k	Conditions	Reference and Experimental Method
1 + 2	7.95	7.9 (33.1)	3.3(375 K)	294-423 K 67-80 torr	[138] L. M. Leyland, J.R.Majer and J.C.Robb, <u>Trans.</u>.<u>Faraday</u> <u>Soc.</u>, <u>66</u>, 904 (1970). For details see $CF_2C\ell$ + Benzene, reference [137].

PREFERRED RATE PARAMETERS

Tentatively as given above

COMMENTS

Although the Arrhenius parameters seem reasonable, the indirect nature of the experimental method casts some doubt on the validity of the rate constants.

REACTION: CHLORODIFLUOROMETHYL RADICAL + 2-HYDROHEPTAFLUOROCYCLOPENTENE

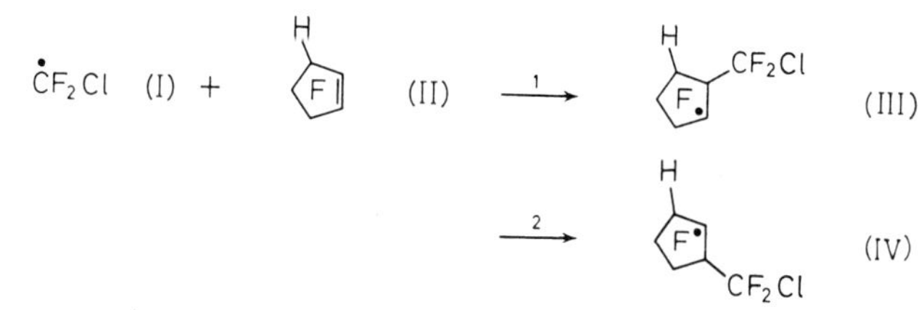

	(I)	(II)	(III)	(IV)		Path 1	Path 2
$\Delta H^\circ_{f(300)}$	-63.7 (-266.6)				ΔH°		
$S^\circ_{(300)}$	67.8 (283.9)				ΔS°		
$C^\circ_{p(300)}$	13.0 (54.4)				ΔC°_p		

Path	log A	E	log k	Conditions	Reference and Experimental Method
1 + 2	8.4	9.3 (38.9)	3.0(375 K)	294-438 K 67-80 torr	[138] L.M.Leyland, J.R.Majer and J.C.Robb, Trans.Faraday Soc., 66, 904 (1970). For details see CF_2Cl + Benzene, reference [137].

PREFERRED RATE PARAMETERS

Tentatively as given above

COMMENTS

Although the Arrhenius parameters seem reasonable, the indirect nature of the experimental method casts some doubt on the validity of the rate constants.

ΔH°_f, ΔH° and E in kcal mol^{-1} (kJ mol^{-1}); S°, C°_p, ΔS° and ΔC°_p in cal mol^{-1} K^{-1} (J mol^{-1} K^{-1}); A and k in l mol^{-1} s^{-1}

REACTION: CHLORODIFLUOROMETHYL RADICAL + 3-HYDROHEPTAFLUOROCYCLOPENTENE

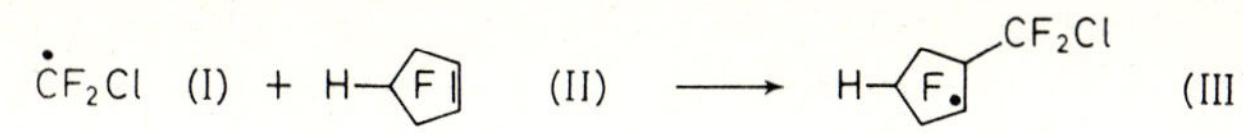

	(I)	(II)	(III)	
$\Delta H^\circ_{f(300)}$	-63.7 (-266.6)			ΔH°
$S^\circ_{(300)}$	67.8 (283.9)			ΔS°
$C^\circ_{p(300)}$	13.0 (54.4)			ΔC°_p

log A	E	log k	Conditions	Reference and Experimental Method
8.0	8.4 (35.2)	3.1(375 K)	293-413 K 67-80 torr	[138] L.M.Leyland, J.R.Majer and J.C.Robb, Trans.Faraday Soc., 66, 904 (1970). For details see CF_2Cl + Benzene, reference [137].

PREFERRED RATE PARAMETERS

Tentatively as given above

COMMENTS

Although the Arrhenius parameters seem reasonable, the indirect nature of the experimental method casts some doubt on the validity of the rate constants.

REACTION: CHLORODIFLUOROMETHYL RADICAL + 1-HYDRONONAFLUOROCYCLOHEXENE

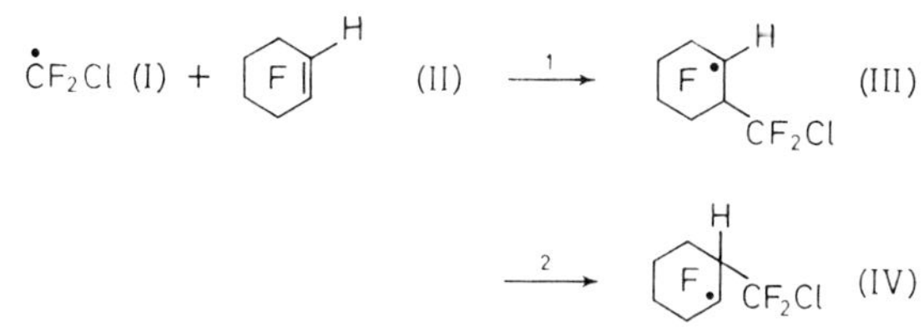

	(I)	(II)	(III)	(IV)		Path 1	Path 2
$\Delta H^\circ_{f(300)}$	-63.7 (-266.6)				ΔH°		
$S^\circ_{(300)}$	67.8 (283.9)				ΔS°		
$C^\circ_{p(300)}$	13.0 (54.4)				ΔC°_p		

Path	log A	E	log k	Conditions	Reference and Experimental Method
1 + 2	9.5	10.3 (43.1)	3.5(375 K)	293-413 K 67-80 torr	[138] L.M.Leyland, J.R.Majer and J.C.Robb, Trans.Faraday Soc., 66, 904 (1970). For details see $CF_2C\ell$ + Benzene, reference [137].

PREFERRED RATE PARAMETERS

log k = [7.8 - 7.4/2.3RT]

COMMENTS

The experimental A-factor is abnormally high. We have calculated $E = 7.4$ kcal mol^{-1} assuming log A = 7.8 and taking the rate constant at 375 K as being correct.

ΔH^o_f, ΔH^o and E in kcal mol^{-1} (kJ mol^{-1}); S^o, C^o_p, ΔS^o and ΔC^o_p in cal mol^{-1} K^{-1} (J mol^{-1} K^{-1}); A and k in l mol^{-1} s^{-1}

REACTION: CHLORODIFLUOROMETHYL RADICAL + 1,2-DIHYDROOCTAFLUOROCYCLOHEXENE

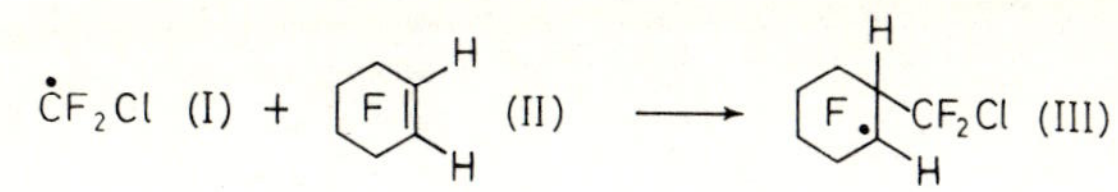

	(I)		(II)		(III)			
$\Delta H^\circ_{f(300)}$	-63.7	(-266.6)	[-369	(-1545)]	-458.8	(-1921.0)	ΔH°	[-26 (-109)]
$S^\circ_{(300)}$	67.8	(283.9)	[109	(456)]	129.2	(541.0)	ΔS°	[-47 (-197)]
$C^\circ_{p(300)}$	13.0	(54.4)	[43.6	(182.6)]	56.6	(237.0)	ΔC°_p	[0

log A	E	log k	Conditions	Reference and Experimental Method
6.9	5.6 (23.4)	3.64(375 K)	294-463 K 67-80 torr	[138] L.M.Leyland, J.R.Majer and J.C.Robb, Trans.Faraday Soc., 66, 904 (1970). For details see CF_2Cl + Benzene, reference [137].

PREFERRED RATE PARAMETERS

log k = [7.8 - 7.2/2.3RT]

COMMENTS

The experimental A-factor seems rather low. We have calculated E = 7.2 by taking log A = 7.8 and assuming the rate constant at 375 K to be correct.

REACTION: CHLORODIFLUOROMETHYL RADICAL + PERFLUOROCYCLOPENTENE

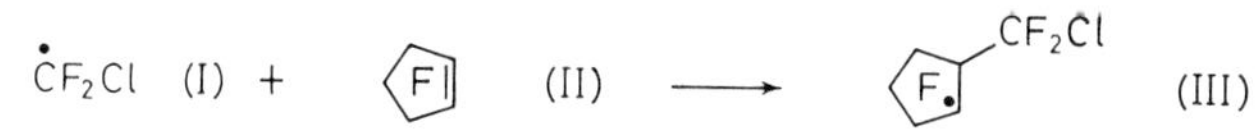

	(I)	(II)	(III)	
$\Delta H^\circ_{f(300)}$	-63.7 (-266.6)			ΔH°
$S^\circ_{(300)}$	67.8 (283.9)			ΔS°
$C^\circ_{p(300)}$	13.0 (54.4)			ΔC°_p

log A	E	log k	Conditions	Reference and Experimental Method
6.7	6.6 (27.6)	2.85(375 K)	293-510 K 67-80 torr	[138] L.M.Leyland, J.R.Majer and J.C.Robb, Trans.Faraday Soc., 66, 904 (1970). For details see $CF_2C\ell$ + Benzene, reference [137].

PREFERRED RATE PARAMETERS

log k = [7.5 - 8.0/2.3RT]

COMMENTS

The experimental A-factor is low in relation to $CF_2C\ell$ additions to the other cyclopentenes. We have calculated E = 8.0 kcal mol^{-1} assuming log A = 7.5 and taking the rate constant at 375 K to be correct. The results are tentative as the rate constants are in doubt owing to the indirect experimental method.

ΔH^o_f, ΔH^o and E in kcal mol^{-1} (kJ mol^{-1}); S^o, C^o_p, ΔS^o and ΔC^o_p in cal mol^{-1} K^{-1} (J mol^{-1} K^{-1}); A and k in l mol^{-1} s^{-1}

REACTION: CHLORODIFLUOROMETHYL RADICAL + PERFLUOROCYCLOHEXENE

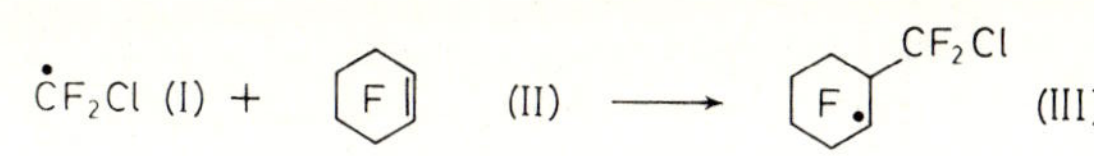

	(I)	(II)	(III)	
$\Delta H^\circ_{f(300)}$	-63.7 (-266.6)	-462.7(-1938.0)		ΔH°
$S^\circ_{(300)}$	67.8 (283.9)			ΔS°
$C^\circ_{p(300)}$	13.0 (54.4)			ΔC°_p

log A	E	log k	Conditions	Reference and Experimental Method
11.3	15.4 (64.5)	2.30(375 K)	293-421 K 67-80 torr	[138] L.M.Leyland, J.R.Majer and J.C.Robb, Trans.Faraday Soc., 66, 904 (1970). For details see $CF_2C\ell$ + Benzene, reference [137].

PREFERRED RATE PARAMETERS

log k = [8.0 - 9.8/2.3RT]

COMMENTS

The experimental Arrhenius parameters are clearly in error since it is physically impossible to have a bimolecular radical reaction with an A-factor near the collision rate and at the same time an activation energy of 15 kcal mol^{-1}. We have calculated E = 9.8 kcal mol^{-1} on the assumption of an A-factor of $10^{8.0}$ ℓ mol^{-1}s^{-1} and taking the measured rate constant at 375 K as being correct.

REACTION: CHLORODIFLUOROMETHYL RADICAL + HEXAFLUOROBENZENE

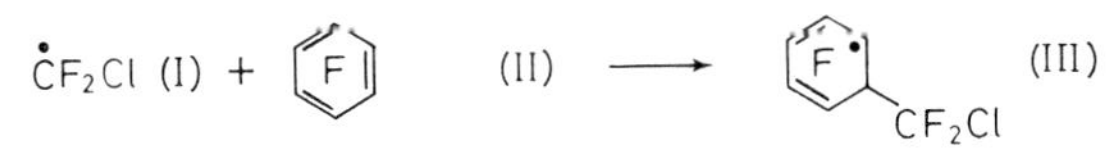

	(I)	(II)	(III)	
$\Delta H^o_{f(300)}$	-63.7 (-266.6)	-226.8 (-949.6)		ΔH^o
$S^o_{(300)}$	67.8 (283.9)	96.6 (404.5)		ΔS^o
$C^o_{p(300)}$	13.0 (54.4)	37.8 (158.3)		ΔC^o_p

log A	E	log k	Conditions	Reference and Experimental Method
5.12	2.4 (10.1)	3.72(375 K)	296-423 K 3.5 torr	[137] J.R.Majer, D.Phillips and J.C.Robb, Trans.Faraday Soc., 61, 110 (1965). For details see CF_2Cl + Benzene.

PREFERRED RATE PARAMETERS

log k = [8.0 - 7.3/2.3RT]

COMMENTS

The experimental A-factor is undoubtedly too low. We have calculated E ≈ 7.3 kcal mol^{-1} by assuming log A ≈ 8.0 and taking the rate constant at 375 K to be correct.

ΔH^o_f, ΔH^o and E in kcal mol^{-1} (kJ mol^{-1}); S^o, C^o_p, ΔS^o and ΔC^o_p in cal mol^{-1} K^{-1} (J mol^{-1} K^{-1}); A and k in l mol^{-1} s^{-1}

REACTION: CHLORODIFLUOROMETHYL RADICAL + 1-CHLOROHEPTAFLUOROCYCLOPENTENE

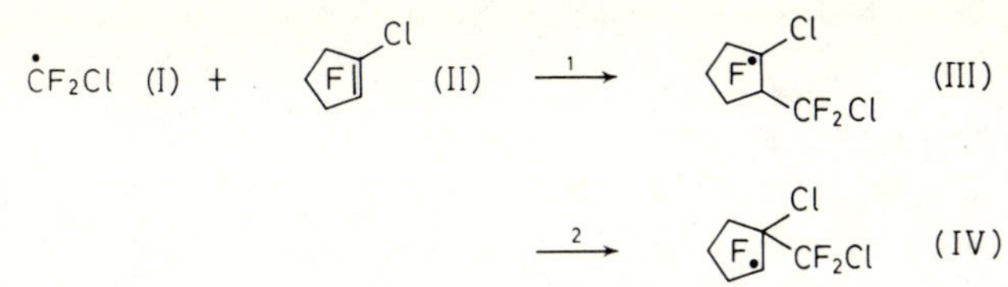

	(I)	(II)	(III)	(IV)		Path 1	Path 2
$\Delta H^\circ_{f(300)}$	-63.7 (-266.6)				ΔH°		
$S^\circ_{(300)}$	67.8 (283.9)				ΔS°		
$C^\circ_{p(300)}$	13.0 (54.4)				ΔC°_p		

Path	log A	E	log k	Conditions	Reference and Experimental Method
1 + 2	5.7	4.7 (19.7)	2.96(375 K)	293-433 K 67-80 torr	[138] L.M.Leyland, J.R.Majer and J.C.Robb, Trans.Faraday Soc., 66, 904 (1970). For details see CF_2Cl + Benzene, reference [137].

PREFERRED RATE PARAMETERS

log k = [7.8 - 8.3/2.3RT]

COMMENTS

The experimental A-factor is abnormally low. We have calculated $E = 8.3$ kcal mol^{-1} by assuming log A = 7.8 and taking the rate constant at 375 K as being correct, but the rate constants are in doubt owing to a suspect experimental method.

REACTION: BROMODIFLUOROMETHYL RADICAL + ETHYLENE $\dot{C}F_2Br(I) + CH_2{=}CH_2(II) \longrightarrow CF_2BrCH_2\dot{C}H_2(III)$

	(I)	(II)	(III)	
$\Delta H^\circ_{f(300)}$		12.5 (52.3)		ΔH°
$S^\circ_{(300)}$	70.7 (296.0)	52.4 (219.4)		ΔS°
$C^\circ_{p(300)}$	13.9 (58.2)	10.3 (43.1)		ΔC°_p

log A	E	log k	Conditions	Reference and Experimental Method
5.20	1.0 (4.2)	4.59(358 K)	278-438 K 150-250 torr	[139] J.M.Tedder and J.C.Walton, Trans.Faraday Soc., 66, 1135 (1970). Static system. CF_2Br radicals from the full medium-pressure Hg arc photolyses of CF_2Br_2. Products analyzed by gas chromatography and mass spectrometry. The following reaction scheme was proposed in which $O\ell$ = olefin:

$$CF_2Br_2 + h\nu \xrightarrow{I} \dot{C}F_2Br + \dot{B}r$$
$$\dot{C}F_2Br + O\ell \xrightarrow{1} CF_2Br\dot{O}\ell$$
$$CF_2Br\dot{O}\ell + CF_2Br_2 \xrightarrow{3} CF_2BrO\ell Br + \dot{C}F_2Br$$
$$2\dot{C}F_2Br \xrightarrow{4} (CF_2Br)_2$$
$$\dot{C}F_2Br + \dot{B}r\ (+M) \xrightarrow{5} CF_2Br_2\ (+M)$$
$$\dot{B}r + \dot{B}r + M \xrightarrow{6} Br_2 + M$$
$$CF_2Br\dot{O}\ell + \dot{C}F_2Br \xrightarrow{7} CF_2BrO\ell CF_2Br$$
$$CF_2Br\dot{O}\ell + \dot{B}r \xrightarrow{8} CF_2BrO\ell Br$$

Under conditions where $[O\ell] \ll [CF_2Br_2]$, it was shown that reactions (7) and (8) could be neglected, and the following rate expression applied:

$$k_1/k_4^{\frac{1}{2}} = R_{CF_2BrO\ell Br}/R_{(CF_2Br)_2}^{\frac{1}{2}}\,[O\ell].$$

Thus, $\log\,(k_1/k_4^{\frac{1}{2}})(\ell^{\frac{1}{2}}mol^{-\frac{1}{2}}s^{-\frac{1}{2}}) = 0.40 - 3.3/2.3RT$, and the above results were calculated taking $\log k_4 = 9.60$, which was estimated from the empirical relation between the rate constant for radical-radical reactions and the molecular weight of the radical [L.Bertrand, G.R.De Maré, G.Huybrechts, J.Olbregts and M.Toth, Chem.Physics Lett., 5, 183 (1970)].

PREFERRED RATE PARAMETERS

log k = [8.3 - 6.1/2.3RT]

ΔH^o_f, ΔH^o and E in kcal mol^{-1} (kJ mol^{-1}); S^o, C^o_p, ΔS^o and ΔC^o_p in cal mol^{-1} K^{-1} (J mol^{-1} K^{-1}); A and k in l mol^{-1} s^{-1}

COMMENTS

The preferred parameters were calculated from the above rate constant at the middle of the experimental range and taking an assumed A-factor of $10^{8.3}$ $\ell\ mol^{-1}s^{-1}$, the experimental A-factor seems much too low. Unfortunately there are no group additivity terms for the group [(C) - (C)(F)$_2$(Br)], but it is clear that any reasonable value of $S°(CF_2BrCH_2\dot{C}H_2)$ and, hence, $\Delta S°$, will lead to a negative entropy of activation for the reverse radical decomposition reaction, on the basis of the above experimental A-factor of the addition reaction. It seems reasonable to assume that the A-factor for CF_2Br addition to C_2H_4 will be similar to those for CH_3, CF_3 and $CC\ell_3$ additions, i.e. $A \sim 10^{8.3}$ $\ell\ mol^{-1}s^{-1}$. The resultant E for CF_2Br addition to C_2H_4 places it intermediate in reactivity between CF_3 and $CC\ell_3$, which is reasonable. We suspect a systematic error in the measurement of the temperature coefficient of $k_1/k_4^{\frac{1}{2}}$ as reported in reference [139].

REACTION: BROMODIFLUOROMETHYL RADICAL + TRIFLUOROETHYLENE

$$\dot{C}F_2Br(I) + CFH{=}CF_2(II) \xrightarrow{1} CF_2BrCFH\dot{C}F_2(III)$$
$$\xrightarrow{2} CF_2BrCF_2\dot{C}FH(IV)$$

	(I)	(II)	(III)	(IV)		Path 1	Path 2
$\Delta H°_{f(300)}$		-117.3 (-491.1)			$\Delta H°$		
$S°_{(300)}$	70.7 (296.0)	70.1 (293.5)			$\Delta S°$		
$C°_{p(300)}$	13.9 (58.2)	16.5 (69.1)			$\Delta C°_p$		

Path	log A	E	log k	Conditions	Reference and Experimental Method
1	6.88	3.3 (13.8)	5.00(384 K)	333-435 K 100-250 torr	[139] J.M.Tedder and J.C.Walton, Trans.Faraday Soc., 66, 1135 (1970). For details see $CF_2Br + CH_2{=}CH_2$. The rate of addition by path 1 was calculated, as for the $CF_2Br + CH_2{=}CH_2$ system, from the rate of formation of the adduct $CF_2BrCFHCF_2Br$, hence log $(k_1(\text{path 1})/k_4^{\frac{1}{2}})$ $(\ell^{\frac{1}{2}}mol^{-\frac{1}{2}}s^{-\frac{1}{2}})$ = 2.08 - 3.3/2.3RT, and similarly the rate of addition by path 2 was calculated from the rate of formation of the adduct $CF_2BrCF_2CF_2HBr$, hence log $(k_2(\text{path 2})/k_4^{\frac{1}{2}})(\ell^{\frac{1}{2}}mol^{-\frac{1}{2}}s^{-\frac{1}{2}})$ = 2.30 - 4.4/2.3RT. The above results are then based on log k_4 = 9.6, which was
2	7.10	4.0 (16.7)	4.82(384 K)	333-435 K 100-250 torr	

Path	log A	E	log k	Conditions	Reference and Experimental Method
					estimated from the empirical relation between rate constant for radical-radical reactions and the molecular weight of the radical [L.Bertrand, G.R.De Maré, G.Huybrechts, J.Olbregts and M.Toth, Chem.Physics Lett., 5, 183 (1970)].

PREFERRED RATE PARAMETERS

$\log k_1 = [8.3 - 5.8/2.3RT]$

$\log k_2 = [8.3 - 6.1/2.3RT]$

COMMENTS

The preferred parameters were calculated from the above rate constants at the middle of the experimental range and taking an assumed A-factor of $10^{8.3}$ ℓ $mol^{-1}s^{-1}$ (see comments on $CF_2Br + CH_2{=}CH_2$ system). The preferred data indicate that $\dot{C}F_2Br$ radicals show little selectivity in additions to $=CH_2$, $=CHF$ or $= CF_2$ in the fluoroethylenes.

ΔH_f^o, ΔH^o and E in kcal mol^{-1} (kJ mol^{-1}); S^o, C_p^o, ΔS^o and ΔC_p^o in cal mol^{-1} K^{-1} (J mol^{-1} K^{-1}); A and k in l mol^{-1} s^{-1}

REACTION: CHLOROMETHYLIDYNE RADICAL + ETHYLENE $CC\ell(I) + CH_2{=}CH_2(II) \longrightarrow [\dot{C}C\ell CH_2CH_2](III)$

	(I)	(II)	(III)		
$\Delta H^\circ_{f(300)}$	106.8 (447.2)	12.5 (52.3)	51.8 (216.9)	ΔH°	-67.5 (-282.6)
$S^\circ_{(300)}$	53.8 (225.3)	52.4 (219.4)	69.9 (292.6)	ΔS°	-36.3 (-152.0)
$C^\circ_{p(300)}$	7.7 (32.2)	10.3 (43.1)	17.0 (71.2)	ΔC°_p	-1.0 (-4.2)

log A	E	log k	Conditions	Reference and Experimental Method
-	-	8.52	298 K 500 torr	[139a] W.J.R.Tyerman, J.Chem.Soc.(A), 2483 (1969). Static system. $CC\ell$ from flash photolysis of $CF_2{=}CC\ell_2$ moderated with N_2. $[CC\ell]$ monitored by absorption at 278 nm using plate densitometry. The optical density of $CC\ell$ $(O.D.)_{CC\ell}$ was shown to be proportional to $[CC\ell]^{\frac{1}{2}}$. Pseudo-first-order rate constants were calculated from $-2d[\ell n(O.D.)_{CC\ell}]/dt = -d\ell n[CC\ell]/dt$, assuming a first-order decay of $CC\ell$, and from the known concentrations of the added olefins, the second-order rate constants for the reaction of $CC\ell$ with the olefins were calculated.

PREFERRED RATE PARAMETERS

Reasonable as given above

COMMENTS

No experimental evidence was obtained for the formation of the cyclic adduct radical, but its presence was inferred from the similar pattern of reactivity of $CC\ell$ with olefins to that for O atom attack, which is known to produce epoxides. It was postulated that the adduct chlorocyclopropyl radicals would probably isomerize to give allyl radicals, e.g.

$$\underline{CH_2-CH_2-\dot{C}C\ell} \longrightarrow CH_2{=}CC\ell{-}\dot{C}H_2 \quad \text{or} \quad \dot{C}H_2{-}CH{=}CHC\ell$$

although it was pointed out that the allyl radicals could also arise from direct insertion of $CC\ell$ into a vinylic C-H bond.

REACTION: CHLOROMETHYLIDYNE RADICAL + PROPENE $\quad CC\ell(I) + CH_2{=}CHCH_3(II) \longrightarrow [\dot{C}C\ell CH_2CHCH_3](III)$

	(I)	(II)	(III)		
$\Delta H^\circ_{f(300)}$	106.8 (447.2)	4.9 (20.5)	44.8 (187.6)	ΔH°	-66.9 (-447.6)
$S^\circ_{(300)}$	53.8 (225.3)	63.8 (267.1)	76.6 (320.7)	ΔS°	-41.0 (-171.7)
$C^\circ_{p(300)}$	7.7 (32.2)	15.3 (64.1)	22.2 (93.0)	ΔC°_p	-0.8 (-3.3)

log A	E	log k	Conditions	Reference and Experimental Method
-	-	9.40	298 K 500 torr	[139a] W.J.R.Tyerman, J.Chem.Soc.(A), 2483 (1969). For details see $CC\ell + CH_2{=}CH_2$.

PREFERRED RATE PARAMETERS

Reasonable as given above

COMMENTS

See comments for $CC\ell + CH_2{=}CH_2$ system.

REACTION: CHLOROMETHYLIDYNE RADICAL + TRANS-BUTENE-2 $\quad CC\ell(I) + \text{trans-}CH_3CH{=}CHCH_3(II) \longrightarrow [\dot{C}C\ell CH(CH_3)CHCH_3](III)$

	(I)	(II)	(III)		
$\Delta H^\circ_{f(300)}$	106.8 (447.2)	-3.0 (-12.6)	37.8 (158.3)	ΔH°	-66.0 (-276.3)
$S^\circ_{(300)}$	53.8 (225.3)	72.4 (303.1)	83.3 (348.8)	ΔS°	-42.9 (-179.6)
$C^\circ_{p(300)}$	7.7 (32.2)	20.7 (86.7)	27.4 (114.7)	ΔC°_p	-1.0 (-4.2)

log A	E	log k	Conditions	Reference and Experimental Method
-	-	9.98	298 K 500 torr	[139a] W.J.R.Tyerman, J.Chem.Soc.(A), 2483 (1969). For details see $CC\ell + CH_2{=}CH_2$.

PREFERRED RATE PARAMETERS

Reasonable as given above

COMMENTS

See comments for $CC\ell + CH_2{=}CH_2$ system.

ΔH°_f, ΔH° and E in kcal mol^{-1} (kJ mol^{-1}); S°, C°_p, ΔS° and ΔC°_p in cal mol^{-1} K^{-1} (J mol^{-1} K^{-1}); A and k in l mol^{-1} s^{-1}

REACTION: CHLOROMETHYLIDYNE RADICAL + 2-METHYLPROPENE (ISOBUTENE) $C\ell(I) + CH_2{=}C(CH_3)_2(II) \longrightarrow [\dot{C}C\ell CH_2\dot{C}(CH_3)_2](III)$

	(I)	(II)	(III)		
$\Delta H^\circ_{f(300)}$	106.8 (447.2)	-3.6 (-15.1)	37.2 (155.8)	ΔH°	-66.0 (-276.3)
$S^\circ_{(300)}$	53.8 (225.3)	71.3 (298.5)	81.8 (342.5)	ΔS°	-43.3 (-181.3)
$C^\circ_{p(300)}$	7.7 (32.2)	21.6 (90.4)	28.2 (118.1)	ΔC°_p	-1.1 (-4.6)

log A	E	log k	Conditions	Reference and Experimental Method
-	-	10.19	298 K 500 torr	[139a] W.J.R.Tyerman, J.Chem.Soc.(A), 2483 (1969). For details see $CC\ell + CH_2{=}CH_2$ system.

PREFERRED RATE PARAMETERS

Reasonable as given above

COMMENTS

See comments for $CC\ell + CH_2{=}CH_2$ system.

REACTION: CHLOROMETHYLIDYNE RADICAL + 1,1-DIFLUORO-2,2-DICHLOROETHYLENE $CC\ell(I) + CF_2{=}CC\ell_2(II) \longrightarrow [\dot{C}C\ell CF_2\dot{C}C\ell_2](III)$

	(I)	(II)	(III)		
$\Delta H^\circ_{f(300)}$	106.8 (447.2)	-79.3 (-332.0)		ΔH°	
$S^\circ_{(300)}$	53.8 (225.3)	79.4 (332.4)		ΔS°	
$C^\circ_{p(300)}$	7.7 (32.2)	21.1 (88.3)		ΔC°_p	

log A	E	log k	Conditions	Reference and Experimental Method
-	-	9.28	298 K 500 torr	[139a] W.J.R.Tyerman, J.Chem.Soc.(A), 2483 (1969). For details see $CC\ell + CH_2{=}CH_2$.

PREFERRED RATE PARAMETERS

Reasonable as given above

COMMENTS

See comments for $CC\ell + CH_2{=}CH_2$ system.

REACTION: CHLOROMETHYLIDYNE RADICAL + TRICHLOROFLUOROETHYLENE

$CC\ell(I) + CFC\ell{=}CC\ell_2(II) \longrightarrow [\dot{C}C\ell CFC\ell CC\ell_2](III)$

	(I)	(II)	(III)	
$\Delta H^\circ_{f(300)}$	106.8 (447.2)			ΔH°
$S^\circ_{(300)}$	53.8 (225.3)	81.9 (342.9)		ΔS°
$C^\circ_{p(300)}$	7.7 (32.2)	21.7 (90.9)		ΔC°_p

log A	E	log k	Conditions	Reference and Experimental Method
-	-	~9.90	298 K 500 torr	[139a] W.J.R.Tyerman, J.Chem.Soc.(A), 2483 (1969). For details see $CC\ell + CH_2{=}CH_2$.

PREFERRED RATE PARAMETERS

Reasonable as given above

COMMENTS

Large experimental errors (±50%) were involved in this measurement. See comments for $CC\ell + CH_2{=}CH_2$ system.

REACTION: CHLOROMETHYLIDYNE RADICAL + TRICHLOROETHYLENE

$CC\ell(I) + CHC\ell{=}CC\ell_2(II) \longrightarrow [\dot{C}C\ell CHC\ell CC\ell_2](III)$

	(I)	(II)	(III)	
$\Delta H^\circ_{f(300)}$	106.8 (447.2)	-2.0 (-8.4)		ΔH°
$S^\circ_{(300)}$	53.8 (225.3)	77.5 (324.5)		ΔS°
$C^\circ_{p(300)}$	7.7 (32.2)	19.3 (80.8)		ΔC°_p

log A	E	log k	Conditions	Reference and Experimental Method
-	-	~9.60	298 K 500 torr	[139a] W.J.R.Tyerman, J.Chem.Soc.(A), 2483 (1969). For details see $CC\ell + CH_2{=}CH_2$.

PREFERRED RATE PARAMETERS

Reasonable as given above

COMMENTS

Large experimental errors (±50%) were involved in this measurement. See comments for $CC\ell + CH_2{=}CH_2$ system.

ΔH^o_f, ΔH^o and E in kcal mol^{-1} (kJ mol^{-1}); S^o, C^o_p, ΔS^o and ΔC^o_p in cal mol^{-1} K^{-1} (J mol^{-1} K^{-1}); A and k in l mol^{-1} s^{-1}

REACTION: CHLOROMETHYLIDYNE RADICAL + TETRACHLOROETHYLENE $CC\ell(I) + CC\ell_2{=}CC\ell_2(II) \longrightarrow [\dot{C}C\ell CC\ell_2CC\ell_2](III)$

	(I)	(II)	(III)	
$\Delta H^\circ_{f(300)}$	106.8 (447.2)	-2.7 (-11.3)		ΔH°
$S^\circ_{(300)}$	53.8 (225.3)	84.2 (352.5)		ΔS°
$C^\circ_{p(300)}$	7.7 (32.2)	22.8 (95.5)		ΔC°_p

log A	E	log k	Conditions	Reference and Experimental Method
-	-	~10.0	298 K 500 torr	[139a] W.J.R.Tyerman, J.Chem.Soc.(A), 2483 (1969). For details see $CC\ell + CH_2{=}CH_2$.

PREFERRED RATE PARAMETERS

Reasonable as given above

COMMENTS

Large experimental errors (±50%) were involved in this measurement. See comments for $CC\ell + CH_2{=}CH_2$ system.

REACTION: CHLOROMETHYLIDYNE RADICAL + ACETYLENE $CC\ell(I) + CH{\equiv}CH(II) \longrightarrow [\dot{C}C\ell CH{=}CH](III)$

	(I)	(II)	(III)	
$\Delta H^\circ_{f(300)}$	106.8 (447.2)	54.2 (226.9)		ΔH°
$S^\circ_{(300)}$	53.8 (225.3)	48.0 (201.0)		ΔS°
$C^\circ_{p(300)}$	7.7 (32.2)	10.5 (44.0)		ΔC°_p

log A	E	log k	Conditions	Reference and Experimental Method
-	-	8.04	298 K 500 torr	[139a] W.J.R.Tyerman, J.Chem.Soc.(A), 2483 (1969). For details see $CC\ell + CH_2{=}CH_2$.

PREFERRED RATE PARAMETERS

Reasonable as given above

COMMENTS

As for the olefin systems, no evidence for the formation of the cyclopropenyl radical adduct was obtained, although it seems the likely product. See also comments for $CC\ell + CH_2{=}CH_2$ system.

REACTION: CHLOROMETHYLIDYNE RADICAL + PROPYNE $CC\ell(I) + CH{\equiv}CCH_3(II) \longrightarrow [\dot{C}C\ell CH{=}CCH_3](III)$

	(I)	(II)	(III)	
$\Delta H^\circ_{f(300)}$	106.8 (447.2)	44.4 (185.9)		ΔH°
$S^\circ_{(300)}$	53.8 (225.3)	59.3 (248.3)		ΔS°
$C^\circ_{p(300)}$	7.7 (32.2)	14.6 (61.2)		ΔC°_p

log A	E	log k	Conditions	Reference and Experimental Method
-	-	9.38	298 K 500 torr	[139a] W.J.R.Tyerman, J.Chem.Soc.(A), 2483 (1969). For details see $CC\ell + CH_2{=}CH_2$.

PREFERRED RATE PARAMETERS

Reasonable as given above

COMMENTS

As for the olefin systems, no evidence for the formation of the cyclopropenyl radical adduct was obtained, although it seems the likely product. See also comments for $CC\ell + CH_2{=}CH_2$ system.

ΔH^o_f, ΔH^o and E in kcal mol^{-1} (kJ mol^{-1}); S^o, C^o_p, ΔS^o and ΔC^o_p in cal mol^{-1} K^{-1} (J mol^{-1} K^{-1}); A and k in l mol^{-1} s^{-1}

REACTION: TRICHLOROMETHYL RADICAL + ETHYLENE $\dot{C}C\ell_3(I) + CH_2{=}CH_2(II) \longrightarrow CC\ell_3CH_3\dot{C}H_2(III)$

	(I)	(II)	(III)		
$\Delta H^\circ_{f(300)}$	18.5 (77.5)	12.5 (52.4)	13.6 (56.9)	ΔH°	-17.4 (-72.9)
$S^\circ_{(300)}$	70.8 (296.4)	52.4 (219.4)	88.8 (371.8)	ΔS°	-34.4 (-144.0)
$C^\circ_{p(300)}$	14.9 (62.4)	10.3 (43.1)	27.8 (116.4)	$\Delta C^\circ_{p(300)}$	2.6 (10.9)
$C^\circ_{p(500)}$	17.3 (72.4)	14.9 (62.4)	35.6 (149.1)	$\Delta C^\circ_{p(500)}$	3.4 (14.2)
				$<\Delta C^\circ_p>$	3.0 (12.6)

log A	E	log k	Conditions	Reference and Experimental Method
6.04	3.24 (13.57)	4.23(391 K)	331-451 K 40-120 torr	[140] J.M.Tedder and J.C.Walton, Trans.Faraday Soc., 60, 1769 (1964).

The Arrhenius parameters were calculated from $E_1 - \frac{1}{2}E_c = 3.24$ kcal mol^{-1} and $\log (A_1/A_c^{\frac{1}{2}})(\ell^{\frac{1}{2}}mol^{-\frac{1}{2}}s^{-\frac{1}{2}}) = 1.22$ and taking $E_c = 0$ kcal and $\log A_c = 9.63$ for the combination of $CC\ell_3$ radicals [J.A.Kerr in Free Radicals, edited by J.K.Kochi, Wiley, New York, to be published].

Static system. $CC\ell_3$ from photolysis ($\lambda > 280$ nm) of $CC\ell_3Br$. Proposed mechanism:

$$CC\ell_3Br + h\nu \xrightarrow{I} \dot{C}C\ell_3 + \dot{B}r$$
$$\dot{C}C\ell_3 + O\ell \xrightarrow{1} CC\ell_3\dot{O}\ell$$
$$CC\ell_3\dot{O}\ell + CC\ell_3Br \xrightarrow{3} CC\ell_3O\ell Br + \dot{C}C\ell_3$$
$$2\dot{C}C\ell_3 \xrightarrow{4} C_2C\ell_6$$
$$\dot{B}r + O\ell \xrightarrow{5} Br\dot{O}\ell$$
$$Br\dot{O}\ell + CC\ell_3Br \xrightarrow{6} BrO\ell Br + \dot{C}C\ell_3$$
$$2CC\ell_3\dot{O}\ell \xrightarrow{7} (CC\ell_3O\ell)_2$$

Under conditions with $[CC\ell_3Br] > [O\ell]$ it was established that reaction (4) was the main chain-terminating step, and the rate constants were calculated from the steady-state expression:

$$[CC\ell_3O\ell Br]/[C_2C\ell_6] = k_1[O\ell]/(k_4\phi I_a)^{\frac{1}{2}} \qquad (I)$$

where the rate of reaction (1) is defined as

$$R_1 = \phi I_a = R_{C_2C\ell_6}$$

The products $CC\ell_3O\ell Br$ and $C_2C\ell_6$ were analyzed by gas chromatography and the Arrhenius parameters determined from the temp. coefficient of equation (I),

log A	E	log k	Conditions	Reference and Experimental Method
				assuming ϕI_a to be temperature independent. With asymmetrical olefins the position of addition of $CC\ell_3$ was identified by analysis of the adduct $CC\ell_3O\ell Br$, and in some cases products were analyzed from addition at both positions.
6.0	3.2 (13.4)	-	-	[141] J.M.Tedder and J.C.Walton, Trans.Faraday Soc., 62, 1859 (1966). As for reference [140].
~8.0	~6.7 (30.1)	~4.2(391 K)	-	[142] J.A.Kerr and M.J.Parsonage, Int.J.Chem.Kinet., in press. A recalculation of the data of reference [140]. It was shown that ϕI_a increases with temp. [H.W.Sidebottom, J.M.Tedder and J.C.Walton, Trans.Faraday Soc., 65, 755 (1969)]. The rate constants for $CC\ell_3$ addition to three of the olefins studied were calculated from the expression: $k_1/k_4^{\frac{1}{2}} = R_{CC\ell_3O\ell Br}/R_{C_2C\ell_6}^{\frac{1}{2}}\,[O\ell]$ which does not depend on ϕI_a. From the detailed results of the three olefins it was concluded that for all the olefins studied by this method, the original E_1 should be increased by 3-4 kcal mol^{-1} and original log A_1 by ~2 log units; hence above results.
(i)7.6 (ii) -	(i)6.3 (ii) -	(i)4.1(391 K) (ii) -	- -	[143] H.W.Sidebottom, J.M.Tedder and J.C.Walton, Int.J.Chem.Kinet., in press. A recalculation of the data of reference [140] according to the reinterpretation given in reference [142]. We have adjusted the reported A-factor in terms log A_c = 9.63 and E_c = 0 kcal mol^{-1} for the combination of $CC\ell_3$ radicals. (i) recalculation based on temperature dependence of ϕI_a given by k_ϕ = $6.1 \times 10^{-3}\exp(-6.3/2.3RT)$ for $[CC\ell_3Br] = 7.7 \times 10^{-4}$ mol ℓ^{-1}. (ii) recalculated as in reference [142]. On the whole, we favour the results calculated by method (ii), which is more direct.

PREFERRED RATE PARAMETERS

log k = 7.6 - 6.3/2.3RT

COMMENTS

The preferred Arrhenius parameters are close to those for CH_3 addition to $CH_2{=}CH_2$. The A-factors for CH_3, CF_3 and $CC\ell_3$ addition to $CH_2{=}CH_2$ are similar, which would be expected.

From the thermodynamic data and the above preferred rate constant, it can be shown that log (k_{-1}/s^{-1}) = 13.3 - 26.5/2.3RT at 298 K for the reverse radical decomposition reaction, a result that seems reasonable compared with the n-C_3H_7 decomposition (see $CH_3 + CH_2{=}CH_2$ system).

ΔH_f^o, ΔH^o and E in kcal mol^{-1} (kJ mol^{-1}); S^o, C_p^o, ΔS^o and ΔC_p^o in cal mol^{-1} K^{-1} (J mol^{-1} K^{-1}); A and k in l mol^{-1} s^{-1}

REACTION: TRICHLOROMETHYL RADICAL + PROPENE

$\dot{C}C\ell_3(I) + CH_2{=}CHCH_3(II) \xrightarrow{1} CC\ell_3CH_2\dot{C}HCH_3(III)$

$\xrightarrow{2} \dot{C}H_2CH(CC\ell_3)CH_3(IV)$

	(I)	(II)	(III)	(IV)		Path 1	Path 2
$\Delta H^o_{f(300)}$	18.5 (77.5)	4.9 (20.5)	5.6 (23.4)	6.5 (27.2)	ΔH^o	-17.8 (-74.5)	-16.9 (-70.8)
$S^o_{(300)}$	70.8 (296.4)	63.8 (267.1)	96.6 (404.5)	95.5 (399.9)	ΔS^o	-38.0 (-159.1)	-39.1 (-163.7)
$C^o_{p(300)}$	14.9 (62.4)	15.3 (64.1)	33.2 (139.0)	33.0 (138.2)	ΔC^o_p	3.0 (12.6)	2.8 (11.7)

Path	log A	E	log k	Conditions	Reference and Experimental Method
1	6.8	3.4 (14.2)	5.98(452 K)	421-483 K 100 torr	[144] J.M.Tedder and J.C.Walton, Trans.Faraday Soc., 63, 2679 (1967). The Arrhenius parameters were calculated from $E_1 - \frac{1}{2}E_C = 3.4$ and $\log (A_1/A_C^{\frac{1}{2}})(\ell^{\frac{1}{2}}mol^{-\frac{1}{2}}s^{-\frac{1}{2}}) = 2.0$, and taking $\log k_C = 9.63$ for the combination of $CC\ell_3$ radicals.
2	-	-	-	421-483 K 100 torr	For details see $CC\ell_3 + CH_2{=}CH_2$, reference [140]. The only product of the addition, $CC\ell_3CH_2CHBrCH_3$.
1	~8.6	~6.9 (31.0)	~5.3(452 K)	421-483 K 100 torr	[142] J.A.Kerr and M.J.Parsonage, Int.J.Chem.Kinet., in press. For details see $CC\ell_3 + CH_2{=}CH_2$. Results calculated from original data for reference [144], adding 3.5 kcal mol^{-1} to E_1 and 2 log units to log A_1.
1	(i)8.9 (ii) -	(i)6.5 (ii) -	(i)5.8(452 K) (ii) -	-	[143] H.W.Sidebottom, J.M.Tedder and J.C.Walton, Int.J.Chem.Kinet., in press A recalculation of the data of reference [144] according to the reinterpretation given in reference [142]. We have adjusted the reported A factor in terms of $\log A_C = 9.63$ and $E_C = 0$ kcal mol^{-1} for the combination of $CC\ell_3$ radicals. For details see $CC\ell_3 + CH_2{=}CH_2$.

PREFERRED RATE PARAMETERS

log k = 8.9 - 6.5/2.3RT

COMMENTS

The preferred Arrhenius parameters are similar to those for CH_3 addition to $CH_3CH{=}CH_2$. It is surprising that $CC\ell_3$ appears to add exclusively to the terminal C atom while there is some CH_3 addition at the central C atom.

REACTION: TRICHLOROMETHYL RADICAL + VINYL FLUORIDE

$\dot{C}Cl_3(I) + CH_2{=}CHF(II) \xrightarrow{1} CCl_3CH_2\dot{C}HF(III)$

$\xrightarrow{2} \dot{C}H_2CHFCCl_3(IV)$

	(I)	(II)	(III)	(IV)		Path 1	Path 2
$\Delta H^\circ_{f(300)}$	18.5 (77.5)	-31.6 (-132.3)	-26.2 (-109.7)	-30.4 (-127.3)	ΔH°	-13.1 (-54.8)	-17.3 (-72.4)
$S^\circ_{(300)}$	70.8 (296.4)	60.4 (252.9)	94.8 (396.9)	[93 (390)]	ΔS°	-36.4 (-152.4)	[-38 (-159)]
$C^\circ_{p(300)}$	14.9 (62.4)	11.9 (49.8)	29.1 (121.8)		ΔC°_p	2.3 (9.6)	

Path	log A	E	log k	Conditions	Reference and Experimental Method
1	6.0	3.3 (13.8)	4.2(398 K)	337-460 K 40 torr	[141] J.M.Tedder and J.C.Walton, Trans.Faraday Soc., 62, 1859 (1966). The Arrhenius parameters were calculated from $E_1 - \frac{1}{2}E_c = 3.3$ (Path 1), $E_2 - \frac{1}{2}E_c = 5.4$ (Path 2), $\log (A_1/A_c^{\frac{1}{2}})(\ell^{\frac{1}{2}}mol^{-\frac{1}{2}}s^{-\frac{1}{2}}) = 1.2$ and $\log (A_2/A_c^{\frac{1}{2}})(\ell^{\frac{1}{2}}mol^{-\frac{1}{2}}s^{-\frac{1}{2}}) = 1.1$, and taking $\log k_c = 9.63$ for combination of CCl_3 radicals. For details see $CCl_3 + CH_2{=}CH_2$, reference [140].
2	5.9	5.4 (22.6)	3.0(398 K)	337-460 K 40 torr	
1	~8.0	~6.8 (28.4)	~4.3(398 K)	-	[142] J.A.Kerr and M.J.Parsonage, Int.J.Chem.Kinet., in press. For details see $CCl_3 + CH_2{=}CH_2$. Results calculated from original data of reference [141] adding 3.5 kcal mol^{-1} to E_1 and E_2 and ~2.0 log units to $\log A_1$ and $\log A_2$.
2	~7.9	~8.9 (37.3)	~3.0(398 K)	-	
1	(i)7.4	(i) 6.4 (26.8)	(i)3.9(398 K)	-	[143] H.W.Sidebottom, J.M.Tedder and J.C.Walton, Int.J.Chem.Kinet., in press. A recalculation of the data of reference [141] according to the reinterpretation given in reference [142]. We have adjusted the reported A factor in terms of $\log A_c = 9.63$ and $E_c = 0$ kcal mol^{-1} for the combination of CCl_3 radicals. For details see $CCl_3 + CH_2{=}CH_2$
	(ii)7.4	(ii)6.2 (25.9)	(ii)4.0(398 K)	-	
2	(i)7.4	(i)8.4 (35.2)	(i)2.8(398 K)	-	
	(ii)7.5	(ii)8.6 (35.9)	(ii)2.8(398 K)	-	

PREFERRED RATE PARAMETERS

$\log k_1 = 7.4 - 6.2/2.3RT$

$\log k_2 = 7.5 - 8.6/2.3RT$

COMMENTS

Path 1 is the favoured route for addition, despite the fact that the enthalpy change for path 2 is probably more favourable.

ΔH^o_f, ΔH^o and E in kcal mol^{-1} (kJ mol^{-1}); S^o, C^o_p, ΔS^o and ΔC^o_p in cal mol^{-1} K^{-1} (J mol^{-1} K^{-1}); A and k in l mol^{-1} s^{-1}

REACTION: TRICHLOROMETHYL RADICAL + 1,1-DIFLUOROETHYLENE

$\dot{C}C\ell_3(I) + CH_2{=}CF_2(II) \xrightarrow{1} CC\ell_3CH_2\dot{C}F_2(III)$

$\xrightarrow{2} \dot{C}H_2CF_2CC\ell_3(IV)$

	(I)	(II)	(III)	(IV)		Path 1	Path 2
$\Delta H^\circ_{f(300)}$	18.5 (77.5)	[-71.5 (-299.4)]	[-77 (-322)]	-78.5 (-328.7)	ΔH°	[-24 (-100)]	-25.5 (-106.8)
$S^\circ_{(300)}$	70.8 (296.4)	64.9 (271.7)	98.0 (410.3)	97.2 (407.0)	ΔS°	-37.7 (-157.8)	-38.5 (-161.2)
$C^\circ_{p(300)}$	14.9 (62.4)	14.8 (62.0)	31.7 (132.7)	32.2 (134.8)	ΔC°_p	2.0 (8.4)	2.5 (10.5)

Path	log A	E	log k	Conditions	Reference and Experimental Method
1	6.1	4.6 (19.3)	3.6(400 K)	344-455 K 40 torr	[141] J.M.Tedder and J.C.Walton, Trans.Faraday Soc., 62, 1859 (1966). The Arrhenius parameters were calculated from $E_1 - \frac{1}{2}E_c = 4.6$ (Path 1), $E_2 - \frac{1}{2}E_c = 8.3$ (Path 2), log $(A_1/A_c^{\frac{1}{2}})(\ell^{\frac{1}{2}}mol^{-\frac{1}{2}}s^{-\frac{1}{2}}) = 1.3$, log $(A_2/A_c^{\frac{1}{2}})(\ell^{\frac{1}{2}}mol^{-\frac{1}{2}}s^{-\frac{1}{2}}) = 1.1$ and taking log $k_c = 9.63$ for combination of $CC\ell_3$ radicals. For details see $CC\ell_3 + CH_2{=}CH_2$, reference [140].
2	5.9	8.3 (34.8)	1.7(435 K)	405-465 K 40 torr	
1	∿8.1	∿8.1 (33.9)	∿3.6(400 K)	-	[142] J.A.Kerr and M.J.Parsonage, Int.J.Chem.Kinet., in press. For details see $CC\ell_3 + CH_2{=}CH_2$. Results calculated from original data of reference [141] adding 3.5 kcal mol^{-1} to E_1 and E_2 and ∿2.0 log units to log A_1 and log A_2.
2	∿7.9	∿11.8 (49.4)	∿1.9(435 K)	-	
1	(i)7.7	(i)7.7 (32.2)	(i)3.5(400 K)	-	[143] H.W.Sidebottom, J.M.Tedder and J.C.Walton, Int.J.Chem.Kinet., in press A recalculation of the data of reference [141] according to the interpretation given in reference [142]. We have adjusted the reported A factor in terms of log $A_C = 9.63$ and $E_C = 0$ kcal mol^{-1} for the combination of $CC\ell_3$ radicals. For details see $CC\ell_3 + CH_2{=}CH_2$.
	(ii)7.7	(ii)7.8 (32.6)	(ii)3.4(400 K)	-	
2	(i)7.7	(i)11.4 (47.7)	(i)2.0(435 K)	-	
	(ii) -	(ii) -	(ii) -	-	

PREFERRED RATE PARAMETERS

$\log k_1 = 7.7 - 7.8/2.3RT$

$\log k_2 = 7.7 - 11.4/2.3RT$

COMMENTS

Path 1 is the favoured route for addition despite the fact that the enthalpy change for path 2 is probably more favourable.

REACTION: TRICHLOROMETHYL RADICAL + TRIFLUOROETHYLENE

$$\dot{C}Cl_3(I) + CHF{=}CF_2(II) \xrightarrow{1} CCl_3CHF\dot{C}F_2(III)$$
$$\xrightarrow{2} CCl_3CF_2\dot{C}HF(IV)$$

	(I)	(II)	(III)	(IV)
$\Delta H^o_{f(300)}$	18.5 (77.5)	-117.3 (-491.9)	[-117 (-490)]	[-118 (-494)]
$S^o_{(300)}$	70.8 (296.4)	70.1 (293.5)	[103 (431)]	102.8 (430.4)
$C^o_{p(300)}$	14.9 (62.4)	16.5 (69.1)		34.0 (142.4)

	Path 1	Path 2
ΔH^o	[-18 (-75)]	[-19 (-80)]
ΔS^o	[-38 (-159)]	-38.1 (-159.5)
ΔC^o_p		2.6 (10.9)

Path	log A	E	log k	Conditions	Reference and Experimental Method
1	6.7	6.1 (25.5)	3.5(416 K)	367-465 K 40 torr	[141] J.M.Tedder and J.C.Walton, Trans.Faraday Soc., 62, 1859 (1966). Arrhenius parameters calculated from $E_1 - \frac{1}{2}E_C = 6.1$ kcal mol^{-1} (Path 1), $E_2 - \frac{1}{2}E_C = 7.1$ (Path 2), $\log (A_1/A_C^{\frac{1}{2}})(\ell^{\frac{1}{2}}mol^{-\frac{1}{2}}s^{-\frac{1}{2}}) = 1.9$, $\log (A_2/A_C^{\frac{1}{2}}) = 2.0$ and taking $\log k_C = 9.63$ for the combination of CCl_3 radicals. For details see $CCl_3 + CH_2{=}CH_2$, reference [140].
2	6.8	7.1 (29.7)	3.1(416 K)	367-465 K 40 torr	
1	∿8.7	∿9.6 (40.2)	∿3.5(416 K)	-	[142] J.A.Kerr and M.J.Parsonage, Int.J.Chem.Kinet., in press. For details see $CCl_3 + CH_2{=}CH_2$. Results calculated from original data of reference [141], adding ∿3.5 kcal mol^{-1} to E_1 and E_2, and ∿2.0 log units to log A_1 and log A_2.
2	∿8.8	∿10.5 (44.4)	∿3.1(416 K)	-	
1	(i)8.4	(i)9.2 (38.5)	(i)3.6(416 K)	-	[143] H.W.Sidebottom, J.M.Tedder and J.C.Walton, Int.J.Chem.Kinet., in press. A recalculation of the data of reference [143] according to the re-interpretation given in reference [142]. We have adjusted the reported A factor in terms of $\log A_C = 9.63$ and $E_C = 0$ kcal mol^{-1} for the combination of CCl_3 radicals. For details see $CCl_3 + CH_2{=}CH_2$
	(ii)7.6	(ii)7.9 (33.1)	(ii)3.5(416 K)	-	
2	(i)8.4	(i)10.2 (42.7)	(i)3.0(416 K)	-	
	(ii) -	(ii) -	(ii) -	-	

PREFERRED RATE PARAMETERS

$\log k_1 = 7.6 - 7.9/2.3RT$

$\log k_2 = 8.4 - 10.2/2.3RT$

ΔH^o_f, ΔH^o and E in kcal mol^{-1} (kJ mol^{-1}); S^o, C^o_p, ΔS^o and ΔC^o_p in cal mol^{-1} K^{-1} (J mol^{-1} K^{-1}); A and k in l mol^{-1} s^{-1}

REACTION: TRICHLOROMETHYL RADICAL + TETRAFLUOROETHYLENE $\dot{C}C\ell_3(I) + CF_2{=}CF_2(II) \longrightarrow CC\ell_3CF_2\dot{C}F_2(III)$

	(I)	(II)	(III)		
$\Delta H^\circ_{f(300)}$	18.5 (77.5)	-157.9 (-661.1)	[-166 (-695)]	ΔH°	[-27 (-113)]
$S^\circ_{(300)}$	70.8 (296.4)	74.6 (312.4)	106.5 (445.9)	ΔS°	-38.9 (-162.9)
$C^\circ_{p(300)}$	14.9 (62.4)	19.4 (81.2)	35.9 (150.3)	ΔC°_p	1.6 (6.7)

log A	E	log k	Conditions	Reference and Experimental Method
7.5	6.1 (25.5)	4.2(398 K)	345-452 K 40 torr	[141] J.M.Tedder and J.C.Walton, Trans.Faraday Soc., 62, 1859 (1966). Arrhenius parameters calculated from $E_1 - \frac{1}{2}E_c$ = 6.1 kcal mol^{-1} and log $(A_1/A_c^{\frac{1}{2}})(\ell^{\frac{1}{2}}mol^{-\frac{1}{2}}s^{-\frac{1}{2}})$ = 2.7 and taking log k_c = 9.63 for the combination of $CC\ell_3$ radicals. For details see $CC\ell_3 + CH_2{=}CH_2$, reference [140].
∿9.5	∿9.6 (40.1)	∿4.2(398 K)	-	[142] J.A.Kerr and M.J.Parsonage, Int.J.Chem.Kinet., in press. For details see $CC\ell_3 + CH_2{=}CH_2$. Results calculated from original data of reference [141], adding 3.5 kcal mol^{-1} to E_1 and ∿2.0 log units to log A_1.
(i)9.0	(i)9.2 (38.5)	(i)3.9(398 K)	-	[143] H.W.Sidebottom, J.M.Tedder and J.C.Walton, Int.J.Chem.Kinet., in press. A recalculation of the data of reference [141] according to the re-interpretation of reference [142]. We have adjusted the reported A-factor in terms of log A_c = 9.63 and E_c = 0 kcal mol^{-1} for the combination of $CC\ell_3$ radicals. For details see $CC\ell_3 + CH_2{=}CH_2$.
(ii)7.7	(ii)6.6 (27.6)	(ii)4.1(398 K)	-	

PREFERRED RATE PARAMETERS

[log k = 7.7 - 6.6/2.3RT]

COMMENTS

The rather large discrepancy in the Arrhenius parameters calculated by methods (i) and (ii) of reference [143] was ascribed to the formation of telomers in the reaction of $CC\ell_3$ with C_2F_4. We have selected the data by method (ii) on the basis that the A-factor seems more reasonable, but obviously they must be assigned considerable errors.

REACTION: TRICHLOROMETHYL RADICAL + 2-FLUOROPROPENE

$\dot{C}Cl_3(I) + CH_2{=}CFCH_3(II) \xrightarrow{1} CCl_3CH_2\dot{C}FCH_3(III)$

$\xrightarrow{2} \dot{C}H_2CF(CCl_3)CH_3(IV)$

	(I)	(II)	(III)	(IV)	
$\Delta H^\circ_{f(300)}$	18.5 (77.5)		[-35.8 (-149.9)]	-35.5 (-148.6)	ΔH°
$S^\circ_{(300)}$	70.8 (296.4)		[-101 (-423)]		ΔS°
$C^\circ_{p(300)}$	14.9 (62.4)				ΔC°_p

Path	log A	E	log k	Conditions	Reference and Experimental Method
1	6.4	3.2 (13.4)	4.7(408 K)	362-454 K 40 torr	[144] J.M.Tedder and J.C.Walton, Trans.Faraday Soc., 63, 2679 (1967). Arrhenius parameters calculated from $E_1 - \frac{1}{2}E_C = 3.2$ kcal mol⁻¹ and $\log (A_1/A_C^{\frac{1}{2}})(l^{\frac{1}{2}}mol^{-\frac{1}{2}}s^{-\frac{1}{2}}) = 1.6$ and taking $\log k_C = 9.63$ for the combination of CCl_3 radicals.
2	∿7.8	6.4 (26.8)	∿4.4(408 K)	362-454 K 40 torr	The major addition product was $CH_3CFBrCH_2CCl_3$, and the minor $CH_3CF(CCl_3)CH_2Br$. For details see $CCl_3 + CH_2{=}CH_2$, reference [140].
1	8.25	6.9 (28.9)	4.68(408 K)	-	[142] J.A.Kerr and M.J.Parsonage, Int.J.Chem.Kinet., in press. For details see $CCl_3 + CH_2{=}CH_2$. Results calculated from data of reference [144] and based on $\log k_C = 9.63$ for combination of CCl_3 radicals.
1	(i)8.1 (ii)7.6	(i)6.3 (26.4) (ii)5.7 (23.8)	(i)4.7(408 K) (ii)4.5(408 K)	-	[143] H.W.Sidebottom, J.M.Tedder and J.C.Walton, Int.J.Chem.Kinet., in press. A recalculation of the data of reference [144] according to the reinterpretation of reference [142]. We have adjusted the reported A factor in terms of $\log A_C = 9.63$ and $E_C = 0$ kcal mol⁻¹ for the combination of CCl_3 radicals. For details see $CCl_3 + CH_2{=}CH_2$.

PREFERRED RATE PARAMETERS

$\log k_1 = 7.6 - 5.7/2.3RT$

COMMENTS

The addition occurs predominantly at the terminal C atom. The product $CH_3CF(CCl_3)CH_2Br$ was observed at 454 and 435 K when the ratio $[CH_3CFBrCH_2CCl_3]/[CH_3CF(CCl_3)CH_2Br]$ was reported as 19.3 and 52.8 respectively. Following the procedure of reference [142], these ratios yield $E_2 \simeq 25$ kcal mol⁻¹, which is much too high. It is not clear how the original value $E_2 = 6.4$ kcal mol⁻¹ was obtained.

ΔH°_f, ΔH° and E in kcal mol⁻¹ (kJ mol⁻¹); S°, C°_p, ΔS° and ΔC°_p in cal mol⁻¹ K⁻¹ (J mol⁻¹ K⁻¹); A and k in l mol⁻¹ s⁻¹

REACTION: TRICHLOROMETHYL RADICAL + 3,3,3-TRIFLUOROPROPENE

$\dot{C}C\ell_3(I) + CH_2{=}CHCF_3(II) \xrightarrow{1} CC\ell_3CH_2\dot{C}HCF_3(III)$

$\xrightarrow{2} \dot{C}H_2CH(CC\ell_3)CF_3(IV)$

	(I)	(II)	(III)	(IV)		Path 1	Path 2
$\Delta H^\circ_{f(300)}$	18.5 (77.5)	-143.6 (-601.3)	-142.7 (-597.5)	-141.8 (-593.7)	ΔH°	-19.4 (-81.2)	-20.3 (-85.0)
$S^\circ_{(300)}$	70.8 (296.4)	75.9 (317.8)	108.7 (455.1)	107.6 (450.5)	ΔS°	-38.0 (-159.1)	-39.1 (-163.7)
$C^\circ_{p(300)}$	14.9 (62.4)	22.0 (92.1)	39.7 (166.2)	39.5 (165.4)	ΔC°_p	2.8 (11.7)	2.6 (10.9)

Path	log A	E	log k	Conditions	Reference and Experimental Method
1	7.4	4.8 (20.1)	4.7(393 K)	313-453 K 40-120 torr	[136] J.Gibb, M.J.Peters, J.M.Tedder, J.C.Walton and K.D.R.Winton, Chem. Comm., 978 (1970). This reference is a preliminary communication with no experimental details, but presumably involved the same method as for other $CC\ell_3$ additions to olefins, see $CC\ell_3 + CH_2{=}CH_2$, reference [140]. The results are taken directly from the paper, but the data for the reference reaction are not given.
1	-	∿8.3 (34.7)	-	-	[142] J.A.Kerr and M.J.Parsonage, Int.J.Chem.Kinet., in press. It is presumed that the results of reference [136] are subject to the same errors of interpretation as for the other $CC\ell_3$ additions to olefins, see $CC\ell_3 + CH_2{=}CH_2$ system. In this case it is only possible to correct the activation energy, as the original data for the A-factor do not include the A-factor taken for the reference reaction.
1	(i)8.4 (ii)7.7	(i)8.0 (33.5) (ii)6.5 (27.1)	(i)4.0(393 K) (ii)4.1(393 K)	- -	[143] H.W.Sidebottom, J.M.Tedder and J.C.Walton, Int.J.Chem.Kinet., in press. A recalculation of the data of reference [136] according to the re-interpretation of reference [142]. We have adjusted the reported A-factor in terms of log A_c = 9.63 and E_c = 0 kcal mol^{-1} for the recombination of $CC\ell_3$ radicals. For details see $CC\ell_3 + CH_2{=}CH_2$.

PREFERRED RATE PARAMETERS

$\log k_1 = 7.7 - 6.5/2.3RT$

COMMENTS

Path 1 was the exclusive mode of addition [Private Communication from the authors].

REACTION: TRICHLOROMETHYL RADICAL + HEXAFLUOROPROPENE

$$\dot{C}Cl_3(I) + CF_2{=}CFCF_3(II) \xrightarrow{1} CCl_3CF_2\dot{C}FCF_3(III)$$
$$\xrightarrow{2} \dot{C}F_2CF(CCl_3)CF_3(IV)$$

	(I)	(II)	(III)	(IV)		Path 1	Path 2
$\Delta H^\circ_{f(300)}$	18.5 (77.5)		-276.2 (-1156.4)	[-271 (1135)]	ΔH°		
$S^\circ_{(300)}$	70.8 (296.4)		[122 (511)]		ΔS°		
$C^\circ_{p(300)}$	14.9 (62.4)				ΔC°_p		

Path	log A	E	log k	Conditions	Reference and Experimental Method
1	6.1	6.2 (26.0)	3.1(450 K)	415-485 K 40 torr	[144] J.M.Tedder and J.C.Walton, Trans.Faraday Soc., 63, 2679 (1967). Arrhenius parameters calculated from $E_1 - E_C/2 = 6.2$ kcal mol^{-1} and $\log (A_1/A_C^{\frac{1}{2}})(l^{\frac{1}{2}}mol^{-\frac{1}{2}}s^{-\frac{1}{2}}) = 1.3$, and taking $\log k_C = 9.63$ for CCl_3 radicals. For details see $CCl_3 + CH_2{=}CH_2$, reference [140].
1	8.34	10.9 (45.6)	3.05(450 K)	-	[142] J.A.Kerr and M.J.Parsonage, Int.J.Chem.Kinet., in press. Recalculation of data of reference [144] above, see $CCl_3 + CH_2{=}CH_2$ system. Results based on $\log k_C = 9.63$ for CCl_3 radicals.
1	(i)8.6	(i)9.3 (38.9)	(i)4.1(450 K)	-	[143] H.W.Sidebottom, J.M.Tedder and J.C.Walton, Int.J.Chem.Kinet., in press.
	(ii)8.3	(ii)10.9 (45.6)	(ii)3.0(450 K)	-	A recalculation of the data of reference [144] according to the reinterpretation given in reference [142]. We have adjusted the reported A factor in terms of $\log A_C = 9.63$ and $E_C = 0$ kcal mol^{-1} for the combination of CCl_3 radicals. For details see $CCl_3 + CH_2{=}CH_2$.

PREFERRED RATE PARAMETERS

$\log k_1 = 8.34 - 10.9/2.3RT$

COMMENTS

The exclusive addition product was $CF_3CFBrCF_2CCl_3$, indicating addition entirely at the terminal C atom.

ΔH°_f, ΔH° and E in kcal mol^{-1} (kJ mol^{-1}); S°, C°_p, ΔS° and ΔC°_p in cal mol^{-1} K^{-1} (J mol^{-1} K^{-1}); A and k in l mol^{-1} s^{-1}

REACTION: TRICHLOROMETHYL RADICAL + 1,1-DIFLUORO-2-CHLOROETHYLENE

$$\dot{C}C\ell_3(I) + CF_2{=}CHC\ell(II) \xrightarrow{1} CC\ell_3CF_2\dot{C}HC\ell(III)$$
$$\xrightarrow{2} \dot{C}F_2CHC\ell CC\ell_3(IV)$$

	(I)	(II)	(III)	(IV)		Path 1	Path 2
$\Delta H^\circ_{f(300)}$	18.5 (77.5)	-78.3 (-327.8)	-85.4 (-357.6)	[-82 (-343)]	ΔH°	-25.6 (-107.2)	[-22 (-92)]
$S^\circ_{(300)}$	70.8 (296.4)	72.7 (304.4)	104.8 (438.8)	106.3 (445.1)	ΔS°	-38.7 (-162.0)	-37.2 (-155.8)
$C^\circ_{p(300)}$	14.9 (62.4)	17.6 (73.7)	35.1 (147.0)	35.2 (147.4)	ΔC°_p	2.6 (10.9)	2.7 (11.3)

Path	log A	E	log k	Conditions	Reference and Experimental Method
1 + 2	-	-	3.46	423 K 40 torr	[145] D.P.Johari, H.W.Sidebottom, J.M.Tedder and J.C.Walton, J.Chem.Soc., (B), 95 (1971). Results calculated from $k(CF_2{=}CHC\ell)/k(CH_2{=}CHC\ell) = 0.05$ and log k = 4.76 at 423 K for $CC\ell_3$ addition to $CH_2{=}CHC\ell$. For details see $CC\ell_3 + CH_2{=}CH_2$. When $CC\ell_3$ adds to an olefin at a C atom containing a $C\ell$ substituent, the adduct radical can undergo a rapid decomposition reaction: $CC\ell_3CXC\ell\dot{C}X_2 \longrightarrow CC\ell_3CX{=}CX_2 + \dot{C}\ell$ (X = H, F, $C\ell$ or Br) as shown by the formation of the propene $CC\ell_3CX{=}CX_2$ in the products, which was the only product from this type of addition to the olefin. Addition at the other C atom yielded the usual $CC\ell_3CX_2CXC\ell Br$ product. Total rates of attack on the olefins $CXC\ell{=}CX_2$ were calculated from the sum of the products $CC\ell_3CX{=}CX_2$ and $CC\ell_3CX_2CXC\ell Br$, relative to addition of $CC\ell_3$ to $CH_2{=}CHC\ell$.

PREFERRED RATE PARAMETERS

Reasonable as given above

COMMENTS

From product analysis it was shown that at 423 K the relative rates of paths 1 : 2 were 25 : 1.

REACTION: TRICHLOROMETHYL RADICAL + 1,1-DIFLUORO-2,2-DICHLOROETHYLENE

$\dot{C}Cl_3(I) + CCl_2{=}CF_2(II) \xrightarrow{1} CCl_3CCl_2\dot{C}F_2(III)$

$\xrightarrow{2} CCl_3CF_2\dot{C}Cl_2(IV)$

	(I)	(II)	(III)	(IV)
$\Delta H^\circ_{f(300)}$	18.5 (77.5)	-79.3 (-332.0)	[-88 (-368)]	-90.3 (-378.1)
$S^\circ_{(300)}$	70.8 (296.4)	79.4 (332.4)		111.1 (465.2)
$C^\circ_{p(300)}$	14.9 (62.4)	21.1 (88.3)		38.0 (159.1)

	Path 1	Path 2
ΔH°	[-27 (-113)]	-29.5 (-123.5)
ΔS°		-39.1 (-163.7)
ΔC°_p		2.0 (8.4)

Path	log A	E	log k	Conditions	Reference and Experimental Method
(1+2)	-	-	3.54	423 K 40 torr	[145] D.P.Johari, H.W.Sidebottom, J.M.Tedder and J.C.Walton, J.Chem.Soc. (B.), 95 (1971). Above result calculated from $k(CF_2{=}CCl_2)/k(CH_2{=}CHCl)$ = 0.06 and log k = 4.76 at 423 K for CCl_3 addition to $CH_2{=}CHCl$. For details see $CCl_3 + CH_2{=}CH_2$ system, reference [140], and $CCl_3 + CF_2{=}CHCl$ system.

PREFERRED RATE PARAMETERS

Reasonable as given above

COMMENTS

From product analysis it was shown that at 423 K the relative rates of paths 1:2 were 1:5.

ΔH^o_f, ΔH^o and E in kcal mol^{-1} (kJ mol^{-1}); S^o, C^o_p, ΔS^o and ΔC^o_p in cal mol^{-1} K^{-1} (J mol^{-1} K^{-1}); A and k in l mol^{-1} s^{-1}

REACTION: TRICHLOROMETHYL RADICAL + BROMOTRIFLUOROETHYLENE

$\dot{C}C\ell_3(I) + CF_2{=}CFBr(II) \xrightarrow{1} CC\ell_3CF_2\dot{C}FBr(III)$

$\xrightarrow{2} \dot{C}F_2CFBrCC\ell_3(IV)$

	(I)	(II)	(III)	(IV)		Path 1	Path 2
$\Delta H^\circ_{f(300)}$	18.5 (77.5)				ΔH°		
$S^\circ_{(300)}$	70.8 (296.4)	79.8 (334.1)			ΔS°		
$C^\circ_{p(300)}$	14.9 (62.4)	20.5 (85.8)			ΔC°_p		

Path	log A	E	log k	Conditions	Reference and Experimental Method
(1+2)	-	-	3.80	423 K 40 torr	[145] D.P.Johari, H.W.Sidebottom, J.M.Tedder and J.C.Walton, J.Chem.Soc. (B), 95 (1971). Rate constant calculated from $k(CF_2{=}CFBr)/k(CH_2{=}CHC\ell) = 0.11$ and $\log k = 4.76$ at 423 K for $CC\ell_3$ addition to $CH_2{=}CHC\ell$. For details see $CC\ell_3 + CF_2{=}CHC\ell$ system.

PREFERRED RATE PARAMETERS

Reasonable as given above

COMMENTS

From product analysis it was shown that at 423 K, the relative rates of paths 1:2 were 33:1.

REACTION: TRICHLOROMETHYL RADICAL + VINYL CHLORIDE

$\dot{C}Cl_3(I) + CH_2{=}CHCl(II) \xrightarrow{1} CCl_3CH_2\dot{C}HCl(III)$

$\xrightarrow{2} \dot{C}H_2CHClCCl_3(IV)$

	(I)	(II)	(III)	(IV)
$\Delta H^\circ_{f(300)}$	18.5 (77.5)	8.6 (36.0)	6.6 (27.6)	5.7 (23.9)
$S^\circ_{(300)}$	70.8 (296.4)	63.0 (263.6)	96.8 (405.3)	97.0 (406.1)
$C^\circ_{p(300)}$	14.9 (62.4)	13.0 (54.4)	30.7 (128.5)	31.3 (131.1)

	Path 1	Path 2
ΔH°	-20.5 (-85.8)	-21.4 (-89.6)
ΔS°	-37.0 (-154.9)	-36.8 (-154.1)
ΔC°_p	2.8 (11.7)	3.4 (14.2)

Path	log A	E	log k	Conditions	Reference and Experimental Method
1	6.4	3.4 (14.2)	4.65(423 K)	361-439 K 40 torr	[145] D.P.Johari, H.W.Sidebottom, J.M.Tedder and J.C.Walton, J.Chem.Soc.,(B), 95 (1971). Arrhenius parameters calculated from $E_1 - \frac{1}{2}E_c = 3.4$ kcal mol^{-1} and $\log (A_1/A_c^{\frac{1}{2}})(l^{\frac{1}{2}}mol^{-\frac{1}{2}}s^{-\frac{1}{2}}) = 1.6$, and taking $\log k_4 = 9.63$ for combination of CCl_3 radicals. For details see $CCl_3 + CH_2{=}CH_2$ system, reference [140].
1	8.70	7.6 (31.8)	4.76(423 K)	-	[142] J.A.Kerr and M.J.Parsonage, Int.J.Chem.Kinet., in press. Recalculation of data of reference [145]. See $CCl_3 + CH_2{=}CH_2$ system. Results based on $\log k_c = 9.63$ for combination of CCl_3 radicals.
1	(i)8.6	(i)6.5 (27.1)	(i)5.2(423 K)	-	[143] H.W.Sidebottom, J.M.Tedder and J.C.Walton, Int.J.Chem.Kinet., in press. A recalculation of the data of reference [145] according to the reinterpretation given in reference [142]. We have adjusted the reported A factor in terms of $\log A_c = 9.63$ and $E_c = 0$ kcal mol^{-1} for the combination of CCl_3 radicals. See $CCl_3 + CH_2{=}CH_2$.
	(ii)8.6	(ii)7.6 (31.8)	(ii)4.7(423 K)	-	

PREFERRED RATE PARAMETERS

$\log k_1 = 8.6 - 7.6/2.3RT$

COMMENTS

Addition occurred exclusively by path 1.

ΔH^o_f, ΔH^o and E in kcal mol^{-1} (kJ mol^{-1}); S^o, C^o_p, ΔS^o and ΔC^o_p in cal mol^{-1} K^{-1} (J mol^{-1} K^{-1}); A and k in l mol^{-1} s^{-1}

REACTION: TRICHLOROMETHYL RADICAL + TRANS-DICHLOROETHYLENE

$\dot{C}C\ell_3$(I) + trans-$CHC\ell{=}CHC\ell$(II) ⟶ $CC\ell_3CHC\ell\dot{C}HC\ell$(III)

	(I)	(II)	(III)		
$\Delta H^\circ_{f(300)}$	18.5 (77.5)	1.2 (5.0)	-1.2 (-5.0)	ΔH°	-20.9 (-87.5)
$S^\circ_{(300)}$	70.8 (296.4)	70.8 (296.4)	105.0 (439.6)	ΔS°	-36.6 (-153.2)
$C^\circ_{p(300)}$	14.9 (62.4)	15.3 (66.2)	34.2 (143.2)	ΔC°_p	3.5 (14.7)

log A	E	log k	Conditions	Reference and Experimental Method
-	-	3.46	423 K 40 torr	[145] D.P.Johari, H.W.Sidebottom, J.M.Tedder and J.C.Walton, J.Chem.Soc.,(B), 95 (1971). Above rate constant calculated from $k(CHC\ell{=}CHC\ell)/k(CH_2{=}CHC\ell)$ = 0.05 and log k = 4.76 at 423 K for $CC\ell_3$ addition to $CH_2{=}CHC\ell$. For details see $CC\ell_3$ + $CF_2{=}CHC\ell$ system.

PREFERRED RATE PARAMETERS

Reasonable as given above

REACTION: TRICHLOROMETHYL RADICAL + TRICHLOROETHYLENE

$\dot{C}Cl_3(I) + CHCl{=}CCl_2(II) \xrightarrow{1} CCl_3CHCl\dot{C}Cl_2(III)$

$\xrightarrow{2} \dot{C}HClCCl_2CCl_3(IV)$

	(I)	(II)	(III)	(IV)		Path 1	Path 2
$\Delta H^\circ_{f(300)}$	18.5 (77.5)	-2.0 (-8.4)	-6.1 (-25.5)	[-8 (-33)]	ΔH°	-22.6 (-94.6)	[-25 (-105)]
$S^\circ_{(300)}$	70.8 (296.4)	77.5 (324.5)	110.9 (464.3)		ΔS°	-37.4 (-156.6)	
$C^\circ_{p(300)}$	14.9 (62.4)	19.3 (80.8)	37.0 (154.9)		ΔC°_p	2.8 (11.7)	

Path	log A	E	log k	Conditions	Reference and Experimental Method
(1+2)	-	-	3.96	423 K 40 torr	[145] D.P.Johari, H.W.Sidebottom, J.M.Tedder and J.C.Walton, J.Chem.Soc.(B), 95 (1971). Above results calculated from $k(CHCl{=}CCl_2)/k(CH_2{=}CHCl) = 0.16$ and $\log k = 4.76$ at 423 K for CCl_3 addition to $CH_2{=}CHCl$. For details see $CCl_3 + CF_2{=}CHCl$ system.

PREFERRED RATE PARAMETERS

Reasonable as given above

COMMENTS

From product analysis it was shown that at 423 K, the relative rates of paths 1:2 were 30:1.

ΔH°_f, ΔH° and E in kcal mol^{-1} (kJ mol^{-1}); S°, C°_p, ΔS° and ΔC°_p in cal mol^{-1} K^{-1} (J mol^{-1} K^{-1}); A and k in l mol^{-1} s^{-1}

REACTION: TRICHLOROMETHYL RADICAL + TETRACHLOROETHYLENE $\dot{C}C\ell_3(I) + CC\ell_2{=}CC\ell_2(II) \longrightarrow CC\ell_3CC\ell_2\dot{C}C\ell_2(III)$

	(I)	(II)	(III)		
$\Delta H^\circ_{f(300)}$	18.5 (77.5)	-2.7 (-11.3)	[-13 (-54)]	ΔH°	[-29 (-121)]
$S^\circ_{(300)}$	70.8 (296.4)	84.2 (352.5)		ΔS°	
$C^\circ_{p(300)}$	14.9 (62.4)	22.8 (95.5)		ΔC°_p	

log A	E	log k	Conditions	Reference and Experimental Method
-	-	<2.24	423 K 40 torr	[145] D.P.Johari, H.W.Sidebottom, J.M.Tedder and J.C.Walton, J.Chem.Soc. (B), 95 (1971). Above rate constant calculated from $k(CC\ell_2{=}CC\ell_2)/k(CH_2{=}CHC\ell) < 0.003$, and $\log k = 4.76$ at 423 K for $CC\ell_3$ addition to $CH_2{=}CHC\ell$. For details see $CC\ell_3 + CF_2{=}CHC\ell$ system.

PREFERRED RATE PARAMETERS

Reasonable as given above

REACTION: TRICHLOROMETHYL RADICAL + ACRYLONITRILE

$\dot{C}C\ell_3(I) + CH_2{=}CHCN(II) \xrightarrow{1} CC\ell_3CH_2\dot{C}HCN(III)$

$\xrightarrow{2} \dot{C}H_2CHCNCC\ell_3(IV)$

	(I)	(II)	(III)	(IV)
$\Delta H^o_{f(300)}$	18.5 (77.5)	43.7 (183.0)	[27 (113)]	40.9 (171.2)
$S^o_{(300)}$	70.8 (296.4)	64.2 (268.8)	99.8 (417.9)	99.2 (415.4)
$C^o_{p(300)}$	14.9 (62.4)	14.9 (62.4)	30.9 (129.4)	33.3 (139.4)

	Path 1	Path 2
ΔH^o	[-35 (-147)]	-21.3 (-89.2)
ΔS^o	-35.2 (-147.4)	-35.8 (-149.9)
ΔC^o_p	1.1 (4.6)	-3.5 (14.7)

Path	log A	E	log k	Conditions	Reference and Experimental Method
1	7.2	3.3 (13.8)	5.4(393 K)	313-473 K 40-120 torr	[136] J.Gibb, M.J.Peters, J.M.Tedder, J.C.Walton and K.D.R.Winton, Chem.Comm., 978 (1970). This reference is a preliminary communication with no experimental details, but presumably involved the same method as for other $CC\ell_3$ additions to olefins (see $CC\ell_3 + CH_2{=}CH_2$, reference [140]). The results are taken directly from the paper, but the data for the reference reaction are not given.
1	-	6.8 (28.4)	-	-	[142] J.A.Kerr and M.J.Parsonage, Int.J.Chem.Kinet., in press. It is presumed that the results of reference [136] are subject to the same errors of interpretation as for the other $CC\ell_3$ additions to olefins (see $CC\ell_3 + CH_2{=}CH_2$ system). In this case, it is only possible to correct the activation energy, as the original data for the A-factor do not include the A-factor taken for the reference reaction.

PREFERRED RATE PARAMETERS

$\log k_1 = [8.6 - 6.8/2.3RT]$

COMMENTS

Path 1 is the exclusive mode of addition [Private Communication from the authors]. The preferred A-factor is based on the results for $CC\ell_3$ addition to $CH_3CH{=}CH_2$.

ΔH^o_f, ΔH^o and E in kcal mol^{-1} (kJ mol^{-1}); S^o, C^o_p, ΔS^o and ΔC^o_p in cal mol^{-1} K^{-1} (J mol^{-1} K^{-1}); A and k in l mol^{-1} s^{-1}

REACTION: HYDROXYL RADICAL + ETHYLENE $\dot{O}H(I) + CH_2{=}CH_2(II) \longrightarrow \dot{C}H_2CH_2OH(III)$

	(I)	(II)	(III)		
$\Delta H^\circ_{f(300)}$	9.4 (39.4)	12.5 (53.3)	-10.6 (-44.4)	ΔH°	-32.5 (-136.1)
$S^\circ_{(300)}$	43.9 (183.8)	52.4 (219.4)	70.6 (295.6)	ΔS°	-25.7 (-107.6)
$C^\circ_{p(300)}$	7.2 (30.1)	10.3 (43.1)	15.3 (64.1)	ΔC°_p	-2.2 (-9.2)

log A	E	log k	Conditions	Reference and Experimental Method
11.60	6.4 (26.8)	8.11(400 K)	350-451 K ? torr	[146] L.I.Avramenko and R.V.Lorentzo, Dokl.Akad.Nauk.SSSR, 67, 867 (1949). Flow system. OH from discharge through H_2O vapour. [OH] measured by U.V. absorption. This system has been shown to be unsuitable for kinetic studies [W.E.Wilson, "A Critical Review of the Gas Phase Reaction Kinetics of Several Bimolecular Reactions of the Hydroxyl Radical", NSRDS-NBS, in press].
-	-	≃9.48(300 K)	300 K <1 torr	[147] W.E.Wilson and A.A.Westenberg, Eleventh Symposium (International) on Combustion, (1967), p.1143. Flow system. OH from reaction $H + NO_2 \longrightarrow OH + NO$ (H from dissociation of H_2) and [OH] measured by esr, and products analyzed by mass spectrometry. Exponential decay of OH measured in presence of large excess of C_2H_4, and rate constants calculated from $\ln([OH]_{t_1}/[OH]_{t_2}) = nk[C_2H_4](t_2 - t_1)$ where k is the bimolecular rate constant of the overall reaction and n is the stoichiometric number, i.e. the number of OH consumed per C_2H_4 reacted. n had to be determined to obtain k since fast secondary reactions of OH with product radicals could not be avoided.
8.88	-0.90 (-3.8)	9.37(398 K)	299-497 K 1-100 torr	[148] N.R.Greiner, J.Chem.Phys., 53, 1284 (1970). Static system. OH from flash photolysis of H_2O vapour/He mixtures. [OH]

ΔH°_f, ΔH° and E in kcal mol^{-1} (kJ mol^{-1}); S°, C°_p, ΔS° and ΔC°_p in cal mol^{-1} K^{-1} (J mol^{-1} K^{-1}); A and k in l mol^{-1} s^{-1}

log A	E	log k	Conditions	Reference and Experimental Method
				determined by kinetic spectroscopy. Rate constants derived in same manner as for reference [147]. The stoichiometric number estimated in this work, to give the calculated rate constant, was based on a single secondary reaction of OH, and it has been suggested that other reactions involving radicals from C_2H_4 photolysis should be considered [149], which would have the effect of reducing the measured rate constant.
-	-	9.04	300 K 1 torr	[149] E.D.Morris,Jr., D.H.Stedman and H.Niki, J.Amer.Chem.Soc., 93, 3570 (1971). Flow system. OH from reaction $H + NO_2 \longrightarrow OH + NO$ (H from microwave discharge in H_2/He mixture). [OH] and [C_2H_4] monitored by mass spectrometer. Rate constants obtained (i) same method as in reference [147] with stoichiometry determined from exponential decay of OH and the amount of RH consumed, and (ii) from the decay of RH in the presence of an excess of OH: $\ell n\,([RH]_{t_1}/[RH]_{t_2}) = k\int_{t_1}^{t_2}[OH]\,dt$ with stoichiometry taken as unity. Product analyses revealed mass spectral peak corresponding to $\dot{C}H_2CH_2OH$, which was enhanced by total pressure increase from 1 to 4 torr.

PREFERRED RATE PARAMETERS

log k = 9.04 at 300 K

COMMENTS

The early data on this reaction are discussed in two recent evaluations [W.E.Wilson, "A Critical Review of the Gas Phase Reaction Kinetics of Several Bimolecular Reactions of the Hydroxyl Radical", NSRDS-NBS, in press; D.D.Drysdale and A.C.Lloyd, "Gas Phase Reactions of the Hydroxyl Radical", Comb.Rev., 4, 157 (1970)]. Much discussion has centred on the question of whether or not the abstraction reaction $OH + CH_2{=}CH_2 \xrightarrow{2} H_2O + CH{=}CH_2$ competes with the addition reaction. Since $\Delta H_1^o - \Delta H_2 \simeq -22$ kcal mol^{-1}, the addition reaction seems likely to predominate at low temperatures, and this is confirmed by the product analyses of reference [149]. The results of reference [149] appear to be reliable and the rate constant can be assigned with confidence to the addition reaction. It should be noted that the experimental errors were estimated to be ± 25%.

REACTION: HYDROXYL RADICAL + PROPENE

$$\dot{O}H(I) + CH_3CH{=}CH_2(II) \xrightarrow{1} CH_3\dot{C}HCH_2OH(III)$$
$$\xrightarrow{2} CH_3CHOH\dot{C}H_2(IV)$$

	(I)	(II)	(III)	(IV)		Path 1	Path 2
$\Delta H^\circ_{f(300)}$	9.4 (39.4)	4.9 (20.5)	[-19 (-80)]	[-19 (-80)]	ΔH°	[-33 (-138)]	[-33 (-138)]
$S^\circ_{(300)}$	43.9 (183.8)	63.8 (267.1)	80.5 (337.1)	77.5 (324.5)	ΔS°	-27.2 (-113.9)	-30.2 (-126.4)
$C^\circ_{p(300)}$	7.2 (30.1)	15.3 (64.0)	20.7 (86.7)	21.3 (89.2)	ΔC°_p	-1.8 (-7.5)	-1.2 (-5.0)

Path	log A	E	log k	Conditions	Reference and Experimental Method
1 + 2	-	-	(i) 9.99 (ii)10.03	300 K 1 torr	[150] E.D.Morris,Jr., D.H.Stedman and H.Niki, J.Amer.Chem.Soc., 93, 3570 (1971). For details see $OH + CH_2{=}CH_2$. The rate constants refer to the total reaction of OH with C_3H_6 as measured by methods (i) and (ii).

PREFERRED RATE PARAMETERS

[$\log k_{1+2} = 9.82$ at 300 K]

COMMENTS

The results of Greiner [J.Chem.Phys., 53, 1070 (1970)] on the reactions of OH with alkanes indicate that the reaction, $\dot{O}H + CH_3CH{=}CH_2 \xrightarrow{3} H_2O + \dot{C}H_2CH{=}CH_2$, should compete with the addition reactions (1) and (2) even at room temperature. Greiner has measured the rate constants of the reactions $\dot{O}H + RH \xrightarrow{4} H_2O + \dot{R}$, where R-H is a primary, secondary and tertiary C-H bond. On the basis that $D(CH_2{:}CHCH_2\text{-}H) = 89$ and $D(R\text{-}H)(\text{tert.C-H}) = 92$ kcal mol^{-1}, we estimate that $k_3 \simeq 2k_4$ (for tert.C-H) at 300 K = $2 \times 1.8 \times 10^9$. It follows from $\log (k_{1+2} + k_3) = 10.01$ that $\log k_{1+2} \simeq 9.82$. It seems likely that path 1 will account for at least 95% of the addition at room temperature. The preferred rate constants for OH additions to C_2H_4 and C_3H_6 yield the relative rate constants $k_{C_3H_6}/k_{C_2H_4} \simeq 6.0$, indicating that OH is an electrophilic species (see section on relative rate constants).

ΔH°_f, ΔH° and E in kcal mol^{-1} (kJ mol^{-1}); S°, C°_p, ΔS° and ΔC°_p in cal mol^{-1} K^{-1} (J mol^{-1} K^{-1}); A and k in l mol^{-1} s^{-1}

REACTION: HYDROXYL RADICAL + ACETYLENE $\dot{O}H(I) + CH{\equiv}CH(II) \longrightarrow \dot{C}H{=}CHOH(III)$

	(I)	(II)	(III)		
$\Delta H^\circ_{(300)}$	9.4 (39.4)	54.2 (226.9)	[33 (138)]	ΔH°	[-31 (-130)]
$S^\circ_{(300)}$	43.9 (183.8)	48.0 (201.0)	[66 (276)]	ΔS°	[-26 (-109)]
$C^\circ_{p(300)}$	7.2 (30.1)	10.5 (44.0)	[13.6 (56.9)]	ΔC°_p	[-4.1 (-17.2)]

log A	E	log k	Conditions	Reference and Experimental Method
11.45	7.3 (30.6)	7.63(417 K)	323-510 K ? torr	[146] L.I.Avramenko and R.V.Lorentzo, Dokl.Akad.Nauk.SSSR, 67, 867 (1949). See $OH + CH_2{=}CH_2$.
9.78	7.0 (29.30)	8.65(1350 K)	1000-1700 K 18 torr	[151] W.G.Browne, R.P.Porter, J.D.Verlin and A.H.Clark, Twelfth Symposium (International) on Combustion, (1969), p.1035. Reaction studied in low pressure laboratory flames. [OH] by absorption spectroscopy and analysis of products by mass spectrometric probe. Authors suggest the abstraction reaction $\dot{O}H + C_2H_2 \xrightarrow{2} H_2O + \dot{C}_2H$ Drysdale and Lloyd (Oxidation and Combustion Reviews, 4, 157 (1970)), suggest the predominance of an abstraction reaction as shown by the work of C.P.Fenimore and G.W.Jones, J.Chem.Phys., 41, 1887 (1964) and Porter et al., Eleventh Symposium (International) on Combustion, (1967), p.907.
-	-	∿7.18(300 K)	300 K 0.40-0.48 torr 2-4 torr	[152] J.N.Bradley and R.S.Tse, Trans.Faraday Soc., 65, 2685 (1969). Flow system. Discharge-flow study of O atoms with C_2H_2. OH generated by $O + CH_2 \longrightarrow OH + CH$, and [OH] followed by e.s.r. A value of $k(OH + C_2H_2 \longrightarrow ?)$ is estimated from the results. This is a very indirect determination of k_1 and there is no indication of the products, although the authors appear to favour the H-transfer reaction, rather than addition.

PREFERRED RATE PARAMETERS

No meaningful estimate possible

COMMENTS

The hydrogen abstraction reaction $\dot{O}H + CH{\equiv}CH \xrightarrow{2} H_2O + \dot{C}{\equiv}CH$ is even less energetically favourable than abstraction of H from $CH_2{=}CH_2$ by OH. Consequently, at room temperature addition should be the major reaction, and on this basis it is difficult to see why the rates of addition to $CH_2{=}CH_2$ and $CH{\equiv}CH$ should be very different. It is possible that the results of reference [151] were measured in the pressure-sensitive region.

COMMENTS:

The hydrogen abstraction reaction $\dot{O}H + CH{\equiv}CH \xrightarrow{2} H_2O + \dot{C}{\equiv}CH$ is even less energetically favourable than abstraction of H from $CH_2{=}CH_2$ by OH. Consequently, at room temperature, addition should be the major reaction, and on this basis it is difficult to see why the rates of addition to $CH_2{=}CH_2$ and $CH{\equiv}CH$ should be very different. It is possible that the results of reference [151] were measured in the pressure-sensitive region.

It has been suggested on the basis of experiments involving crossed molecular beams, [M.Gehring, K.Hoyermann, H.G.Wagner and J.Wolfrum, Zeit.Naturforschung, 25, 675 (1970)], that the primary reaction between HO and $CH{\equiv}CH$ is as follows: $H\dot{O} + CH{\equiv}CH \longrightarrow CO + \dot{C}H_3$. It is difficult to see how such a reaction can be elementary.

REACTION: HYDROPEROXYL RADICAL + ETHYLENE $\quad H\dot{O}_2(I) + CH_2{=}CH_2(II) \longrightarrow \dot{C}H_2CH_2OOH(III)$

	(I)	(II)	(III)		
$\Delta H^\circ_{f(300)}$	5.0 (20.9)	12.5 (52.3)	[6.6 (27.6)]	ΔH°	[-11 (-46)]
$S^\circ_{(300)}$	54.4 (227.8)	52.4 (219.4)	[78.8 (329.9)]	ΔS°	[-28.0 (-117.2)]
$C^\circ_{p(300)}$	8.3 (34.8)	10.3 (43.1)	[19.9 (83.3)]	ΔC°_p	[1.3 (5.4)]

log A	E	log k	Conditions	Reference and Experimental Method
8.0	1.1 (4.6)	7.2(315 K)	295-333 K 70 torr	[153] L.I.Avramenko, L.M.Evlashkina and R.V.Kolesnikova, Izv.Akad.Nauk.Ser.Khim., 259 (1967). Static system. HO_2 radicals from reaction of H (from discharged H_2) and O_2. Reaction rate followed by product analysis and rate constants based on rate of formation of formaldehyde according to proposed mechanism:

$$H\dot{O}_2 + CH_2{=}CH_2 \longrightarrow \dot{C}H_2CH_2O_2H$$
$$\dot{C}H_2CH_2O_2H \longrightarrow CH_2OHCH_2\dot{O}$$
$$CH_2OHCH_2\dot{O} \longrightarrow \dot{C}H_2OH + CH_2O$$
$$\dot{C}H_2OH + O_2 \longrightarrow CH_2O + HO_2$$
$$2\,\dot{C}H_2OH \longrightarrow CH_2O + CH_3OH$$

ΔH^o_f, ΔH^o and E in kcal mol^{-1} (kJ mol^{-1}); S^o, C^o_p, ΔS^o and ΔC^o_p in cal mol^{-1} K^{-1} (J mol^{-1} K^{-1}); A and k in l mol^{-1} s^{-1}

[9.58]	[11.5] (48.2)	[1.2](300 K)	-	[154] A.Fish, Advan.Chem., 76, 69 (1968). Estimated parameters.
-	-	∿3.69(734 K)	734-798 K 70-186 torr	[155] D.E.Hoare and M.Patel, Trans.Faraday Soc., 65, 1325 (1969). Results calculated from $k_1/k_3 = 0.4$ at 734 K, taking $\log k_3 = 9.0 - 16.5/2.3RT$ [A.C.Lloyd, Int.J.Chem.Kinet., in press] where reaction (3) refers to $H\dot{O}_2 + CO \xrightarrow{3} CO_2 + \dot{O}H$. Static system. HO_2 produced in combustion of CH_4, C_2H_6 and C_2H_4 mixtures thus: $CO + \dot{O}H \longrightarrow CO_2 + \dot{H}$; $\dot{H} + O_2 + M \longrightarrow H\dot{O}_2 + M$ and reaction rate followed by product analysis using gas chromatography.
[8.70]	[5.0] (20.9)	[5.06](300 K)	-	[156] A.C.Lloyd, Int.J.Chem.Kinet., in press. Estimated parameters.

PREFERRED RATE PARAMETERS

$\log k = [8.70 - 5.0/2.3RT]$

COMMENTS

For an extensive discussion of this reaction, the reader should consult the evaluation of Lloyd [156]. It has been pointed out that both of the above experimental values refer to the total reaction of HO_2 with $CH_2{=}CH_2$, although the addition reaction should be very much faster than the abstraction reaction $H\dot{O}_2 + CH_2{=}CH_2 \xrightarrow{4} H_2O_2 + \dot{C}H{=}CH_2$ at room temperature, since $\Delta H_4^o \simeq +18$ kcal mol^{-1}.

We favour the data estimated by Lloyd [156]. The rate constant at 300 K is estimated on the basis that the data of reference [153] are ∿10^2 too high owing to the additional reaction of O atoms with ethylene. Thus, $\log k_1 \simeq 5.06$ at 300 K, and assuming $\log A_1 = 8.7$ leads to the above value of E_1. The data of Hoare and Patel [155] were not considered reliable owing to the complexity of the reaction scheme involved and the uncertainty in the rate constant of the reference reaction.

REACTION: HYDROPEROXYL RADICAL + 2-METHYLPROPENE (ISOBUTENE)

$$H\dot{O}_2(I) + CH_2{=}C(CH_3)_2(II) \xrightarrow{1} HOOCH_2\dot{C}(CH_3)_2(III)$$
$$\xrightarrow{2} \dot{C}H_2C(O_2H)(CH_3)_2(IV)$$

	(I)	(II)	(III)	(IV)
$\Delta H^\circ_{f(300)}$	5.0 (20.9)	-3.6 (-14.9)	[-12.4 (-51.9)]	[-11.7 (-49.0)]
$S^\circ_{(300)}$	54.4 (227.8)	71.3 (298.4)	[93.2 (390.2)]	[91.3 (382.3)]
$C^\circ_{p(300)}$	8.3 (34.8)	21.6 (90.4)	[30.3 (126.9)]	[31.6 (132.3)]

	Path 1	Path 2
ΔH°	[-14 (-59)]	[-13 (-55)]
ΔS°	[-32.5 (-136.1)]	[-34.4 (-144.0)]
ΔC°_p	[0.4 (1.7)]	[1.7 (7.1)]

Path	log A	E	log k	Conditions	Reference and Experimental Method
1 + 2	9.48	1.7 (7.1)	8.31(315 K)	295-333 K 70 torr	[153] L.I.Avramenko, L.M.Evlashkina and R.V.Kolesnikova, Izv.Akad.Nauk.Ser. Khim., 259 (1967). For details see $HO_2 + CH_2{=}CH_2$.

PREFERRED RATE PARAMETERS

Insufficient data to make a meaningful estimate

COMMENTS

Lloyd has pointed out in his evaluation [156] that the reliability of the above data is unlikely to be high since the reaction was complex and could well have involved reactions of other radicals. The above experimental A-factor and rate constants would certainly appear to be over-estimates in relation to the recommended data for HO_2 addition to $CH_2{=}CH_2$.

ΔH^o_f, ΔH^o and E in kcal mol^{-1} (kJ mol^{-1}); S^o, C^o_p, ΔS^o and ΔC^o_p in cal mol^{-1} K^{-1} (J mol^{-1} K^{-1}); A and k in l mol^{-1} s^{-1}

REACTION: METHANETHIYL RADICAL + ETHYLENE $CH_3\dot{S}(I) + CH_2{=}CH_2(II) \longrightarrow CH_3SCH_2\dot{C}H_2(III)$

	(I)	(II)	(III)		
$\Delta H^\circ_{f(300)}$	[32.5 (136)]	12.5 (52.3)	[32 (134)]	ΔH°	[-13 (-54)]
$S^\circ_{(300)}$	[59 (247)]	52.4 (219.4)	[85 (356)]	ΔS°	[-26 (-109)]
$C^\circ_{p(300)}$	[9.0 (37.7)]	10.3 (43.1)	[22.6 (94.6)]	ΔC°_p	[3.3 (13.8)]

log A	E	log k	Conditions	Reference and Experimental Method
-	-	5.68(298 K)	298-333 K 50-130 torr	[157] D.M.Graham and J.F.Soltys, Can.J.Chem., 47, 2529 (1969).

Static system. Photolysis of CH_3SH (λ = 254-265 nm) in presence of an olefin. Reaction rate determined by rate of decrease of pressure using either a wide-bore Hg manometer or a differential manometer or a pressure transducer. Proposed mechanism: (Oℓ = olefin)

$$CH_3SH + h\nu \xrightarrow{I} CH_3\dot{S} + \dot{H}$$
$$CH_3\dot{S} + O\ell \xrightarrow{1} CH_3S\dot{O}\ell$$
$$CH_3S\dot{O}\ell \xrightarrow{2} CH_3\dot{S} + O\ell$$
$$CH_3S\dot{O}\ell + CH_3SH \xrightarrow{3} CH_3SO\ell H + CH_3\dot{S}$$
$$2CH_3\dot{S} \xrightarrow{4} CH_3SSCH_3$$

and from a steady-state treatment the rate law is

$$\frac{-d[O\ell]}{dt} = \frac{R_I^{\frac{1}{2}}}{(2k_4)^{\frac{1}{2}}} \frac{k_1[O\ell]k_3[CH_3SH]}{k_2 + k_3[CH_3SH]}$$

Mechanism established by light intensity and [Oℓ] changes. Rate constants obtained from plots of $R_I/(-d[O\ell]/dt)$ versus $1/[CH_3SH]$. The value of k_1 for ethylene given above is that calculated by Graham and Soltys [157] from the data of Graham et al.[Can.J.Chem., 42, 2250 (1964)] taking a revised value of log k_4 = 9.40.

PREFERRED RATE PARAMETERS

Reasonable as given above

COMMENTS

The above rate constant indicates that CH_3S radicals are quite reactive in addition reactions to olefins.

REACTION: METHANETHIYL RADICAL + PROPENE

$$CH_3\dot{S}(I) + CH_2{=}CHCH_3(II) \xrightarrow{1} CH_3SCH_2\dot{C}HCH_3(III)$$
$$\xrightarrow{2} \dot{C}H_2CH(SCH_3)CH_3(IV)$$

	(I)	(II)	(III)	(IV)		Path 1	Path 2
$\Delta H^\circ_{f(300)}$	[32.5 (136)]	4.9 (20.5)	23.2 (97.1)	24.5 (102.6)	ΔH°	[-14 (-59)]	[-13 (-54)]
$S^\circ_{(300)}$	[59 (247)]	63.8 (267.0)	90.2 (377.7)	89.5 (374.7)	ΔS°	[-33 (-138)]	[-33 (-138)]
$C^\circ_{p(300)}$	[9.0 (37.7)]	15.3 (64.0)	27.9 (116.8)	28.2 (118.1)	ΔC°_p	[3.6 (15.1)]	[3.9 (16.3)]

Path	log A	E	log k	Conditions	Reference and Experimental Method
1 + 2	-	-	-	298-333 K 50-130 torr	[158] D.M.Graham, R.L.Mieville, R.H.Pallen and C.Sivertz, Can.J.Chem., 42, 2250 (1964). For details see $CH_3\dot{S} + CH_2{=}CH_2$ system, reference [157].

PREFERRED RATE PARAMETERS

No data obtained on addition reaction

ΔH°_f, ΔH° and E in kcal mol^{-1} (kJ mol^{-1}); S°, C°_p, ΔS° and ΔC°_p in cal mol^{-1} K^{-1} (J mol^{-1} K^{-1}); A and k in l mol^{-1} s^{-1}

REACTION: METHANETHIYL RADICAL + BUTENE-2 $CH_3\dot{S} + CH_3CH{=}CHCH_3(II) \longrightarrow CH_3CH(CH_3S)\dot{C}HCH_3(III)$

	(I)	(II)	(III)		
$\Delta H^\circ_{f(300)}$	[32.5 (136)]	-1.9 (-7.9)	[16 (67)]	ΔH°	[-15 (-62)]
$S^\circ_{(300)}$	[59 (247)]	73.6 (308.2)	[97 (406)]	ΔS°	[-36 (-151)]
$C^\circ_{p(300)}$	[9.0 (37.7)]	19.4 (81.2)	[34 (142)]	ΔC°_p	[5.6 (23.4)]

log A	E	log k	Conditions	Reference and Experimental Method
-	-	6.2(298 K)*	298-333 K 50-130 torr	[157] D.M.Graham and J.F.Soltys, Can.J.Chem., 47, 2529 (1969). For details see $CH_3\dot{S} + CH_2{=}CH_2$ system.

* cis-trans equilibrium weighted rate constant.

PREFERRED RATE PARAMETERS

Reasonable as given above

REACTION: METHANETHIYL RADICAL + ACETYLENE $CH_3\dot{S}(I) + CH{\equiv}CH(II) \longrightarrow CH_3SCH{=}\dot{C}H(III)$

	(I)	(II)	(III)		
$\Delta H^\circ_{f(300)}$	[32.5 (136)]	54.2 (226.9)	70.7 (296.0)	ΔH°	[-16 (-67)]
$S^\circ_{(300)}$	[59 (247)]	48.0 (201.0)		ΔS°	
$C^\circ_{p(300)}$	[9.0 (37.7)]	10.5 (44.0)		ΔC°_p	

log A	E	log k	Conditions	Reference and Experimental Method
-	-	4.90(298 K)	298-333 K 50-130 torr	[157] D.M.Graham and J.F.Soltys, Can.J.Chem., 47, 2529 (1969). For details see $CH_3\dot{S} + CH_2{=}CH_2$ system.

PREFERRED RATE PARAMETERS

Reasonable as given above

ΔH^o_f, ΔH^o and E in kcal mol^{-1} (kJ mol^{-1}); S^o, C^o_p, ΔS^o and ΔC^o_p in cal mol^{-1} K^{-1} (J mol^{-1} K^{-1}); A and k in l mol^{-1} s^{-1}

REACTION: DIFFLUOROAMINO RADICAL + ETHYLENE $\dot{N}F_2(I) + CH_2{=}CH_2(II) \longrightarrow NF_2CH_2\dot{C}H_2(III)$

	(I)	(II)	(III)		
$\Delta H^\circ_{f(300)}$	8.5 (35.6)	12.5 (52.3)	21.4 (89.6)	ΔH°	0.4 (1.7)
$S^\circ_{(300)}$	59.7 (250.0)	52.4 (219.4)		ΔS°	
$C^\circ_{p(300)}$	9.8 (41.0)	10.3 (43.1)		ΔC°_p	

log A	E	log k	Conditions	Reference and Experimental Method
7.6	15.5 (64.9)	-2.1(350 K)	351-428 K 40-550 torr	[159] A.J.Dijkstra, J.A.Kerr and A.F.Trotman-Dickenson, J.Chem.Soc.(A), 105 (1967).

Static system. NF_2 from equilibrium dissociation $N_2F_4 \rightleftharpoons 2NF_2$.
Reaction rate measured by rate of formation of bisdifluoroamine adduct.
$O\ell(NF_2)_2$ analyzed by gas chromatography and identified by molecular weight determination (gas density balance). Proposed mechanism: ($O\ell$ = olefin, * vibrationally excited).

$$\dot{N}F_2 + O\ell \xrightarrow{1} NF_2\dot{O}\ell^*$$
$$NF_2\dot{O}\ell^* \xrightarrow{-1^*} \dot{N}F_2 + O\ell$$
$$NF_2\dot{O}\ell^* + M \xrightarrow{3} NF_2\dot{O}\ell + M$$
$$NF_2\dot{O}\ell \xrightarrow{-1} \dot{N}F_2 + O\ell$$
$$NF_2\dot{O}\ell + \dot{N}F_2 \xrightarrow{4} O\ell(NF_2)_2$$

Steady-state treatment yields

$$\frac{1}{k_{obs}} = \frac{1}{k_1}\left\{1 + \frac{k_{-1*}}{k_3}\frac{1}{[M]}\right\}\left\{1 + \frac{k_{-1}}{k_4}\frac{1}{[NF_2]}\right\} \quad \text{where}$$

M = total gas pressure, $d[O\ell(NF_2)_2]/dt = k_{obs}[O\ell][NF_2]$, and $[NF_2]$ is calculated from the equilibrium constant for $N_2F_4 \rightleftharpoons 2NF_2$ and the pressure. With [M] constant, $1/k_{obs.}$ was plotted against $1/[NF_2]$ and $(1/k_1)(1 + k_{-1}*/k_3[M])$ obtained from intercept, and k_{-1}/k_4 obtained from ratio of slope to intercept. With $[NF_2]$ constant, $1/k_{obs.}$ was plotted against $1/[M]$ and k_{-1}/k_3 similarly obtained. k_1 was derived from both sets of values. Above parameters obtained from Arrhenius plot of derived k_1 values. Several tests confirmed the validity of the mechanism.

PREFERRED RATE PARAMETERS

Reasonable as given above

COMMENTS

The above E_1 and the value of ΔH_1° yield E_{-1} = 14.5 kcal mol^{-1} for the reverse radical decomposition. It is not possible to make a meaningful estimate of ΔS°, and therefore A_{-1} cannot be calculated from A_1. The above estimate of E_{-1} compares well with the result $\log (k_{-1}/s^{-1}) = 12.7 - 13.7/2.3RT$ which can be obtained from $\log (k_{-1}/k_4)(\ell^{-1}mol) = 2.5 - 13.7/2.3RT$ reported in reference [159], and assuming $\log k_4 = 10.2$ (based on the empirical relation of Bertrand et al. [Chem.Phys.Lett., 5, 183 (1970)]). It should be noted that the low experimental value of $\log (A_{-1}/s^{-1}) = 12.7$ probably indicates that the Arrhenius parameters of reaction (-1) have been slightly under-estimated, although they are also subject to any error in the assigned value of k_4.

ΔH_f^o, ΔH^o and E in kcal mol^{-1} (kJ mol^{-1}); S^o, C_p^o, ΔS^o and ΔC_p^o in cal mol^{-1} K^{-1} (J mol^{-1} K^{-1}); A and k in l mol^{-1} s^{-1}

REACTION: DIFLUOROAMINO RADICAL + PROPENE

$$\dot{N}F_2(I) + CH_2{=}CHCH_3(II) \xrightarrow{1} NF_2CH_2\dot{C}HCH_3(III)$$
$$\xrightarrow{2} \dot{C}H_2CH(NF_2)CH_3(IV)$$

	(I)	(II)	(III)	(IV)		Path 1	Path 2
$\Delta H^\circ_{f(300)}$	8.5 (35.6)	4.9 (20.5)	13.0 (54.4)	12.7 (53.2)	ΔH°	-0.4 (-1.7)	-0.7 (-2.9)
$S^\circ_{(300)}$	59.7 (250.0)	63.8 (267.1)			ΔS°		
$C^\circ_{p(300)}$	9.8 (41.0)	15.3 (64.1)			ΔC°_p		

Path	log A	E	log k	Conditions	Reference and Experimental Method
1 + 2	7.2	13.7 (57.4)	-1.4(350 K)	334-391 K 27-725 torr	[160] A.J.Dijkstra, J.A.Kerr and A.F.Trotman-Dickenson, J.Chem.Soc.(A), 582 (1966). For details see $NF_2 + CH_2{=}CH_2$ system, reference [159].

PREFERRED RATE PARAMETERS

Reasonable as given above

COMMENTS

See comments for $NF_2 + CH_2{=}CH_2$ system. From $E_1 = 13.7$ and $\Delta H^\circ_1 = -0.4$, it follows that $E_{-1} = 13.5$ kcal mol^{-1}. This calculated value of E_{-1} is in excellent agreement with $\log (k_{-1}/s^{-1}) = 13.1 - 13.5/2.3RT$ derived from $\log (k_{-1}/k_4)(\ell^{-1}mol) = 3.0 - 13.5/2.3RT$ from reference [160], taking $\log k_4 = 10.1$ (Bertrand et al. [Chem.Phys. Lett., 5, 183 (1970)]). It is not possible to tell from the experiments which addition path is favoured, although the enthalpy changes indicate that path 2 is slightly more favourable. There is probably an error in the thermochemical estimates, since path 1 should be favoured.

REACTION: DIFLUOROAMINO RADICAL + BUTENE-1

$$\dot{N}F_2(I) + CH_2{=}CHCH_2CH_3(II) \xrightarrow{1} NF_2CH_2\dot{C}HCH_2CH_3(III)$$
$$\xrightarrow{2} \dot{C}H_2CH(NF_2)CH_2CH_3(IV)$$

	(I)	(II)	(III)	(IV)		Path 1	Path 2
$\Delta H^\circ_{f(300)}$	8.5 (35.6)	0.0 (0.0)	8.0 (33.5)	7.8 (32.7)	ΔH°	-0.5 (-2.1)	-0.7 (-2.9)
$S^\circ_{(300)}$	59.7 (250.0)	73.6 (308.0)			ΔS°		
$C^\circ_{p(300)}$	9.8 (41.0)	20.6 (86.2)			ΔC°_p		

Path	log A	E	log k	Conditions	Reference and Experimental Method
1 + 2	7.1	13.6 (56.9)	-1.4(350 K)	334-391 K 40-550 torr	[159] A.J.Dijkstra, J.A.Kerr and A.F.Trotman-Dickenson, J.Chem.Soc.(A), 105 (1967). For details see $NF_2 + CH_2{=}CH_2$.

PREFERRED RATE PARAMETERS

Reasonable as given above

COMMENTS

See comments for $NF_2 + CH_2{=}CH_2$ system. From $E_1 = 13.6$ and $\Delta H^\circ_1 = -0.5$, it follows that $E_{-1} = 13.5$ kcal mol^{-1}. This calculated value of E_{-1} is higher than from $\log (k_{-1}/s^{-1}) = 12.1 - 10.5/2.3RT$, calculated from $\log (k_{-1}/k_4)(\ell^{-1}mol) = 2.1 - 10.3/2.3RT$ from reference [159], taking $\log k_4 = 10.0$, Bertrand et al. [Chem.Phys.Lett., 5, 183 (1970)].

It appears that E_{-1} and A_{-1} have been under-estimated. It is not possible to tell from the experiments which addition path is favoured, although the enthalpy changes indicate that path 2 should be slightly more favourable than path 1. There is probably an error in the thermochemical estimates, since path 1 should be preferred.

ΔH^o_f, ΔH^o and E in kcal mol^{-1} (kJ mol^{-1}); S^o, C^o_p, ΔS^o and ΔC^o_p in cal mol^{-1} K^{-1} (J mol^{-1} K^{-1}); A and k in l mol^{-1} s^{-1}

REACTION: DIFLUOROAMINO RADICAL + *cis*-BUTENE-2 $\dot{N}F_2$(I) + *cis*-$CH_3CH{=}CHCH_3$(II) ⟶ $CH_3CH(NF_2)\dot{C}HCH_3$(III)

	(I)	(II)	(III)		
$\Delta H^\circ_{f(300)}$	8.5 (35.6)	-1.9 (-7.9)	4.3 (18.0)	ΔH°	-2.3 (-9.6)
$S^\circ_{(300)}$	59.7 (250.0)	73.6 (308.2)		ΔS°	
$C^\circ_{p(300)}$	9.8 (41.0)	19.4 (81.2)		ΔC°_p	

log A	E	log k	Conditions	Reference and Experimental Method
6.5	11.9 (49.8)	-0.9(350 K)	334-391 K 26-542 torr	[160] A.J.Dijkstra, J.A.Kerr and A.F.Trotman-Dickenson, J.Chem.Soc.(A), 582 (1966). For details see NF_2 + $CH_2{=}CH_2$ system, reference [159].

PREFERRED RATE PARAMETERS

Reasonable as given above

COMMENTS

See comments for NF_2 + $CH_2{=}CH_2$ system. From E_1 = 11.9 and ΔH°_1 = -2.3, it follows that E_{-1}= 13.6 kcal mol^{-1}. This calculated value of E_{-1} is in good agreement with log (k_{-1}/s^{-1}) = 13.6 - 13.6/2.3RT derived from log $(k_{-1}/k_4)(\ell^{-1} mol)$ = 3.6 - 13.6/2.3RT from reference [160], taking log k_4 = 10.0, Bertrand et al. [Chem.Phys. Lett., 5, 183 (1970)].

REACTION: DIFLUOROAMINO RADICAL + TRANS-BUTENE-2 $\dot{N}F_2$(I) + trans-$CH_3CH=CHCH_3$(II) ⟶ $CH_3CH(NF_2)\dot{C}HCH_3$(III)

	(I)	(II)	(III)		
$\Delta H^\circ_{f(300)}$	8.5 (35.6)	-2.9 (-12.1)	4.3 (18.0)	ΔH°	-1.3 (-5.4)
$S^\circ_{(300)}$	59.7 (250.0)	72.4 (302.8)		ΔS°	
$C^\circ_{p(300)}$	9.8 (41.0)	20.7 (86.6)		ΔC°_p	

log A	E	log k	Conditions	Reference and Experimental Method
6.5	11.9 (49.8)	-0.9(350 K)	334-391 K 22-561 torr	[160] A.J.Dijkstra, J.A.Kerr and A.F.Trotman-Dickenson, J.Chem.Soc.(A), 582 (1966). For details see $NF_2 + CH_2=CH_2$ system, reference [159].

PREFERRED RATE PARAMETERS

Reasonable as given above

COMMENTS

See comments for $NF_2 + CH_2=CH_2$ system. From E_1 = 11.9 and ΔH°_1 = -1.3, it follows that E_{-1} = 12.6 kcal mol^{-1}. This calculated value of E_{-1} is in agreement with $\log (k_{-1}/s^{-1}) = 13.6 - 13.6/2.3RT$ derived from $\log (k_{-1}/k_4)(\ell^{-1}mol) = 3.6 - 13.6/2.3RT$ from reference [160], taking $\log k_4 = 10.0$, Bertrand et al. [Chem. Phys.Lett., 5, 183 (1970)].

ΔH^o_f, ΔH^o and E in kcal mol^{-1} (kJ mol^{-1}); S^o, C^o_p, ΔS^o and ΔC^o_p in cal mol^{-1} K^{-1} (J mol^{-1} K^{-1}); A and k in l mol^{-1} s^{-1}

REACTION: DIFLUOROAMINO RADICAL + 2-METHYLPROPENE (ISOBUTENE)

$\dot{N}F_2(I) + (CH_3)_2C{=}CH_2(II) \xrightarrow{1} (CH_3)_2\dot{C}CH_2NF_2(III)$

$\xrightarrow{2} (CH_3)_2C(NF_2)\dot{C}H_2(IV)$

	(I)	(II)	(III)	(IV)		Path 1	Path 2
$\Delta H^\circ_{f(300)}$	8.5 (35.6)	-3.6 (-14.9)	2.4 (10.0)	4.7 (19.7)	ΔH°	-2.5 (-10.5)	-0.2 (-0.8)
$S^\circ_{(300)}$	59.7 (250.0)	71.3 (298.4)			ΔS°		
$C^\circ_{p(300)}$	9.8 (41.0)	21.6 (90.4)			ΔC°_p		

Path	log A	E	log k	Conditions	Reference and Experimental Method
1 + 2	6.8	11.8 (49.4)	-0.6(350 K)	314-373 K 25-550 torr	[159] A.J.Dijkstra, J.A.Kerr and A.F.Trotman-Dickenson, J.Chem.Soc.(A), 105 (1967). For details see $NF_2 + CH_2{=}CH_2$.

PREFERRED RATE PARAMETERS

Reasonable as given above

COMMENTS

See comments for $NF_2 + CH_2{=}CH_2$ system. From E_1 = 11.8 and ΔH°_1 = -2.5, it follows that E_{-1} = 13.7 kcal mol^{-1}. No experimental value of E_{-1} is available. No information on which path for addition is favoured, but path 1 seems most likely, although the A-factor tends to suggest path 2 is favoured. Probably both paths 1 and 2 are occurring.

REACTION: DIFLUOROAMINO RADICAL + 2-METHYLBUTENE-2

$\dot{N}F_2(I) + (CH_3)_2C{=}CHCH_3(II) \xrightarrow{1} (CH_3)_2\dot{C}CH(NF_2)CH_3(III)$

$\xrightarrow{2} (CH_3)_2C(NF_2)\dot{C}HCH_3(IV)$

	(I)	(II)	(III)	(IV)		Path 1	Path 2
$\Delta H^\circ_{f(300)}$	8.5 (35.6)	-11.3 (-47.3)	-6.2 (-26.0)	-3.8 (-15.9)	ΔH°	-3.4 (-14.2)	-1.0 (-4.2)
$S^\circ_{(300)}$	59.7 (250.0)	79.9 (334.4)			ΔS°		
$C^\circ_{p(300)}$	9.8 (41.0)	26.8 (112.1)			ΔC°_p		

Path	log A	E	log k	Conditions	Reference and Experimental Method
1 + 2	6.0	10.1 (42.3)	-0.3(350 K)	314-373 K 40-550 torr	[159] A.J.Dijkstra, J.A.Kerr and A.F.Trotman-Dickenson, J.Chem.Soc.(A), 105 (1967). For details see $NF_2 + CH_2{=}CH_2$.

PREFERRED RATE PARAMETERS

Tentatively as given above

COMMENTS

See comments for $NF_2 + CH_2{=}CH_2$ system. From $E_1 = 10.1$ and $\Delta H^\circ_1 = -3.4$, it follows that $E_{-1} = 12.9$ kcal mol^{-1}. This calculated value is higher than from $\log (k_{-1}/s^{-1}) = 12.8 - 9.7/2.3RT$, derived from $\log (k_{-1}/k_4)(\ell^{-1}mol) = 2.9 - 9.7/2.3RT$, from reference [159], taking $\log k_4 = 9.9$, Bertrand et al. [Chem. Phys.Lett., 5, 183 (1970)]. E_{-1} and A_{-1} appear to have been under-estimated. It is also probable that the Arrhenius parameters for the addition reaction (1) have also been slightly under-estimated, as indicated by the very low value of A_1, although this should be compensated for by the larger negative ΔS°, since addition is to a crowded olefin. Log A_1 would appear to be about 0.5 log units too low. From the experiments it is not possible to tell which addition path is favoured, although path 1 seems the more likely.

ΔH°_f, ΔH° and E in kcal mol^{-1} (kJ mol^{-1}); S°, C°_p, ΔS° and ΔC°_p in cal mol^{-1} K^{-1} (J mol^{-1} K^{-1}); A and k in l mol^{-1} s^{-1}

REACTION: DIFLUOROAMINO RADICAL + 2,3-DIMETHYLBUTENE-2

$\dot{N}F_2(I) + (CH_3)_2C{=}C(CH_3)_2(II) \longrightarrow (CH_3)_2C(NF_2)\dot{C}(CH_3)_2(III)$

	(I)	(II)	(III)		
$\Delta H^\circ_{f(300)}$	8.5 (35.6)	-19.6 (-81.9)	-14.3 (-59.9)	ΔH°	-3.2 (-13.4)
$S^\circ_{(300)}$	59.7 (250.0)	87.4 (365.3)		ΔS°	
$C^\circ_{p(300)}$	9.8 (41.0)	33.0 (137.9)		ΔC°_p	

log A	E	log k	Conditions	Reference and Experimental Method
5.3	8.3 (34.8)	0.1(350 K)	314-373 K 40-550 torr	[159] A.J.Dijkstra, J.A.Kerr and A.F.Trotman-Dickenson, J.Chem.Soc.(A), 105 (1967). For details see $NF_2 + CH_2{=}CH_2$.

PREFERRED RATE PARAMETERS

Tentatively as given above

COMMENTS

See comments for $NF_2 + CH_2{=}CH_2$ system. From $E_1 = 8.3$ and $\Delta H^\circ_1 = -3.2$, it follows that $E_{-1} = 10.9$ kcal mol^{-1}. This calculated value of E_{-1} is higher than from $\log (k_{-1}/s^{-1}) = 11.5 - 8.3/2.3RT$, calculated from $\log (k_{-1}/k_4)(\ell^{-1}mol) = 1.7 - 8.3/2.3RT$ from reference [159], taking $\log k_4 = 9.8$, Bertrand et al. [Chem.Phys. Lett., 5, 183 (1970)]. E_{-1} and A_{-1} appear to have been under-estimated, and it is also possible that the Arrhenius parameters for the addition reaction (1) have also been slightly under-estimated, as indicated by the very low value of A_1, although this should be compensated for by the larger negative ΔS°, since addition is to a crowded olefin. Log A_1 would appear to be about 0.8 log units too low.

REACTION: DIFLUOROAMINO RADICAL + VINYL FLUORIDE

$$\dot{N}F_2(I) + CH_2{=}CHF(II) \xrightarrow{1} NF_2CH_2\dot{C}HF(III)$$
$$\xrightarrow{2} \dot{C}H_2CHFNF_2(IV)$$

	(I)	(II)	(III)	(IV)		Path 1	Path 2
$\Delta H^\circ_{f(300)}$	8.5 (35.6)	-31.6 (-132.3)	-18.3 (-76.5)		ΔH°	4.8 (20.1)	
$S^\circ_{(300)}$	59.7 (250.0)	60.4 (252.9)			ΔS°		
$C^\circ_{p(300)}$	9.8 (41.0)	11.9 (49.8)			ΔC°_p		

Path	log A	E	log k	Conditions	Reference and Experimental Method
1 + 2	-	-	(-0.99)	463 K 65 torr	[161] A.J.Dijkstra, J.A.Kerr and A.E.Trotman-Dickenson, J.Chem.Soc.(A), 864 (1967). For details see $NF_2 + CH_2{=}CH_2$, reference [159].

PREFERRED RATE PARAMETERS

Tentatively as given above

COMMENTS

The above value is only approximate, since vinyl fluoride was so unreactive that the reaction required a very high temperature for a measurable rate, and the results were thought to be complicated by side-reactions.

ΔH^o_f, ΔH^o and E in kcal mol^{-1} (kJ mol^{-1}); S^o, C^o_p, ΔS^o and ΔC^o_p in cal mol^{-1} K^{-1} (J mol^{-1} K^{-1}); A and k in l mol^{-1} s^{-1}

REACTION: DIFLUOROAMINO RADICAL + VINYL CHLORIDE

$\dot{N}F_2(I) + CH_2{=}CHCl(II) \xrightarrow{1} NF_2CH_2\dot{C}HCl(III)$

$\xrightarrow{2} \dot{C}H_2CH(NF_2)Cl(IV)$

	(I)	(II)	(III)	(IV)		Path 1	Path 2
$\Delta H^\circ_{f(300)}$	8.5 (35.6)	8.6 (36.0)	14.1 (59.0)		ΔH°	-3.0 (-12.5)	
$S^\circ_{(300)}$	59.7 (250.0)	63.0 (263.8)			ΔS°		
$C^\circ_{p(300)}$	9.8 (41.0)	13.0 (54.4)			ΔC°_p		

Path	log A	E	log k	Conditions	Reference and Experimental Method
1 + 2	6.4	12.9 (54.0)	-1.7(350 K)	351-405 K 40-550 torr(?)	[161] A.J.Dijkstra, J.A.Kerr and A.F.Trotman-Dickenson, J.Chem.Soc.(A), 864 (1967). For details see $NF_2 + CH_2{=}CH_2$ system, reference [159].

PREFERRED RATE PARAMETERS

Reasonable as given above

COMMENTS

See comments for $NF_2 + CH_2{=}CH_2$ system. From $E_1 = 12.9$ and $\Delta H^\circ_1 = -3.0$, it follows that $E_{-1} = 15.3$ kcal mol^{-1}. The above value of E_{-1} is considerably higher than the value from $\log (k_{-1}/s^{-1}) = 10.8 - 9.2/2.3RT$, calculated from $\log (k_{-1}/k_4)(\ell^{-1}mol) = 0.5 - 9.2/2.3RT$ (reference [161]), and taking $\log k_4 = 10.3$, Bertrand et al. [Chem.Phys.Lett., 5, 183 (1970)]. It appears that E_{-1} and A_{-1} have been under-estimated. The experiments gave no indication of which path addition follows.

REACTION: DIFLUOROAMINO RADICAL + VINYL BROMIDE

$\dot{N}F_2(I) + CH_2{=}CHBr(II) \xrightarrow{1} NF_2CH_2\dot{C}HBr(III)$

$\xrightarrow{2} \dot{C}H_2CH(NF_2)Br(IV)$

	(I)	(II)	(III)	(IV)		Path 1	Path 2
$\Delta H^\circ_{f(300)}$	8.5 (35.6)	19.0 (79.6)			ΔH°		
$S^\circ_{(300)}$	59.7 (250.0)	65.9 (275.9)			ΔS°		
$C^\circ_{p(300)}$	9.8 (41.0)	13.2 (55.3)			ΔC°_p		

Path	log A	E	log k	Conditions	Reference and Experimental Method
1 + 2	6.6	13.2 (55.3)	-1.6(350 K)	351-405 K 40-550 torr	[161] A.J.Dijkstra, J.A.Kerr and A.F.Trotman-Dickenson, J.Chem.Soc.(A), 864 (1967). For details see $NF_2 + CH_2{=}CH_2$ system, reference [159].

PREFERRED RATE PARAMETERS

Reasonable as given above

COMMENTS

See comments for $NF_2 + CH_2{=}CH_2$ system. No estimates of the kinetic parameters for the reverse radical decomposition can be made, since the thermodynamic properties of the adduct radicals are unknown. From reference [161], $\log (k_{-1}/s^{-1}) = 12.0 - 11.4/2.3RT$, calculated from $\log (k_{-1}/k_4)(\ell^{-1}mol) = 1.9 - 11.4/2.3RT$, and taking $\log k_4 = 10.1$, Bertrand et al. [Chem.Phys.Lett., 5, 183 (1970)].

ΔH^o_f, ΔH^o and E in kcal mol^{-1} (kJ mol^{-1}); S^o, C^o_p, ΔS^o and ΔC^o_p in cal mol^{-1} K^{-1} (J mol^{-1} K^{-1}); A and k in l mol^{-1} s^{-1}

REACTION: DIFLUOROAMINO RADICAL + CYCLOPENTENE $\dot{N}F_2$ (I) +

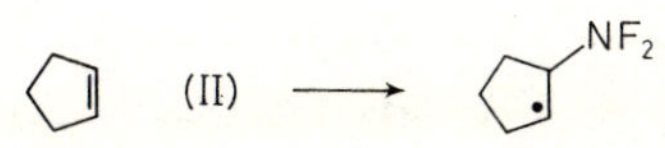

(III)

	(I)	(II)	(III)		
$\Delta H^\circ_{f(300)}$	8.5 (35.6)	8.6 (36.0)	15.9 (66.5)	ΔH°	-1.2 (-5.0)
$S^\circ_{(300)}$	59.7 (250.0)	70.8 (296.4)		ΔS°	
$C^\circ_{p(300)}$	9.8 (41.0)	18.1 (75.8)		ΔC°_p	

log A	E	log k	Conditions	Reference and Experimental Method
5.9	11.0 (46.1)	-1.0(350 K)	334-391 K 23-454 torr	[159] A.J.Dijkstra, J.A.Kerr and A.F.Trotman-Dickenson, J.Chem.Soc.(A), 105 (1967). For details see $NF_2 + CH_2{=}CH_2$.

PREFERRED RATE PARAMETERS

Tentatively as given above

COMMENTS

See comments for $NF_2 + CH_2{=}CH_2$ system. From $E_1 = 11.0$ and $\Delta H^\circ_1 = -1.2$, it follows that $E_{-1} = 11.6$ kcal mol^{-1}. The above calculated E_{-1} is higher than the value from $\log (k_{-1}/s^{-1}) = 10.9 - 9.8/2.3RT$ calculated from $\log (k_{-1}/k_4)(\ell^{-1}mol) = 1.0 - 9.8/2.3RT$ (reference [159]) and taking $\log k_4 = 9.9$, Bertrand et al. [Chem.Phys.Lett., 5, 183 (1970)]. It appears that E_{-1} and A_{-1} have been under-estimated, and the low value of A_1, for the addition reaction, also suggests a slight under-estimation of E_1 and A_1.

REACTION: SULPHUR PENTAFLUORIDE RADICAL + ETHYLENE $\dot{S}F_5(I) + CH_2{=}CH_2(II) \longrightarrow SF_5CH_2\dot{C}H_2(III)$

	(I)	(II)	(III)	
$\Delta H^o_{f(300)}$		12.5 (52.3)		ΔH^o
$S^o_{(300)}$		52.4 (219.4)		ΔS^o
$C^o_{p(300)}$		10.3 (43.1)		ΔC^o_p

log A	E	log k	Conditions	Reference and Experimental Method
5.44	1.9 (8.0)	4.31(367 K)	313-421 K 120-154 torr	[162] H.W.Sidebottom, J.M.Tedder and J.C.Walton, Trans.Faraday Soc., 65, 2103 (1969).

Static system. SF_5 from photolysis of $SF_5C\ell$ with a medium-pressure Hg arc filtered with 1% $CuSO_4$ solution. Products were analyzed by gas chromatography and identified by mass spectrometry to be $CH_2C\ell CH_2C\ell$, $CH_2C\ell CH_2SF_5$ and S_2F_{10}. The proposed mechanism was:

$$SF_5C\ell + h\nu \xrightarrow{I} \dot{S}F_5 + \dot{C}\ell$$
$$\dot{S}F_5 + O\ell \xrightarrow{1} SF_5\dot{O}\ell$$
$$SF_5\dot{O}\ell \xrightarrow{-1} \dot{S}F_5 + O\ell$$
$$SF_5\dot{O}\ell + SF_5C\ell \xrightarrow{3} SF_5O\ell C\ell + \dot{S}F_5$$
$$\dot{C}\ell + O\ell \xrightarrow{4} C\ell\dot{O}\ell$$
$$C\ell\dot{O}\ell + SF_5C\ell \xrightarrow{5} C\ell O\ell C\ell + \dot{S}F_5$$
$$2\dot{S}F_5 \xrightarrow{6} S_2F_{10}$$

where $O\ell$ = olefin, and with $CH_2{=}CH_2$, the products are as given above. The reversibility of reaction (1) was confirmed by the fact that rate of formation of the adduct, $SF_5O\ell C\ell$, decreased with increasing temperature. A steady-state treatment gave

$$\frac{d[SF_5O\ell C\ell]/dt}{d[C\ell O\ell C\ell]/dt} = \frac{k_1[O\ell]}{(\phi I_a k_6)^{\frac{1}{2}}}\left\{\frac{1}{1 + k_{-1}/(k_3[SF_5C\ell])}\right\}$$

where the rate of reaction (I) = ϕI_a. This becomes

$$\frac{[C\ell O\ell C\ell]}{[SF_5O\ell C\ell][SF_5C\ell]^{\frac{1}{2}}} = \frac{(Ck_6)^{\frac{1}{2}}}{k_1[O\ell]}\left\{1 + \frac{k_{-1}}{k_3[SF_5C\ell]}\right\}$$

ΔH^o_f, ΔH^o and E in kcal mol^{-1} (kJ mol^{-1}); S^o, C^o_p, ΔS^o and ΔC^o_p in cal mol^{-1} K^{-1} (J mol^{-1} K^{-1}); A and k in l mol^{-1} s^{-1}

log A	E	log k	Conditions	Reference and Experimental Method
				where $C = \phi I_0 2.3\varepsilon\ell$ and it is assumed that $I_a = I_0 2.3\varepsilon\ell[SF_5C\ell]$. Plots of $[C\ell O\ell C\ell]/[SF_5O\ell C\ell][SF_5C\ell]^{\frac{1}{2}}$ against $1/[SF_5C\ell]$ were linear, and values of $k_1/k_6^{\frac{1}{2}}$ and k_{-1}/k_3 were calculated from the intercepts and slopes. An Arrhenius plot yielded: $\log (A_1/A_6^{\frac{1}{2}})(\ell^{\frac{1}{2}}mol^{-\frac{1}{2}}s^{-\frac{1}{2}}) = 0.59$ and $E_1 - \frac{1}{2}E_6 = 1.9$ kcal mol^{-1}. The experimental results above were calculated assuming $\log k_6 = 9.7$ [L.Bertrand, G.R.De Maré, G.Huybrechts, J.Olbregts and M.Toth, Chem.Phys.Lett., 5, 183 (1970)]. Note also results for radical decomposition: $\log (A_{-1}/A_3)(\ell^{-1}mol) = 2.07$ and $E_{-1} - E_3 = 6.6$ kcal mol^{-1} and assuming a reasonable value of $\log k_3 \simeq 8 - 7/2.3RT$ yields $\log (k_{-1}/s^{-1}) \simeq 10 - 14/2.3RT$.

PREFERRED RATE PARAMETERS

Tentatively, $\log k = [7.5 - 5.4/2.3RT]$

COMMENTS

The experimental A-factor is abnormally low. Although it is not possible to make meaningful estimates of the entropies of SF_5 and the adduct radical, there is no reason why they should be abnormal, and consequently $\Delta S°$ should fall within the range 30 to 37 cal mol^{-1}K^{-1}. In conjunction with $\log A_1 = 5.44$, this corresponds to $\log (A_{-1}/s^{-1}) = 10.2$ to 11.7, which is incompatible with a bond breaking process. The rate constants for the SF_5 addition reaction, which are probably reliable, indicate that SF_5 is about as reactive as CH_3, and consequently we have tentatively assigned an A-factor of $10^{7.5}$ which yields $E_1 = 5.4$ kcal mol^{-1} from the rate constant at 367 K. Although the kinetic treatment of this system is thorough and the results seem self-consistent, it is possible that errors have arisen in the extrapolations which were necessary to calculate k_1. Further indication of a systematic error comes from the results for the reverse radical decomposition (k_{-1}) derived in this study. From the experimental rate constant ratio, a reasonable estimate of A_{-1} is 10^{10} s^{-1} (see experimental section). This is an abnormally low A-factor for a bond breaking process and suggests that the Arrhenius parameters have been under-estimated.

REACTION: SULPHUR PENTAFLUORIDE RADICAL + 1,1-DIFLUOROETHYLENE

$$\dot{S}F_5(I) + CH_2{=}CF_2(II) \xrightarrow{1} SF_5CH_2\dot{C}F_2(III)$$
$$\xrightarrow{2} SF_5CF_2\dot{C}H_2(IV)$$

	(I)	(II)	(III)	(IV)
$\Delta H^\circ_{f(300)}$		-71.5 (-299.4)		
$S^\circ_{(300)}$		64.9 (271.7)		
$C^\circ_{p(300)}$		14.8 (62.0)		

	Path 1	Path 2
ΔH°		
ΔS°		
ΔC°_p		

Path	log A	E	log k	Conditions	Reference and Experimental Method
1	5.4	2.0 (8.4)	4.25(357 K)	296-418 K 120-154 torr(?)	[163] H.W.Sidebottom, J.M.Tedder and J.C.Walton, Trans.Faraday Soc., 66, 2038 (1970). For details see $SF_5 + CH_2{=}CH_2$, reference [162]. Major adduct identified as $SF_5CH_2CF_2Cl$ by mass spectrometry. Only traces of $SF_5CF_2CH_2Cl$ were detected; hence path 1 is the principal route for addition. Experimental results calculated from log $(A_1/A_6^{\frac{1}{2}})(l^{\frac{1}{2}}mol^{-\frac{1}{2}}s^{-\frac{1}{2}}) = 0.62$ and $E_1 - \frac{1}{2}E_6 = 2.0$ kcal mol^{-1}, assuming log $k_6 = 9.7$.
2	-	-	-		

PREFERRED RATE PARAMETERS

Tentatively, log k_1 = [7.5 - 5.4/2.3RT]

COMMENTS

See comments for $SF_5 + CH_2{=}CH_2$ system. Here again the experimental Arrhenius parameters appear to have been under-estimated, and we have recalculated, assigning an A-factor of $10^{7.5}$. This conclusion is supported by the abnormally low A-factor, $(\log(A_{-1}/s^{-1}) \simeq 11)$, deduced in this paper for the reverse radical decomposition.

ΔH°_f, ΔH° and E in kcal mol^{-1} (kJ mol^{-1}); S°, C°_p, ΔS° and ΔC°_p in cal mol^{-1} K^{-1} (J mol^{-1} K^{-1}); A and k in l mol^{-1} s^{-1}

REACTION: SULPHUR PENTAFLUORIDE RADICAL + TRIFLUOROETHYLENE

$\dot{S}F_5(I) + CFH{=}CF_2(II) \xrightarrow{1} SF_5CFH\dot{C}F_2(III)$

$\xrightarrow{2} SF_5CF_2\dot{C}FH(IV)$

	(I)	(II)	(III)	(IV)		Path 1	Path 2
$\Delta H^\circ_{f(300)}$		-117.3 (-491.1)			ΔH°		
$S^\circ_{(300)}$		70.1 (293.5)			ΔS°		
$C^\circ_{p(300)}$		16.5 (69.1)			ΔC°_p		

Path	log A	E	log k	Conditions	Reference and Experimental Method
1	6.25	3.4 (14.2)	4.21(364 K)	296-433 K 120-154 torr	[163] H.W.Sidebottom, J.M.Tedder and J.C.Walton, Trans.Faraday Soc., 66, 2038 (1970). For details see $SF_5 + CH_2{=}CH_2$ system, reference [162]. Adduct identified as SF_5CFHCF_2Cl by mass spectrometry. Only small amounts of SF_5CF_2CFHCl detected, ($SF_5CFHCF_2Cl/SF_5CF_2CFHCl$ varied from 11 to 36 between 356 and 433 K, [H.W.Sidebottom, J.M.Tedder and J.C.Walton, Chem.Comm., 253 (1970)]), and hence path 1 is the major route for addition. Experimental results calculated from log $(A_1/A_6^{\frac{1}{2}})$ $(l^{\frac{1}{2}}mol^{-\frac{1}{2}}s^{-\frac{1}{2}})$ = 1.40 and $E_1 - \frac{1}{2}E_6$ = 3.4 kcal mol^{-1}, assuming log k_6 = 9.7.
2	-	-	-		

PREFERRED RATE PARAMETERS

Tentatively, log k_1 = [8.0 - 6.3/2.3RT]

COMMENTS

See comments for $SF_5 + CH_2{=}CH_2$ system. Preferred parameters calculated from assumed log A = 8.0 and k at 364 K. Under-estimation of experimental Arrhenius parameters supported by abnormally low A-factor (log $(A_{-1}/s^{-1}) \simeq 10$) for reverse radical decomposition, reported in this study.

REACTION: SULPHUR PENTAFLUORIDE RADICAL + TETRAFLUOROETHYLENE $\dot{S}F_5(I) + CF_2{=}CF_2(II) \longrightarrow SF_5CF_2\dot{C}F_2(III)$

	(I)	(II)	(III)	
$\Delta H^\circ_{f(300)}$		-157.9 (-661.1)		ΔH°
$S^\circ_{(300)}$		74.6 (312.4)		ΔS°
$C^\circ_{p(300)}$		19.4 (81.2)		ΔC°_p

log A	E	log k	Conditions	Reference and Experimental Method
-	-	-	301-418 K 120-154 torr(?)	[163] H.W.Sidebottom, J.M.Tedder and J.C.Walton, Trans.Faraday Soc., 66, 2038 (1970). For details see $SF_5 + CH_2{=}CH_2$ system, reference [162]. Telomer molecules containing two and three C_2F_4 units were major products as well as the expected $SF_5CF_2CF_2Cl$ adduct. It was not possible to analyze for S_2F_{10} as well as the telomers, and no kinetic data on the addition reaction could be derived.

PREFERRED RATE PARAMETERS

No determination possible from this data

ΔH^o_f, ΔH^o and E in kcal mol^{-1} (kJ mol^{-1}); S^o, C^o_p, ΔS^o and ΔC^o_p in cal mol^{-1} K^{-1} (J mol^{-1} K^{-1}); A and k in l mol^{-1} s^{-1}

3. RELATIVE RATE DATA

RELATIVE RATE CONSTANTS FOR ADDITION OF H TO ALKENES AND FLUOROALKENES

Substrate	Relative k at room temperature				
	Ref.[5]	Ref.[17]	Ref.[18]	Preferred	Ref.[21]
$CH_2=CH_2$	(0.56)	0.67	0.64	0.83	1.00
$CH_2=CHCH_3$	1.00	1.00	1.00	1.00	-
$CH_2=CHCH_2CH_3$	1.08	0.99	-	1.03	-
cis-$CH_3CH=CHCH_3$	0.47	-	-	0.47	-
trans-$CH_3CH=CHCH_3$	0.59	-	-	0.59	-
$CH_2=C(CH_3)_2$	2.46	2.56	2.55	2.52	-
$(CH_3)_2C=CHCH_3$	-	-	1.03	1.03	-
$(CH_3)_2C=C(CH_3)_2$	0.81	-	0.87	0.84	-
$CH_2=CHCH_2CH_2CH_3$	-	-	0.89	0.89	-
cis-$CH_3CH=CHCH_2CH_3$	-	-	0.39	0.39	-
trans-$CH_3CH=CHCH_2CH_3$	-	-	0.44	0.44	-
$CH_2=CHCH=CH_2$	4.8	5.0	(5.3)	4.9	-
$CH_2=CHF$	-	-	-	-	0.79
$CH_2=CF_2$	-	-	-	-	1.45
cis-$CHF=CHF$	-	-	-	-	0.70
trans-$CHF=CHF$	-	-	-	-	1.16
$CHF=CF_2$	-	-	-	-	1.65
$CF_2=CF_2$	-	-	-	-	1.69

References

[5] K.R.Jennings and R.J.Cvetanović, J.Chem.Phys., 35, 1233 (1961).
See H + C_2H_4 system.

[17] G.R.Woolley and R.J.Cvetanović, J.Chem.Phys., 50, 4697 (1969).
See H + C_2H_4 system.

[18] R.J.Cvetanović and L.C.Doyle, J.Chem.Phys., 50, 4705 (1969).
See H + C_2H_4 system.

[21] R.D.Penzhorn and H.L.Sandoval, J.Phys.Chem., 74, 2065 (1970).
See H + C_2H_4 system.

COMMENTS

The preferred relative rate constants in column 5 of the Table are the mean values from references [5,17,18] as given in reference [17], but based on propene as the standard and with the exception of the value for ethylene. This latter relative rate constant has been obtained from the preferred absolute rate constants $k_{C_2H_4}$ and $k_{C_3H_6}$. It has been pointed out that the data for ethylene from references

[5,17,18] lack infinite pressure extrapolations and consequently are probably less reliable than the result from the absolute measurements [M.J.Kurylo, N.C.Peterson and W.Braun, J.Chem.Phys., 53, 2776 (1970)]. This criticism does not apply to any of the data on the relative rate constants for olefins higher in the series than ethylene, but in calculating absolute rate constants from these data we have made the calculations relative to propene as the standard, rather than ethylene.

The relative rate constants for H addition to fluoro-alkenes listed in column 6 of the Table are probably reliable with the qualification that the data for C_2H_4 and C_2H_3F may have been measured slightly below the high-pressure limiting conditions.

RELATIVE RATE CONSTANTS FOR ADDITION OF N TO ALKENES

	Relative k at 435 K
Substrate	Ref.[43]
CH_2=CH_2	1.0
CH_2=$CHCH_3$	2.8
CH_2=$CHCH_2CH_3$	3.4
cis-CH_3CH=$CHCH_3$	2.4
trans-CH_3CH=$CHCH_3$	3.0
CH_2=$C(CH_3)_2$	4.1
$(CH_3)_2C$=$CHCH_3$	3.5
$(CH_3)_2C$=$C(CH_3)_2$	3.5

Reference

[43] J.T.Herron and R.E.Huie, J.Phys.Chem., 72, 2538 (1968).
See N + CH_2=CH_2 system.

Preferred Data

As given above

COMMENTS

See comments for N + CH_2=CH_2 system.

RELATIVE RATE CONSTANTS FOR ADDITION OF N TO ALKYNES

Substrate	Relative k at 435 K Ref.[43]
$CH\equiv CH$	1.0
$CH\equiv CCH_3$	6.5
$CH\equiv CCH_2CH_3$	13
$CH_3C\equiv CCH_3$	12
$CH\equiv CCH_2CH_2CH_3$	14
$CH\equiv CCH_2CH_2CH_2CH_3$	14
$CH_3CH_2C\equiv CCH_2CH_3$	14

Reference

[43] J.T.Herron and R.E.Huie, J.Phys.Chem., 72, 2538 (1968).
See N + CH_2=CH_2 system.

Preferred Data

As given above

COMMENTS

See comments for N + CH_2=CH_2 system.

RELATIVE RATE CONSTANTS FOR ADDITION OF $S(^3P)$ TO ALKENES

	Relative k at 298 K
Substrate	Ref.[49]
$CH_2=CH_2$	1.00
$CH_2=CHCH_3$	6.9
$CH_2=CHCH_2CH_3$	10
cis-$CH_3CH=CHCH_3$	16
trans-$CH_3CH=CHCH_3$	20
$CH_2=C(CH_3)_2$	50
$CH_2=CHCH_2CH_2CH_3$	10
$(CH_3)_2C=CHCH_3$	56
cyclo-C_5H_8	18
$CH_2=CHC\ell$	1.4

Reference

[49] E.M.Lown, E.L.Dedio, O.P.Strausz and H.E.Gunning, J.Amer.Chem.Soc., 89, 1056 (1967). See S + $CH_2=CH_2$ system.

Preferred Data

As given above

RELATIVE RATE CONSTANTS FOR ADDITION OF $S(^1D)$ TO ALKENES

Substrate	Relative k at room temperature Ref.[49]
$CH_2=CH_2$	1.00
$CH_2=CHCH_3$	1.7
$CH_2=C(CH_3)_2$	3.5

Reference

[49] E.M.Lown, E.L.Dedio, O.P.Strausz and H.E.Gunning, J.Amer.Chem.Soc., 89, 1056 (1967).

Preferred Data

As given above

COMMENTS

See experimental section of S + $CH_2=CH_2$ system. The reaction of $S(^1D)$ with alkenes gives rise to mercaptans as well as episulfides, but the detailed mechanisms are not understood.

RELATIVE RATE CONSTANTS FOR ADDITION OF Se(^{3}P) TO ALKENES

Substrate	Relative k at room temperature Ref.[77]
CH_2=CH_2	1.00
CH_2=$CHCH_3$	2.6
CH_2=$CHCH_2CH_3$	7.1
cis-CH_3CH=$CHCH_3$	24
trans-CH_3CH=$CHCH_3$	56
CH_2=$C(CH_3)_2$	45
CH_2=$CHCH_2CH_2CH_3$	5.0
CH_2=CHCN	3.8
CH_2=CHCℓ	1.3
CH_2=$CHCH_2$Cℓ	2.3
CH_2=CHCH=CH_2	114

Reference

[77] A.B.Callear and W.J.R.Tyerman, Trans.Faraday Soc., 62, 2760 (1966).
See Se + CH_2=CH_2 system.

Preferred Data

As given above

RELATIVE RATE CONSTANTS FOR ADDITION OF Br(^{2}P) TO ALKENES

Substrate	Relative k at room temperature Ref.[79]
$CH_2=CH_2$	1.00
$CH_2=CHCH_3$	17.7
$CH_2=CHCH_2CH_3$	22.6
cis-$CH_3CH=CHCH_3$	95.0
trans-$CH_3CH=CHCH_3$	98.5
$CH_2=C(CH_3)_2$	384

Reference

[79] P.I.Abell, *Trans.Faraday Soc.*, *60*, 2214 (1964).
See Br + C_2H_4 system.

Preferred Data

As given above

COMMENTS

The above relative rate constants, which seem to be reliable, indicate that Br atoms exhibit marked electrophilic behaviour in their reactions with olefins. (See summary of relative rate constants p.370).

RELATIVE RATE CONSTANTS FOR ADDITION OF $^1CH_2$ AND $^3CH_2$ TO ALKENES

Substrate	Relative k at 297 K Ref.[164] $^3CH_2$	 Ref.[164] $^1CH_2$	 Ref.[164] $^1CH_2$†	 Ref.[165]
$CH_2{=}CH_2$	1.0	1.0	1.0	1.0
$CH_2{=}CHCH_3$	1.0	1.27	1.31	-
$CH_2{=}CHCH_2CH_3$	1.6	1.63	1.65	-
cis-$CH_3CH{=}CHCH_3$	0.94	1.37	1.44	-
trans-$CH_3CH{=}CHCH_3$	0.89	1.39	1.46	-
$CH_2{=}C(CH_3)_2$	2.86	1.96	1.85	-
$(CH_3)_2C{=}CHCH_3$	1.83	2.12	2.17	-
$(CH_3)_2C{=}C(CH_3)_2$	2.74	2.16	2.09	-
$CH_2{=}CHCHCH_2$	18.97	3.94	3.20	-
$CH_2{=}CHF$	-	-	-	0.60
$CH_2{=}CF_2$	-	-	-	0.33
$CHF{=}CF_2$	-	-	-	0.16
$CF_2{=}CF_2$	-	-	-	0.10

† $^1CH_2$ but assuming 10% $^3CH_2$

References

[164] S.Krzyzanowski and R.J.Cvetanović, Can.J.Chem., 45, 665 (1967).

[165] F.Casas, J.A.Kerr and A.F.Trotman-Dickenson, J.Chem.Soc., 1141 (1965).

Preferred Data

Reasonable as given above

RELATIVE RATE CONSTANTS FOR ADDITION OF CH_3 TO ALKENES

Substrate	Relative k at 453 K: Ref.[93]	Relative k at 453 K: Calculated from Preferred Absolute Rate Constants
$CH_2=CH_2$	1.0	1.0
$CH_2=CHCH_3$	0.72	0.7
$CH_2=CHCH_2CH_3$	0.7	0.6
cis-$CH_3CH=CHCH_3$	0.2	0.1
trans-$CH_3CH=CHCH_3$	0.4	0.3
$CH_2=C(CH_3)_2$	1.1	1.0
$(CH_3)_2C=CHCH_3$	0.4	0.3
$(CH_3)_2C=C(CH_3)_2$	0.2	0.07
$CH_2=CHCH=CH_2$	12	13

Reference

[93] R.J.Cvetanović and R.S.Irwin, J.Chem.Phys., 46, 1694 (1967).
See CH_3 + $CH_2=CH_2$ system.

COMMENTS

The data of reference [93], which are in good agreement with the relative rate constants shown in the next column and calculated from the preferred absolute rate constants of Section 2, are reliable.

RELATIVE RATE CONSTANTS FOR ADDITION OF CF_3 TO ALKENES

Substrate	Relative k at 338 K	
	Refs.[126,129,166]	Calculated from Preferred Absolute Rate Constants
$CH_2=CH_2$	1.0	1.0
$CH_2=CHCH_3$	1.2	4.7
$CH_2=CHCH_2CH_3$	1.2 [a]	-
cis-$CH_3CH=CHCH_3$	0.9	3.2
trans-$CH_3CH=CHCH_3$	1.0	3.2
$CH_2=C(CH_3)_2$	3.7	13
$(CH_3)_2C=CHCH_3$	-	-
$(CH_3)_2C=C(CH_3)_2$	1.3	6.3
$CH_2=CHCH=CH_2$	8.7	7.2

[a] measured in 2,3-dimethylbutane solution.

Reference

[166] G.E.Owen, J.M.Pearson and M.Szwarc, Trans.Faraday Soc., 60, 564 (1964).

COMMENTS

There is a considerable discrepancy between the two series of relative rate constants listed above. This arises from the choice of ethylene as a standard since the evaluated absolute rate parameters for ethylene are substantially different to the data which can be obtained from the system for the relative rate constants. For the purposes of comparison, the relative data obtained directly from references [126,129,166] are preferable.

RELATIVE RATE CONSTANTS FOR ADDITION OF CCl_3 TO HALOGENATED ALKENES

	Relative k at 423 K
Substrate	Ref.[145]
CF_2=CHCl	0.05
CF_2=CCl_2	0.06
CF_2=CFBr	0.11
CH_2=CHCl	1.0
CHCl=CHCl	0.05
CHCl=CCl_2	0.16
CCl_2=CCl_2	<0.003

Reference

[145] D.P.Johari, H.W.Sidebottom, J.M.Tedder and J.C.Walton, J.Chem.Soc.(B), 95 (1971).
See CCl_3 + CH_2=CH_2 system.

Preferred Data

As given above

COMMENTS

The absolute rate constants for the above substrates obtained in Section 2 are calculated from the relative rate constants of reference [145].

RELATIVE RATE CONSTANTS FOR ADDITION OF CCO TO UNSATURATES

Substrate	Ref.[167] $CCO(a^1\Delta)$ 273 K	Ref.[167] $CCO(X^3\Sigma)$ 273 K	Ref.[168] $CCO(X^3\Sigma)$ 296 K	Ref.[169] $CCO(X^3\Sigma)$ 298 K	Ref.[169] $CCO(a^1\Delta)$ 303.5 K	Preferred value $CCO(X^3\Sigma)$
$CH_2=CH_2$	1.00	1.00	1.00	1.0	1.0	1.0
$CH_2=CHCH_3$	0.313	0.287	5.3	6.1	1.2	5.7
$CH_2=CHCH_2CH_3$	0.101	0.072	-	7.0	-	7.0
cis-$CH_3CH=CHCH_3$	0.155	-	8.0	10.1	1.9	9.1
trans-$CH_3CH=CHCH_3$	0.091	-	9.7	11.4	-	10.6
$CH_2=C(CH_3)_2$	0.096	-	42	58	-	50
$(CH_3)_2C=CHCH_3$	0.040	-	-	100	-	100
$(CH_3)_2C=C(CH_3)_2$	0.020	-	-	250	2.1	250
$CH_2=CHCH=CH_2$	-	-	-	210	2.4	210
$(CH_3)_2C=C=C(CH_3)_2$	-	-	-	$113^\dagger$	-	$113^\dagger$
cyclo-C_5H_8	-	-	-	19	-	19
C_6H_6	-	-	-	<1	-	<1
$CH{\equiv}CH$	-	-	-	$0.3^\dagger$	-	$0.3^\dagger$
$CH_3C{\equiv}CCH_3$	-	-	-	$8.5^\dagger$	-	$8.5^\dagger$

$C_2O(X^3\Sigma)$ marked thus † are at 303.5 K.

References

[167] R.T.K.Baker, J.A.Kerr and A.F.Trotman-Dickenson, J.Chem.Soc.(A), 975 (1966).

[168] C.Willis and K.D.Bayes, J.Amer.Chem.Soc., 88, 3203 (1966).

[169] D.G.Williamson and K.D.Bayes, J.Amer.Chem.Soc., 90, 1957 (1968).

Preferred Data

1) $(a^1\Delta)$ Reasonable as given in reference [169]
2) $(X^3\Sigma)$ The preferred data is that given in reference [169] except that, where possible, the mean value of references [168] and [169] is taken

COMMENTS

The data of references [168,169] are preferred on the basis that they have been confirmed by two independent methods, and it seems implausible that the radical CCO should display nucleophilic properties, as indicated by the data of reference [167]. Nevertheless, it is difficult to explain the discrepancies between the sets of data, and the results must be considered tentative in the meantime.

RELATIVE RATE CONSTANTS FOR ADDITION OF NF_2 TO ALKENES

Substrate	Relative k at 373 K	
	Ref.[159]	Ref.[161]
$CH_2=CH_2$	1.0	1.0
$CH_2=CHCH_3$	4.4	-
$CH_2=CHCH_2CH_3$	4.3	-
cis-$CH_3CH=CHCH_3$	10.3	-
trans-$CH_3CH=CHCH_3$	10.8	-
$CH_2=C(CH_3)_2$	19.6	-
$(CH_3)_2C=CHCH_3$	33.6	-
$(CH_3)_2C=C(CH_3)_2$	87.6	-
$CH_2=CHF$	-	<0.001
$CH_2=CHCl$	-	2.2
$CH_2=CHBr$	-	2.4

References

[159] A.J.Dijkstra, J.A.Kerr and A.F.Trotman-Dickenson, J.Chem.Soc.(A), 105 (1967).
See NF_2 + $CH_2=CH_2$ system.

[161] A.J.Dijkstra, J.A.Kerr and A.F.Trotman-Dickenson, J.Chem.Soc.(A), 864 (1967).

Preferred Data

Reasonable as given above

TABLE VII. SUMMARY OF RELATIVE RATE CONSTANTS FOR ADDITION OF ATOMS AND RADICALS TO ALKENES

Olefin	$Br(^2P)$ (298 K)	$O(^3P)$ (298 K)	$S(^3P)$ (298 K)	$Se(^3P)$ (298 K)	NF_2 (373 K)	$CH_2(^1A_1)$ (297 K)	H (298 K)	CF_3 (338 K)	$CCO(^3\Sigma)$ (296 K)	$CH_2(^3\Sigma g)$ (297 K)	CH_3 (453 K)
	1.00	1.0	1.00	1.00	1.0	1.00	1.00	1.0	1.00	1.00	1.0
	17.7	5.8	6.9	2.6	4.4	1.27	1.20	1.2	5.7	1.0	0.72
	22.6	5.8	10	7.1	4.3	1.63	1.24	1.2	7.0	1.6	0.7
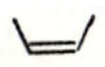	95.0	24.0	16	24	10.3	1.37	0.57	0.9	9.1	0.94	0.2
	98.5	28.0	20	56	10.8	1.39	0.71	1.0	10.6	0.89	0.4
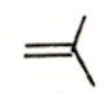	38.4	25.0	50	45	19.6	1.96	3.03	3.7	50	2.86	1.1
	-	79.0	56	-	33.6	2.12	1.24	-	100	1.83	0.4
	-	101.0	-	-	87.6	2.16	1.01	1.3	250	2.74	0.2
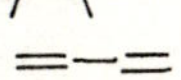	-	-	-	114	-	-	5.9	8.7	210	-	12

COMMENTS

The above summary of relative rate constants are the preferred data from Section 3, with the inclusion of data for $O(^3P)$ atoms for comparison [R.J.Cvetanović, Can.J.Chem., 38, 1678 (1960)]. It is worth repeating that the accuracies of many of the relative rate constants are much better than those of the absolute rate constants. Following the example of Cvetanović and Irwin [93], the relative rate constants above are arranged in decreasing order of electrophilicity of the attacking species from left to right along the Table.

Thus the atoms Br, O, S, Se and NF_2 radicals display typical electrophilic trends, the rates of addition increasing progressively with double bond substitution. At the other end of the scale CH_3 radicals display much less selectivity, although the reactions are very much slower, and here the governing factors appear to be the stability of the adduct radical and steric effects. The position of H atoms, just to the left of CF_3 radicals, is interesting and perhaps a little surprising. The relative positions of H atoms and CH_3 radicals in the Table might be interpreted as indicating that CH_3 radicals are displaying nucleophilic properties.

4. INDEXES

Author Index *

* Numbers in parentheses following a page number signify the frequency of reference to the given author on that page.

Subject Index